# Lecture Notes in Physics

**For information about Vols. 1–151, please contact your bookseller or Springer-Verlag.**

Vol. 152: Physics of Narrow Gap Semiconductors. Proceedings, 1981. Edited by E. Gornik, H. Heinrich and L. Palmetshofer. XIII, 485 pages. 1982.

Vol. 153: Mathematical Problems in Theoretical Physics. Proceedings, 1981. Edited by R. Schrader, R. Seiler, and D.A. Uhlenbrock. XII, 429 pages. 1982.

Vol. 154: Macroscopic Properties of Disordered Media. Proceedings, 1981. Edited by R. Burridge, S. Childress, and G. Papanicolaou. VII, 307 pages. 1982.

Vol. 155: Quantum Optics. Proceedings, 1981. Edited by C.A. Engelbrecht. VIII, 329 pages. 1982.

Vol. 156: Resonances in Heavy Ion Reactions. Proceedings, 1981. Edited by K.A. Eberhard. XII, 448 pages. 1982.

Vol. 157: P. Niyogi, Integral Equation Method in Transonic Flow. XI, 189 pages. 1982.

Vol. 158: Dynamics of Nuclear Fission and Related Collective Phenomena. Proceedings, 1981. Edited by P. David, T. Mayer-Kuckuk, and A. van der Woude. X, 462 pages. 1982.

Vol. 159: E. Seiler, Gauge Theories as a Problem of Constructive Quantum Field Theory and Statistical Mechanics. V, 192 pages. 1982.

Vol. 160: Unified Theories of Elementary Particles. Critical Assessment and Prospects. Proceedings, 1981. Edited by P. Breitenlohner and H.P. Dürr. VI, 217 pages. 1982.

Vol. 161: Interacting Bosons in Nuclei. Proceedings, 1981. Edited by J.S. Dehesa, J.M.G. Gomez, and J. Ros. V, 209 pages. 1982.

Vol. 162: Relativistic Action at a Distance: Classical and Quantum Aspects. Proceedings, 1981. Edited by J. Llosa. X, 263 pages. 1982.

Vol. 163: J.S. Darrozes, C. Francois, Mécanique des Fluides Incompressibles. XIX, 459 pages. 1982.

Vol. 164: Stability of Thermodynamic Systems. Proceedings, 1981. Edited by J. Casas-Vázquez and G. Lebon. VII, 321 pages. 1982.

Vol. 165: N. Mukunda, H. van Dam, L.C. Biedenharn, Relativistic Models of Extended Hadrons Obeying a Mass-Spin Trajectory Constraint. Edited by A. Böhm and J.D. Dollard. VI, 163 pages. 1982.

Vol. 166: Computer Simulation of Solids. Edited by C.R.A. Catlow and W.C. Mackrodt. XII, 320 pages. 1982.

Vol. 167: G. Fieck, Symmetry of Polycentric Systems. VI, 137 pages, 1982.

Vol. 168: Heavy-Ion Collisions. Proceedings, 1982. Edited by G. Madurga and M. Lozano. VI, 429 pages. 1982.

Vol. 169: K. Sundermeyer, Constrained Dynamics. IV, 318 pages. 1982.

Vol. 170: Eighth International Conference on Numerical Methods in Fluid Dynamics. Proceedings, 1982. Edited by E. Krause. X, 569 pages. 1982.

Vol. 171: Time-Dependent Hartree-Fock and Beyond. Proceedings, 1982. Edited by K. Goeke and P.-G. Reinhard. VIII, 426 pages. 1982.

Vol. 172: Ionic Liquids, Molten Salts and Polyelectrolytes. Proceedings, 1982. Edited by K.-H. Bennemann, F. Brouers, and D. Quitmann. VII, 253 pages. 1982.

Vol. 173: Stochastic Processes in Quantum Theory and Statistical Physics. Proceedings, 1981. Edited by S. Albeverio, Ph. Combe, and M. Sirugue-Collin. VIII, 337 pages. 1982.

Vol. 174: A. Kadić, D.G.B. Edelen, A Gauge Theory of Dislocations and Disclinations. VII, 290 pages. 1983.

Vol. 175: Defect Complexes in Semiconductor Structures. Proceedings, 1982. Edited by J. Giber, F. Beleznay, J.C. Szép, and J. László. VI, 308 pages. 1983.

Vol. 176: Gauge Theory and Gravitation. Proceedings, 1982. Edited by K. Kikkawa, N. Nakanishi, and H. Nariai. X, 316 pages. 1983.

Vol. 177: Application of High Magnetic Fields in Semiconductor Physics. Proceedings, 1982. Edited by G. Landwehr. XII, 552 pages. 1983.

Vol. 178: Detectors in Heavy-Ion Reactions. Proceedings, 1982. Edited by W. von Oertzen. VIII, 258 pages. 1983.

Vol. 179: Dynamical Systems and Chaos. Proceedings, 1982. Edited by L. Garrido. XIV, 298 pages. 1983.

Vol. 180: Group Theoretical Methods in Physics. Proceedings, 1982. Edited by M. Serdaroğlu and E. İnönü. XI, 569 pages. 1983.

Vol. 181: Gauge Theories of the Eighties. Proceedings, 1982. Edited by R. Raitio and J. Lindfors. V, 644 pages. 1983.

Vol. 182: Laser Physics. Proceedings, 1983. Edited by J.D. Harvey and D.F. Walls. V, 263 pages. 1983.

Vol. 183: J.D. Gunton, M. Droz, Introduction to the Theory of Metastable and Unstable States. VI, 140 pages. 1983.

Vol. 184: Stochastic Processes – Formalism and Applications. Proceedings, 1982. Edited by G.S. Agarwal and S. Dattagupta. VI, 324 pages. 1983.

Vol. 185: H.N. Shirer, R. Wells, Mathematical Structure of the Singularities at the Transitions between Steady States in Hydrodynamic Systems. XI, 276 pages. 1983.

Vol. 186: Critical Phenomena. Proceedings, 1982. Edited by F.J.W. Hahne. VII, 353 pages. 1983.

Vol. 187: Density Functional Theory. Edited by J. Keller and J.L. Gázquez. V, 301 pages. 1983.

Vol. 188: A.P. Balachandran, G. Marmo, B.-S. Skagerstam, A. Stern, Gauge Symmetries and Fibre Bundles. IV, 140 pages. 1983.

Vol. 189: Nonlinear Phenomena. Proceedings, 1982. Edited by K.B. Wolf. XII, 453 pages. 1983.

Vol. 190: K. Kraus, States, Effects, and Operations. Edited by A. Böhm, J.W. Dollard and W.H. Wootters. IX, 151 pages. 1983.

Vol. 191: Photon Photon Collisions. Proceedings, 1983. Edited by Ch. Berger. V, 417 pages. 1983.

Vol. 192: Heidelberg Colloquium on Spin Glasses. Proceedings, 1983. Edited by J.L. van Hemmen and I. Morgenstern. VII, 356 pages. 1983.

Vol. 193: Cool Stars, Stellar Systems, and the Sun. Proceedings, 1983. Edited by S.L. Balliunas and L. Hartmann. VII, 364 pages. 1984.

Vol. 194: P. Pascual, R. Tarrach, QCD: Renormalization for the Practitioner. V, 277 pages. 1984.

# Lecture Notes in Physics



214

Hendrik Moraal

# Classical, Discrete Spin Models:

## Symmetry, Duality and Renormalization

Springer-Verlag
Berlin Heidelberg GmbH 1984

**Author**

Hendrik Moraal
Institut für Theoretische Physik, Universität zu Köln
Zülpicher Straße 77, D-5000 Köln 41, FRG

ISBN 978-3-540-13896-9 ISBN 978-3-540-39108-1 (eBook)
DOI 10.1007/978-3-540-39108-1


Originally published by Springer-Verlag Berlin Heidelberg New York Tokyo in 1984

2153/3140-543210

Nadie puede escribir un libro.
(Borges)

# PREFACE.

The present book is an outgrowth of a series of lectures held for students at the graduate level at the Institute for Theoretical Physics of the University of Cologne over the past five years. The purpose of the book is to present three aspects of the statistical mechanical theory of discrete, classical spin models in a coherent and essentially self-contained fashion. These three aspects correspond to the three parts of the book and are:

(i) The theory of the possible (finite) symmetry groups of spin models and their classification. This is an essentially mathematical subject and is treated as such in the first six chapters. Nearly all of the theory of finite groups needed is developed in the text.

(ii) The theory of the duality transformation is introduced for several reasons; it not only gives exact information on the phase diagrams of spin models, but also allows for a simple introduction of lattice gauge theories. These are treated exclusively from a statistical mechanics point of view.

(iii) The theory of the renormalization of fields and interactions on recursively defined sequences of graphs gives a qualitative overview of possible symmetry-breaking patterns and other aspects of the phase diagrams of spin and gauge systems.

In order to interrupt the flow of reasoning as little as possible, the number of references in the text has been kept to a minimum. For further reading, general references are given at the ends of some of the chapters. These are mostly to review articles, which, in turn, allow for easy access to the original literature.

The numbering of equations, theorems, figures, etc., is by section. Reference is made to these as follows: suppose the reader finds, in Section 3 of Chapter 5: eq. (5); this is the fifth equation of the same section of the same chapter; Lemma 2.1; this is the first lemma in the second section of the same chapter; Fig. 3.5.1; this is the first figure of the fifth section of the third chapter, etc.

The author thanks Professor J. Hajdu for pressing him not only to start, but also to finish the present work. Thanks are also due to Mrs. A. Schneider for expertly drawing the many figures.

Cologne, September 1984. H. Moraal

## PREFACE

Cologne, September 1984

# Table of Contents

# PART A : THEORY OF THE SYMMETRY GROUPS OF SPIN MODELS.

## 1. SPIN MODELS AND THEIR SYMMETRY GROUPS.

### 1.1. Statistical mechanics of spin models on graphs.

In the following, the statistical mechanics of an assembly of spins, which for the purpose of these lectures are classical variables taking on a finite number of values (this number will always be denoted by M in what follows) will be considered. If these spins interact with each other pairwise only, as will always be assumed to be the case, it is expedient to consider them as attached to the vertices of a graph, two vertices of which are connected by an edge if and only if (iff) the corresponding spins have a nonzero interaction. The mathematical definition of a graph is ([1]):

Definition 1. A graph $G$ is a triple $(V,E,I)$ consisting of the two disjoint sets $V$ (of vertices) and $E$ (of edges) and of the incidence function $I$ which maps every element $e\in E$ on an unordered pair of vertices $v_1$ and $v_2$ from $V$ :

$$I(e)=\{v_1,v_2\}. \tag{1}$$

Graphs on which spin systems are considered will, in general, be such that $v_1\neq v_2$ holds in eq. (1) and that $e_1\neq e_2$ implies $I(e_1)\neq I(e_2)$. These two conditions are easily seen to imply that (i) the graph has no loops, i.e., there are no edges beginning and ending at the same vertex, and (ii) there are no multiple edges connecting the same pair of vertices. In graph theory, the class of graphs satisfying conditions (i) and (ii) is called the class of simple graphs. For a pictorial representation, see Fig. 1 below.

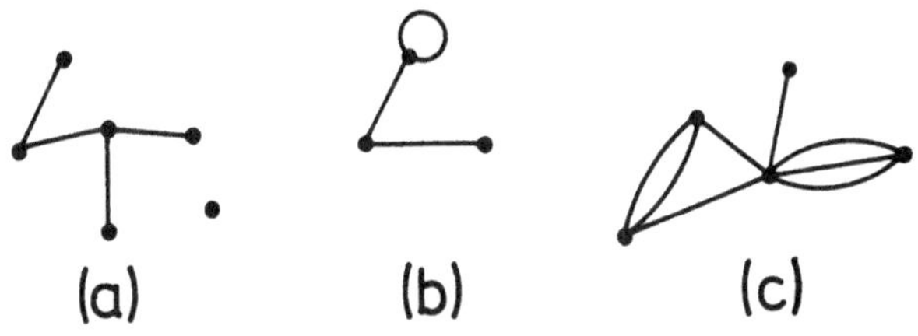


Fig. 1. (a) A simple graph. (b) A graph with a loop. (c) A graph with multiple edges.

Given a graph G, the pair interaction energy of the assembly of spins on the vertices can be calculated once the spin-spin energy function $E(i,j)$ is known for all $i,j\epsilon S$, where $S$ is the set of integers $\{1,2,..,M\}$. The total pair interaction energy is

$$E_{pair} = \sum_{\substack{e\epsilon E \\ I(e)=\{v_1,v_2\}}} E_e(i_{v_1}, i_{v_2}), \tag{2}$$

where provision has been made for a possible edge-dependence of the energy function. In order to calculate the total energy, the interaction with external fields $F(i)$ should also be taken into account:

$$E_{field} = \sum_{v\epsilon V} F_v(i_v). \tag{3}$$

Now the partition function of the spin model on the graph G can be written down as

$$Z(G) = \sum_{\{i_v\}} \prod_{e\epsilon E} \Omega_e(i_{v_1}, i_{v_2}) \prod_{v\epsilon V} A_v(i_v), \tag{4}$$

where the abbreviations ($\beta=1/k_BT$, $k_B$ the Boltzmann constant)

$$\Omega(i,j) = \exp-\beta E(i,j) \tag{5}$$

and

$$A(i) = \exp-\beta F(i) \tag{6}$$

have been introduced. The sum in eq. (4) is over the values from S for all spin variables.

Interesting properties of spin systems on graphs can, of course, only be expected in the thermodynamic limit, i.e., eq. (4) must be evaluated for a sequence $\{G_n\}$ of graphs such that the number of vertices goes to infinity for $n\to\infty$. Special classes of such sequences, which are obtained from recursive prescriptions, are the subject of Chapter 8 of this work. In the following sections, the very generally defined interaction energy $E(i,j)$ will be studied from a group-theoretical point of view after certain restrictions have been introduced.

## 1.2. Interactions.

The M×M matrix $\underline{E}$ of the interaction energies $E(i,j)$ will now be

taken such that it satisfies the following three requirements:
(i) The pair interaction is invariant with respect to the interchange of the spins:

$$E(i,j)=E(j,i) \text{ for all } i,j\varepsilon S. \tag{1}$$

(ii) Let $g$ be a permutation of the set $S$ ; the energy function $E(i,j)$ is invariant with respect to $g$ if

$$E(i,j)=E(g(i),g(j)) \tag{2}$$

holds for all $i,j\varepsilon S$. The set of all permutations of $S$ which leave the energy function invariant form a group $\mathcal{G}$, which is a subgroup of the group $\mathcal{S}(M)$ of <u>all</u> permutations of the $M$ elements of $S$. The group $\mathcal{S}(M)$ is called the <u>symmetric</u> group on $M$ objects and has $M!$ elements, denoted as

$$|\mathcal{S}(M)|= M! \, . \tag{3}$$

The second requirement is now that the symmetry group $\mathcal{G}$ of the pair interaction must be <u>transitive</u>, which means that

$$\{g(i)\,|\,g\varepsilon\mathcal{G}\}=S \tag{4}$$

holds for all $i\varepsilon S$. This means that for any pair $i,j\varepsilon S$, there is at least one permutation $g\varepsilon\mathcal{G}$ which maps $i$ on $j$ : $g(i)=j$. Referral to eq. (1.2) shows that the transitivity of $\mathcal{G}$ implies that a certain value of the total pair interaction energy can be obtained in at least $M$ ways: a particular spin in state $i$ can be transformed into any other state $j$ by a permutation $g\varepsilon\mathcal{G}$ ; if all other spins are also subjected to the same permutation, the total pair energy must stay invariant. In particular, the ground state energy in the absence of external fields will be at least $M$-fold degenerate, so that one may expect symmetry-breaking phase transitions to occur at sufficiently low temperatures (in a suitable thermodynamic limit). This problem is considered at length in Chapter 9.
(iii) The transitivity requirement (ii) implies that all "diagonal" energies $E(i,i)$ are equal for all $i$; this energy is chosen as zero:

$$E(i,i)=E(1,1)=0 \text{ for all } i\varepsilon S. \tag{5}$$

The above three requirements allow for a characterization of all possible pair interactions in a graph-theoretical way. Let $E_k$ be a value that $E(i,j)$ can take for $i \neq j$. Then the number of times that $E_k$ occurs in this matrix for a fixed value of $i$ is

$$N(E_k) = \sum_{j \neq i} \delta\big(E(i,j), E_k\big), \tag{6}$$

where $\delta(a,b)=1$ for $a=b$ and $\delta(a,b)=0$ otherwise. The transitivity of the group $\mathcal{G}$ implies that this number does not depend on the index $i$. This, together with the symmetry $E(i,j)=E(j,i)$, shows that a particular value $E_k$ shows up $N(E_k)$ times in every row and column of $\underline{\underline{E}}$. The set $E_k$ of different values that $E(i,j)$ can take on for $i \neq j$ then has at most $M-1$ elements, since one has

$$\sum_{E_k} N(E_k) = \sum_{j \neq i} 1 = M-1. \tag{7}$$

This shows that $E(i,j)$ can be written as

$$E(i,j) = \sum_{k=1}^{s} E_k \, M_k(i,j), \tag{8}$$

where $s \leqq M-1$ holds and the $\underline{\underline{M}}_k$ are given as

$$M_k(i,j) = \begin{cases} 1 \text{ for } E(i,j)=E_k, \\ 0 \text{ otherwise.} \end{cases} \tag{9}$$

Now $\underline{\underline{M}}_k$ can be read off from a (simple) graph $G_k$ with $M$ vertices, pairs of which are connected by edges iff the corresponding entries of $M_k(i,j)$ are nonzero. The simplicity of $G_k$ follows from the fact that the diagonal elements of the energy matrix are zero. Further, since $\underline{\underline{M}}_k$ has exactly $N(E_k)$ entries equal to 1 in each row and column, the graph $G_k$ is <u>regular</u>, i.e., every vertex has $z_k=N(E_k)$ edges emanating from it; the number of edges of $G_k$ is then $z_k M/2$.

The energy function of eq. (8) is, therefore, completely defined once the $s$ graphs $G_k$ corresponding to the $s$ matrices $\underline{\underline{M}}_k$ are given. These graphs all have the same vertex set, but their edge sets are disjoint. By definition, the union of these edge sets must contain all edges which can be drawn between $M$ vertices, so that the $G_k$ "add up" to the <u>complete</u> graph $K(M)$ on $M$ vertices:

$$K(M) = \sum_{k=1}^{s} G_k . \qquad (10)$$

(Technically, the sum is in the sense of the homology group modulo 2, see, e.g., [2].) Eq. (10) is illustrated for two examples in Fig. 1.

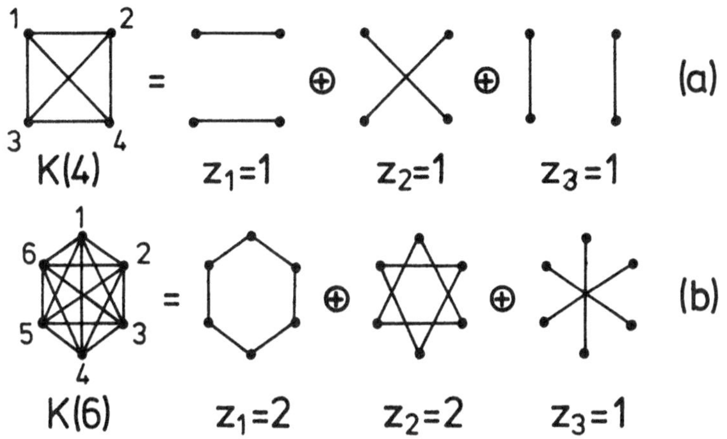


Fig. 1. Two decompositions of complete graphs into regular graphs. (a) M=4, s=3. (b) M=6, s=3.

As shown above, an energy function $E(i,j)$ satisfying requirements (i), (ii) and (iii) above leads, for an M-component spin model, to a decomposition of $K(M)$ into regular graphs with disjoint edge sets. On the other hand, not every such decomposition defines a spin model with the correct properties. Given a particular decomposition, let $\mathcal{G}(G_k)$ be the subgroup of $\mathcal{S}(M)$ which leaves the k-th graph $G_k$ invariant, i.e., $\mathcal{G}(G_k)$ consists of those permutations from $\mathcal{S}(M)$ which map the edges of $G_k$ onto each other; then the symmetry group $\mathcal{G}$ of the interaction matrix $\underline{\underline{E}}$ consists precisely of those permutations which are common to all $\mathcal{G}(G_k)$:

$$\mathcal{G} = \bigcap_{k=1}^{s} \mathcal{G}(G_k) . \qquad (11)$$

A decomposition of $K(M)$ into regular graphs $G_k$ will be called an <u>interaction</u> if the group $\mathcal{G}$ defined by eq. (11) is transitive. Since $\mathcal{G}$ is a subgroup of $\mathcal{G}(G_k)$, notation $\mathcal{G} \leqq \mathcal{G}(G_k)$, this implies that all $\mathcal{G}(G_k)$ must be transitive. There is clearly a one-to-one correspondence between energy functions satisfying requirements (i), (ii) and (iii) on the one hand and interactions on the other hand. Formal definitions are:

Definition 1. A graph $G_k$ with $M$ vertices is called hyperregular if its automorphism group $\mathcal{G}(G_k)$ is transitive (on the vertex set).

Definition 2. A decomposition of the complete graph $K(M)$ into $s$ hyperregular graphs $G_k$ is an interaction if the intersection of the $s$ automorphism groups $\mathcal{G}(G_k)$ is a transitive group $\mathcal{G}$, eq. (11).

The following lemmas follow directly from these definitions.

Lemma 1. The complete graph $K(M)$ is hyperregular with automorphism group $\mathcal{S}(M)$.

Proof. Since every pair of vertices of $K(M)$ is connected by an edge, every permutation of the vertices leaves $K(M)$ invariant, so that its automorphism group is $\mathcal{S}(M)$. Further, $\mathcal{S}(M)$ is clearly transitive since it contains all transpositions $(ij)$, i.e., all interchanges of pairs of vertices. ¶ (This symbol always denotes the end of a proof.)

Definition 3. Let $G$ be a graph with $M$ vertices. The complement $\bar{G}$ of $G$ is obtained from $K(M)$ by deleting the edges of $G$ from it, i.e., $G$ and $\bar{G}$ have disjoint edge sets and their sum is $K(M)$.

Lemma 2. If $G$ is hyperregular with automorphism group $\mathcal{G}(G)$, then $\bar{G}$ is hyperregular with the same automorphism group.

Proof. Clear from Definition 3 and Lemma 1. ¶

Lemma 3. The group $\mathcal{G}$ of an interaction is already obtained by taking the intersection of any $s-1$ of the $s$ groups $\mathcal{G}(G_k)$.

Proof. Eq. (11) can be written as

$$\mathcal{G}=\mathcal{G}(G_a)\cap \{ \bigcap_{\substack{k=1 \\ k\neq a}}^{s} \mathcal{G}(G_k) \}. \tag{12}$$

On the other hand, eq. (10) and Definition 3 imply

$$\sum_{\substack{k=1 \\ k\neq a}}^{s} G_k=\bar{G}_a. \tag{13}$$

Eq. (13) and Lemma 2 now yield

$$\mathcal{G}(G_a)=\mathcal{G}(\bar{G}_a)\geqq \bigcap_{\substack{k=1 \\ k\neq a}}^{s} \mathcal{G}(G_k). \tag{14}$$

Eqs. (12) and (14) imply the lemma. ¶

The next lemma is a statement of some simple facts concerning transitive permutation groups; since the proof is very simple, it is omitted.

Lemma 4. Let $\mathcal{G}$ be a transitive permutation group on M letters. (The set of states, vertices, etc. on which the permutations operate is traditionally called the set of letters if the exact nature of these objects is unimportant.) Then the subgroups $H_j$ which keep the letter j fixed all have equally many elements:

$$|H_j| = |\mathcal{G}|/M, \text{ for all } j=1,2,..,M. \tag{15}$$

Further, let $g_j$ be some element of $\mathcal{G}$ such that $g_j(1)=j$ holds. Then the set of all such elements is $g_jH_1$. This defines a (disjoint) decomposition in right cosets:

$$\mathcal{G}= \bigcup_{m=1}^{M} g_mH_1, \tag{16}$$

where $g_1=e$, the unit element of $\mathcal{G}$. $H_j$ can be obtained from $H_1$ by

$$H_j = g_jH_1g_j^{-1}. \tag{17}$$

If $\mathcal{G}$ is the group of an interaction with graph groups $\mathcal{G}(G_k)$, then eq. (11) implies that one can choose the $g_j$ equal for $\mathcal{G}$ and for the $\mathcal{G}(G_k)$, so that, if $H_1^{(k)}$ is the letter-1-fixing subgroup of $\mathcal{G}(G_k)$, eq. (11) can also be interpreted as

$$H_1 = \bigcap_{k=1}^{s} H_1^{(k)} . \tag{18}$$

## 1.3. Maximal interactions and permissible groups.

In the previous section, it was shown that there is a one-to-one correspondence between energy functions with transitive symmetry groups and special decompositions of complete graphs. It is the purpose of the present section to show that the transitive symmetry groups occurring are of a special type (called permissible groups) and that all interactions can be expressed in terms of certain maximal interactions associated with these permissible groups.

Let $\mathcal{G}$ be a transitive permutation group on M letters. This group defines a unique interaction by the following construction:
Draw one edge from the vertex "1" to an arbitrary other vertex in the edge-empty graph O(M) consisting of M vertices only; operate on this

edge with all elements $g \in \mathcal{G}$ ; the result will be a hyperregular graph $G_1$; if $G_1 \neq K(M)$, repeat the construction starting with a second edge with vertex "1" as one of its endpoints and not yet contained in $G_1$ ; this gives a hyperregular graph $G_2$; if the sum of $G_1$ and $G_2$ is still not equal to $K(M)$, repeat the construction, etc. until after $s \leq M-1$ steps the sum of the graphs is $K(M)$.
Since by construction $\mathcal{G}(G_k) \geq \mathcal{G}$ , the intersection of all $\mathcal{G}(G_k)$, called $\mathcal{G}^{(p)}$ here, also satisfies $\mathcal{G}^{(p)} \geq \mathcal{G}$ . Now since $\mathcal{G}$ is transitive, so is $\mathcal{G}^{(p)}$, so that the sequence of graphs constructed above is an interaction by Definition 2.2. It is clear that the above construction yields the same graphs, and, therefore, also $\mathcal{G}^{(p)}$ as intersection of the automorphism groups of these graphs, if one starts out with the group $\mathcal{G}^{(p)}$ instead of $\mathcal{G}$. This suggests the following definitions:

Definition 1. The group $\mathcal{G}^{(p)}$ obtained by the construction outlined above is called the permissible group corresponding to the transitive permutation group $\mathcal{G}$.

Definition 2. A transitive permutation group $\mathcal{G}$ is called permissible if $\mathcal{G}=\mathcal{G}^{(p)}$ holds.

Definition 3. The interaction constructed above is called the maximal interaction or MI of the groups $\mathcal{G}$ and $\mathcal{G}^{(p)}$.

The above Definitions 1 and 2 make sense only if not every transitive permutation group is permissible. It will be shown below that there do exist nonpermissible transitive groups. The reason for Definition 3 is clear: no graph $G_k$ of a MI can be further decomposed without altering the symmetry group of the interaction. The next lemma is a trivial consequence of the basic construction:

Lemma 1. An interaction with symmetry group $\mathcal{G}$ is the maximal interaction corresponding to $\mathcal{G}$ iff $\mathcal{G}$ is transitive on the edge sets of all graphs $G_k$.

Now follows a basic theorem concerning interactions:

Theorem 1. The permutation groups which can occur as the symmetry groups of an interaction are exactly the permissible groups.
Proof. Let

$$K(M) = G_1 + G_2 + \ldots + G_s \tag{1}$$

be an interaction with symmetry group $\mathcal{G}$. Let the maximal interaction corresponding to $\mathcal{G}$ be

$$K(M) = G_1' + G_2' + \ldots + G_t'. \tag{2}$$

Since $\mathcal{G}$ is transitive on the edges of the $G_k'$ and since $\mathcal{G}$ leaves the $G_k$ invariant, these latter graphs must be sums of graphs from eq. (2):

$$G_k = \sum_{j\varepsilon I_k} G_j', \quad \bigcup_{k=1}^{s} I_k = \{1,2,..,t\}, \quad I_k \cap I_m = \phi \text{ for } k \neq m. \tag{3}$$

Now suppose that $\mathcal{G}$ is not permissible, i.e., one has $\mathcal{G} < \mathcal{G}^{(p)}$ and

$$\mathcal{G} = \bigcap_{k=1}^{s} \mathcal{G}(G_k), \quad \mathcal{G}^{(p)} = \bigcap_{m=1}^{t} \mathcal{G}(G_m'). \tag{4}$$

But eq. (3) and the second of eqs. (4) imply

$$\mathcal{G}(G_k) \geqq \bigcap_{j\varepsilon I_k} \mathcal{G}(G_j') \geqq \mathcal{G}^{(p)}, \tag{5}$$

so that the first of eqs. (4) yields $\mathcal{G} \geqq \mathcal{G}^{(p)}$, which is a contradiction, so that $\mathcal{G}$ is permissible. This shows that a symmetry group occurring as the group of an interaction is permissible. Since a permissible group always occurs as the symmetry group of its MI, this proves the theorem.

Now some nonpermissible transitive permutation groups will be exhibited. To this end, two simple lemmas are proved.

Lemma 2. The symmetric group $\mathcal{S}(M)$ is permissible. Its MI consists of $K(M)$ only, the corresponding energy function is the one of the Potts model.

Proof. Since the automorphism group of $K(M)$ is $\mathcal{S}(M)$, the construction of the maximal interaction ends after one step yielding $G_1 = K(M)$, so that $\mathcal{S}(M)$ is certainly permissible. The corresponding energy function is

$$E(i,j) = E_1\{1-\delta(i,j)\}, \tag{6}$$

which has two possible values, 0 for $i=j$ and $E_1$ for $i \neq j$. This is (up to the arbitrary zero of energy) exactly the energy function of the M-state Potts model [3]. For $M=2$, this reduces to the Ising model. ¶

Lemma 3. Let $\mathcal{G}$ be a doubly transitive permutation group on $M \geqq 3$ letters, i.e., for every pair of ordered pairs $(i,j)$ and $(k,m)$ there is a $g \in \mathcal{G}$ with $g(i)=k$ and $g(j)=m$. Then $\mathcal{G}^{(p)}=\mathcal{S}(M)$.

Proof. Clearly, the double transitivity implies that the elements of $\mathcal{G}$ map a given edge onto any other edge of $K(M)$. Therefore, the construction of the maximal interaction of $\mathcal{G}$ yields only one graph $G_1=K(M)$, so that $\mathcal{G}^{(p)}=\mathcal{S}(M)$ follows. ¶

Since there are many doubly transitive groups which are proper subgroups of $\mathcal{S}(M)$ ([4]), the above lemma implies that not all transitive permutation groups are permissible, thus giving a first justification of Definitions 1 and 2.

As a first application of Theorem 1, it is shown below that many permissible groups can be obtained by studying graphs:

Theorem 2. Let $G$ be a hyperregular graph with $M$ vertices. Then its automorphism group is a permissible group $\mathcal{G}(G)$ on $M$ letters.

Proof. By Theorem 1, it is only necessary to exhibit an interaction with $\mathcal{G}(G)$ as symmetry group. To this end, a sequence of graphs is constructed from $G$ by means of the distance function $d_G(e)$ defined on the edges of the complete graph $K(M)$ by the prescription that $d_G(e)=k$ iff the minimum number of edges of $G$ in a path connecting the end points of $e$ is $k$. If these end points are not connected in $G$, one sets $d_G(e)=\infty$. One now defines $G^{(k)}$ as that subgraph of $K(M)$ for which $d_G(e)=k$ holds, so that $G^{(1)}=G$, $G^{(2)}$ contains edges between vertices which are connected by a minimal path of length 2 in $G$, etc., whereas $G^{(\infty)}$ contains all edges between vertices which are not connected in $G$. Clearly, $K(M)$ can be written as a (finite) sum

$$K(M) = \sum_{k=1}^{\infty} G^{(k)}. \tag{7}$$

By construction, the automorphism groups $\mathcal{G}(G^{(k)})$ for $k \geq 2$ all contain $\mathcal{G}(G)$, so that the symmetry group of the decomposition of eq. (7) is indeed $\mathcal{G}(G)$. Since $G$ is hyperregular, eq. (7) is an interaction; it is, however, not necessarily the MI corresponding to $\mathcal{G}(G)$, but this is not necessary for the proof of the theorem. ¶

Example 1. Consider the decomposition of $K(6)$ shown in Fig. 2.1.(b). If the hexagon is denoted by $G$, then the other two graphs shown are $G^{(2)}$ and $G^{(3)}$, respectively. Since the symmetry group of the hexagon is the dihedral group $\mathcal{D}(6)$ generated by the $60°$ rotation $(123456)$ and by the reflection $(1)(25)(36)(4)$, it is certainly hyperregular; therefore, Fig. 2.1.(b) shows an interaction with permissible group $\mathcal{D}(6)$. In fact, the construction at the beginning of this section shows it to be the MI corresponding to $\mathcal{D}(6)$.

Example 2. The decomposition of K(4) shown in Fig. 2.1.(a) has been obtained by construction as the MI of $\mathcal{K}(4)$, the Klein group with elements e, (12)(34), (13)(24) and (14)(23). Clearly, the symmetry groups of all three graphs are much larger than this, but their intersection is really $\mathcal{K}(4)$, so that this is an example of a permissible group not given by a hyperregular graph.

## 1.4. Completely permissible groups and P-algebras.

Let the MI corresponding to a permissible group $\mathcal{G}$ consist of s graphs $G_k$ which are regular with $z_k$ edges emanating from every vertex. The numbers $z_k$ are related to the double cosets ([5]) of $\mathcal{G}$ with respect to its letter-1-fixing subgroup $\mathcal{H}_1$:

Lemma 1. Let $T_k$ be a transitivity set (or orbit) of $\mathcal{H}_1$, i.e., a maximal set of letters on which $\mathcal{H}_1$ acts transitively; then, if $g_k$ maps 1 on $k \varepsilon T_k$, every element of $\mathcal{H}_1 g_k \mathcal{H}_1$ also maps 1 onto an element of $T_k$. The set $\mathcal{H}_1 g_k \mathcal{H}_1$ is a double coset of $\mathcal{G}$. A transitivity set $T_k$ is called self-conjugate if $\mathcal{H}_1 g_k \mathcal{H}_1 = \mathcal{H}_1 g_k^{-1} \mathcal{H}_1$ holds; in this case, $z_k$ for the graph $G_k$ generated from $\{1,k\}$ satisfies

$$z_k = |\mathcal{H}_1| / |\mathcal{H}_1 \cap \mathcal{H}_k| . \tag{1}$$

If $T_{k'}$ is not self-conjugate, then the graphs generated by $\{1,k'\}$ and by $\{1,\bar{k}'\}$ with $\bar{k}' \varepsilon T_{\bar{k}'}$, $g_{\bar{k}'} \varepsilon \mathcal{H}_1 g_{k'}^{-1} \mathcal{H}_1$ , are identical and one has

$$z_{k'} = 2|\mathcal{H}_1| / |\mathcal{H}_1 \cap \mathcal{H}_{k'}| . \tag{2}$$

Proof. This is immediate as soon as it is realized that the following proposition is true: A set $T_k$ is self-conjugate iff there is an element $\sigma_k \varepsilon \mathcal{G}$ with $\sigma_k(1)=k$ and $\sigma_k(k)=1$. To see this, let first $T_k$ be self-conjugate; then there exist elements $h_1$, $h_2 \varepsilon \mathcal{H}_1$ such that $h_1 g_k h_2 = g_k^{-1}$ holds. Consider $g_k h_1$; one has

$$g_k h_1(1) = g_k(1) = k, \tag{3a}$$

$$g_k h_1(k) = g_k h_1 g_k(1) = g_k h_1 g_k h_2(1) = g_k g_k^{-1}(1) = 1, \tag{3b}$$

so that $g_k h_1$ can play the role of $\sigma_k$. On the other hand, if there is such a $\sigma_k \varepsilon G$, then $\sigma_k{}^2 \varepsilon H_1$ and $\sigma_k$ is of even order. Therefore, one has $h_1 g_k h_2 = g_k{}^{-1}$ for $g_k = \sigma_k$, $h_1 = \sigma_k{}^{-2}$ and $h_1 = e$.

Now the number of different edges that $G$ produces from $\{1,k\}$ is clearly $|G|/|H_1 \cap H_k|$ for the self-conjugate case and else twice as many. Setting these numbers equal to $z_k M/2$ yields eqs. (1) and (2) above. ¶

Lemma 2. If the MI of a permissible group $G$ consists of $s$ graphs and $H_1$ has $t$ self-conjugate transitivity sets (not counting the single vertex 1), then $G$ has 2s-t+1 different double cosets.
Proof. This follows immediately from Lemma 1: the double cosets of are (i) the subgroup $H_1$ itself, (ii) $t$ self-conjugate double cosets $H_1 g_k H_1 = H_1 g_k{}^{-1} H_1$ and (iii) $2(s-t)$ double cosets in pairs of inverses.

Lemma 3. Any $G$-invariant $M \times M$ matrix $\underline{Q}$ is a linear combination of (i) the identity, (ii) the $s$ symmetric $\underline{\underline{M}}_k$ matrices associated with the graphs from the MI of $G$ and (iii) the $s-t$ antisymmetric matrices $\underline{\underline{S}}_{k'}$, defined only for pairs of non-self-conjugate transitivity sets by:

$$S_{k'}(i,j) = \begin{cases} +1 & \text{if there is a } g \varepsilon G \text{ with } g(1)=i,\ g(k')=j, \\ -1 & \text{if there is a } g' \varepsilon G \text{ with } g'(1)=j,\ g'(k')=i, \\ 0 & \text{otherwise.} \end{cases} \qquad (4)$$

Proof. The invariance of the $\underline{\underline{S}}_{k'}$ matrices is evident by construction, and there are clearly $s-t$ of them by Lemma 2. ¶

Remark. The above Lemmas 1,2 and 3 are also valid for a general transitive permutation group $G$. If this group is nonpermissible, then $G^{(p)}$ has at least as many self-conjugate transitivity sets as $G$.

It is obvious from the above, that in general not every invariant matrix can be read off from the MI of a permissible group $G$. For an important subclass of permissible groups, this is the case, however:

Definition 1. A permissible group $G$ is completely permissible if the most general $M \times M$ matrix invariant with respect to all $g \varepsilon G$ is necessarily symmetric.

From the above three lemmas, the following corollary is obtained:

Corollary 1. The following statements are all equivalent for $G$ a permissible group with $s$ graphs in its MI:
(a) $G$ is completely permissible.
(b) For every pair of letters $(1,k)$ there is a $\sigma_k \varepsilon G$ with $\sigma_k(1)=k$,

$\sigma_k(k)=1$.

(c) $\mathcal{G}$ has exactly $s+1$ double cosets.

(d) Every transitivity set of $\mathcal{H}_1$ is self-conjugate.

The Potts model groups $\mathcal{S}(M)$ are completely permissible by criterion (b), since these are doubly transitive, see Lemmas 3.2 and 3.3.

Completely permissible groups have a number of pleasing properties. One of these is related to the concept of a P-algebra:

Definition 2. A permissible group has a P-algebra if the $\underline{\underline{M}}_k$ matrices belonging to the graphs of its MI all commute pairwise.

Lemma 4. A completely permissible group has a P-algebra.

Proof. Since the $\underline{\underline{M}}_k$ are all symmetric, one has that the product $\underline{\underline{M}}_k\underline{\underline{M}}_m$ equals $\underline{\underline{M}}_m\underline{\underline{M}}_k$ iff $\underline{\underline{M}}_k\underline{\underline{M}}_m$ is symmetric. But since this product is certainly invariant with respect to all $g \in \mathcal{G}$, the complete permissibility of $\mathcal{G}$ guarantees this symmetry. ¶

The inverse of the statement of Lemma 4 cannot be proved generally; in some cases, complete permissibility and the property of having a P-algebra are equivalent, see Corollary 5.2 below.

The existence of a P-algebra has a simple graph-theoretical interpretation:

Lemma 5. Two incidence matrices $\underline{\underline{M}}_i$ and $\underline{\underline{M}}_j$ commute with each other iff the number of ways to reach a vertex $k$ from the vertex $1$ by using first an edge from $G_i$ and then one from $G_j$ equals the number of ways to do this by first using an edge from $G_j$ and then a second one from $G_i$. Therefore, $\mathcal{G}$ has a P-algebra iff every pair of graphs from its MI satisfies this requirement.

Proof. It is easily seen, that the requirement reads, in terms of the matrices $\underline{\underline{M}}_i$ and $\underline{\underline{M}}_j$ :

$$\sum_m M_i(1,m)\; M_j(m,k) \;=\; \sum_m M_j(1,m)\; M_i(m,k). \tag{5}$$

This implies that the two matrices commute by their invariance with respect to the permutations of the transitive group $\mathcal{G}$ . ¶

A useful criterion for the existence of a P-algebra makes use of the distance graphs defined in the proof of Theorem 3.2:

Lemma 6. If the MI of a permissible group consists of a graph $G_1=G^{(1)}$ and of the distance graphs $G^{(2)}, G^{(3)},.., G^{(\infty)}$, then it has a P-algebra.

Proof. Let the incidence matrix of $G_1$ be $\underline{\underline{M}}_1$; then the k-th power of this matrix must be expressible as a linear combination of the identity $\underline{\underline{I}}=\underline{\underline{M}}_0$ and of the matrices $\underline{\underline{M}}_m$ corresponding to the $G_m$ with $m\leq k$. This follows immediately from the definition of the distance graphs. Now if $G_1$ is connected, this means that the powers $\underline{\underline{M}}_1{}^k$ can all be expressed as

$$\underline{\underline{M}}_1{}^k = \sum_{m=0}^{k} \alpha_{km} \underline{\underline{M}}_m \quad , \text{ with } \alpha_{mm}>0. \tag{6}$$

Clearly, the matrix of the $\alpha_{km}$ is invertible, so that the $\underline{\underline{M}}_k$ can also be expressed as a linear combination of the powers $\underline{\underline{M}}_1{}^k$; therefore, the $\underline{\underline{M}}_k$ all commute with each other. If $G_1$ is not connected, all $\underline{\underline{M}}_k$ matrices not corresponding to $G^{(\infty)}$ commute with each other by the same argument. The incidence matrix of $G^{(\infty)}$ is given by

$$\underline{\underline{M}}_\infty = \underline{\underline{E}} - \sum_{k=1}^{s-1} \underline{\underline{M}}_k \quad , \tag{7}$$

with $\underline{\underline{E}}$ the incidence matrix of $K(M)$:

$$E(i,j) = \begin{cases} 0 \text{ for } i=j, \\ \\ 1 \text{ for } i\neq j. \end{cases} \tag{8}$$

$\underline{\underline{E}}$ is easily seen to commute with the incidence matrix of any regular graph, so that $\underline{\underline{M}}_\infty$ also commutes with all $\underline{\underline{M}}_k$ with $k$ finite. ¶

## 1.5. External fields and representation theory.

In Section 1 of this chapter, an external field acting on a spin in state $i$ was defined simply as an arbitrary function $F(i)$ defined on S. In many applications, it will be necessary to define different types of fields for a spin model described by a permissible symmetry group $\mathcal{G}$. One possible definition of this makes use of the fact that the group $\mathcal{G}$ is given in a special representation; if one associates with the letter $k\varepsilon S=\{1,2,..,M\}$ the k-th basis vector $\underline{e}_k$, defined as

$$(\underline{e}_k)_i = \delta(i,k) \quad , \tag{1}$$

of an M-dimensional vector space $V_M$, then the permutation $g\varepsilon\mathcal{G}$ is represented by an $M\times M$ matrix $\underline{\underline{D}}(g)$:

$$D_{ij}(g) = \delta\{i,g(j)\} \ , \tag{2}$$

since one has

$$\underline{\underline{D}}\ \underline{e}_k = \underline{e}_m \quad \text{iff} \quad m=g(k). \tag{3}$$

This special representation $\theta_{per}$ of $\mathcal{G}$ is, in general, reducible, i.e., there exist several invariant subspaces of $V_M$. The real irreducible representations of the group $\mathcal{G}$ will be denoted by $\theta_i$ for $i=1,2,..,q$. It is well-known ([6]), that a change of basis in $V_M$ will bring all matrices $\underline{\underline{D}}(g)$ in block-diagonal form $\tilde{\underline{\underline{D}}}(g)$, so that these blocks form irredicible representations of $\mathcal{G}$ operating in invariant minimal subspaces of $V_M$ :

$$\tilde{\underline{\underline{D}}}(g) = \begin{pmatrix} \underline{\underline{D}}_1(g) & & & & \\ & \underline{\underline{D}}_2(g) & & & \\ & & \cdot & & \\ & & & \cdot & \\ & & & & \underline{\underline{D}}_r(g) \end{pmatrix}. \tag{4}$$

Let the i-th irreducible representation $\theta_i$ of $\mathcal{G}$ occur $c_i$ times ($c_i$ can be zero) in the reduced form of eq. (4); one them writes

$$\theta_{per} = \sum_{i=1}^{q} c_i\theta_i. \tag{5}$$

If $\theta_i$ consists of $n_i{\times}n_i$ matrices, one has the following decomposition of $V_M$ in mutually orthogonal invariant subspaces:

$$V_M = \sum_{i=1}^{q} c_iV_{n_i} \ , \qquad M= \sum_{i=1}^{q} c_in_i \ , \tag{6}$$

where the first sum is a direct one. This decomposition is, of course, independent of the special basis in which the $\underline{\underline{D}}(g)$ have the form of eq. (4). Therefore, the following definition is possible:

<u>Definition 1</u>. Let $V_M$ be decomposed in invariant subspaces as in eq. (6), but in the original basis of eq. (1); then the external field $\underline{F}$ can be decomposed in its projections on the invariant subspaces $c_iV_{n_i}$. The lengths of these projections are the different field parameters,

called invariant field parameters.

The above definition of the invariant field parameters has the advantage that it does not depend on the detailed interactions present in in the system. For many applications (see, e.g., the Cayley branch with a field on the boundary in Chapter 9), however, the interactions effectively define a different set of field parameters as follows. Suppose all spin-spin interactions on a graph have the same energy function $E(i,j)$; then it is clear from eq. (1.4) that the matrix of Boltzmann factors $\Omega(i,j)=\exp-\beta E(i,j)$ will play a prominent role in the statistical mechanics of the model. The matrix $\underline{\underline{\Omega}}$ can be expressed in terms of the $\underline{\underline{M}}_k$ matrices of the MI of $\mathcal{G}$ by $(\underline{\underline{M}}_0=\underline{\underline{I}})$

$$\underline{\underline{\Omega}} = \underline{\underline{M}}_0 + \sum_{k=1}^{s} \omega_k \underline{\underline{M}}_k \ , \quad \omega_k=\exp-\beta E_k . \tag{7}$$

Since $\underline{\underline{\Omega}}$ is symmetric, it has a complete set of orthonormal eigenvectors (corresponding to real eigenvalues), which span $V_M$. This suggests a second definition of field parameters:

Definition 2. Let $V_M$ be decomposed into mutually orthogonal subspaces, each of which is spanned by the eigenvectors of $\underline{\underline{\Omega}}$ belonging to one of its (in general degenerate) eigenvalues. Then the interaction field parameters (with respect to $\underline{\underline{\Omega}}$) are defined as the lengths of the projections of $\underline{F}$ on these subspaces.

For a general permissible group $\mathcal{G}$, the interaction field parameters will change upon taking different $E_k$-values in eq. (7). This is, however, not the case if $\mathcal{G}$ has a P-algebra, since then every symmetric matrix $\underline{\underline{P}}$ invariant with respect to $\mathcal{G}$ commutes with every other such matrix, so that all matrices of the form

$$\underline{\underline{P}} = \alpha_0 \underline{\underline{M}}_0 + \sum_{k=1}^{s} \alpha_k \underline{\underline{M}}_k \tag{8}$$

can be diagonalized simultaneously.

Theorem 1. If the permissible group $\mathcal{G}$ has a P-algebra, then the interaction field parameters defined with respect to a general matrix of the form of eq. (8) coincide with the invariant field parameters. For such a group, one has $c_i=0$ or $1$ in eq. (5), and, conversely, this guarantees that $\mathcal{G}$ has a P-algebra.

Proof. First consider the second part of the theorem. If the $\underline{\underline{D}}_i(g)$ are the matrices of a real irreducible representation $\theta_i$ of $\mathcal{G}$ (consisting of $n_i \times n_i$ matrices), then $a\underline{\underline{I}}_i$, with $\underline{\underline{I}}_i$ the $n_i \times n_i$ identity, commutes with all $\underline{\underline{D}}_i(g)$ trivially. If $\theta_i$ occurs more than once in $\theta_{per}$, this implies that there are also nontrivial matrices $\underline{\underline{N}}$ commuting with all $\underline{\underline{D}}_i(g)$; an example for $\theta_i$ occurring twice in $\theta_{per}$ is shown in eq. (9):

$$\underline{\underline{D}}(g) = \begin{pmatrix} \cdot & & & & \\ & \cdot & & & \\ & & \underline{\underline{D}}_i(g) & & \\ & & & \underline{\underline{D}}_i(g) & \\ & & & & \cdot \\ & & & & & \cdot \end{pmatrix} , \quad \underline{\underline{N}} = \begin{pmatrix} \cdot & & & \\ & \cdot & & \\ & & \begin{array}{|ccc|ccc|} \hline a & & & b & & \\ & \ddots & & & \ddots & \\ & & a & & & b \\ \hline b & & & c & & \\ & \ddots & & & \ddots & \\ & & b & & & c \\ \hline \end{array} & \\ & & & \cdot \\ & & & & \cdot \end{pmatrix} . \quad (9)$$

Now the matrices $\underline{\underline{N}}$ are symmetric in the basis in which the $\underline{\underline{D}}(g)$ have the form of eq. (4), but then they are also symmetric in the original basis, since different bases are connected by an orthogonal transformation. This means that $\underline{\underline{N}}$ is of the form of eq. (8) in the original basis. But clearly, different N's do not all commute, so that $\mathcal{G}$ can not have a P-algebra for the case that any of the $c_i$ is larger than 1. Now let all $c_i$ equal 0 or 1. Real irreducible representations are of three different types ([7]): (i) $\theta_i$ is of the first kind if it is absolutely irreducible; this means that the invariant subspace which carries this representation stays invariant upon extension of the coefficients of the basis vectors from the real to the complex field (c-extension); (ii) $\theta_i$ is of the second kind, if its invariant subspace splits up into two invariant subspaces upon c-extension and if the two absolutely irreducible representations $\theta_{i1}$ and $\theta_{i2}$ carried by these subspaces are equivalent; (iii) $\theta_i$ is of the third kind if $\theta_{i1}$ and $\theta_{i2}$, which exist as in case (ii), are not equivalent. The most general real matrices commuting with all matrices of a real irreducible representation are described by the following. If $\theta_i$ is of the first kind, these are the multiples $a\underline{\underline{I}}_i$ ($a$ real) by Schur's Lemma; this set of matrices is obviously isomorphic to the real number field. If $\theta_i$ is of the third kind, the commuting matrices are isomorphic to the complex number field; the real unit is $\underline{\underline{I}}_i$, whereas the imaginary unit must be represented by an antisymmetric matrix, whose square is $-\underline{\underline{I}}_i$. Finally, if $\theta_i$ is of the second kind, then the commuting matrices are isomorphic to the field of quaternions. The real unit is again $\underline{\underline{I}}_i$, whereas the three imaginary quaternion units are represented by three mutually anticommuting antisymmetric matrices with square $-\underline{\underline{I}}_i$. From this it follows that all symmetric matrices which commute with all $\underline{\underline{D}}_i(g)$ are always given by $a\underline{\underline{I}}_i$, independent of the kind of $\theta_i$. Since $c_i = 0$ or 1, this implies that

$\mathcal{G}$ has a P-algebra. The equality of the two types of field parameters is now clear if the interaction field parameters are defined with respect to a general matrix of the form of eq. (8). (For some special matrix of this type, there could be accidental degeneracies.) ¶

Remark. The trivial one-dimensional irreducible representation $\theta_0$ in which each group element is represented by the number 1 is clearly of the first kind. It is easy to see that $\theta_0$ occurs exactly once in $\theta_{per}$, since $\mathcal{G}$ is transitive [8].

From the proof of the above theorem follow a number of interesting corollaries:

Corollary 1. Let the permissible group $\mathcal{G}$ have a P-algebra. Let $\theta_{per}$ contain $s_1$ irreducible representations of the first kind, $s_2$ of the second kind and $s_3$ of the third kind. Then there are $s_1+s_2+s_3-1$ graphs in its MI and there are $3s_2+s_3$ linearly independent antisymmetric matrices invariant with respect to $\mathcal{G}$; one has $2s_2 \leqq s_1-1$. (Note that $s_1 \geqq 1$ by the remark above.)

Proof. Since every irreducible representation that occurs at all can occur only once, $s_1+s_2+s_3$ is the number of free parameters of a general symmetric matrix commuting with all $\underline{\underline{D}}(g)$, so that a comparison with eq. (8) yields the desired result $s=s_1+s_2+s_3$. Further, every $\theta_i$ of the second kind yields three antisymmetric matrices, whereas every $\theta_i$ of the third kind yields one. The inequality follows from the fact that there can be no more antisymmetric matrices than there are symmetric ones, see Lemma 4.3. ¶

Corollary 2. Let the permissible group $\mathcal{G}$ have a P-algebra. Then $\mathcal{G}$ is completely permissible iff all $\theta_i$ occurring in $\theta_{per}$ are of the first kind. In particular, if $\mathcal{G}'$ is an abstract group such that all its absolutely irreducible representations are real, then a permutation representation of $\mathcal{G}'$ that is permissible is either completely permissible or it has no P-algebra. The symmetric groups $\mathcal{S}(M)$ are, as abstract groups, in this class [9].

Proof. The first assertion follows immediately from Corollary 1 and the definition of a completely permissible group. The second assertion is then obvious. ¶

In the proofs of Theorem 1 and its corollaries, no direct use is made of the permissibility of the groups. That this property does play a not unimportant role is shown by the next lemmas.

Lemma 1. Let $\mathcal{G}$ be a permissible group with a P-algebra. Suppose that a linear combination of the (symmetric, invariant) $\underline{\underline{M}}$-matrices and of the (antisymmetric, invariant) $\underline{\underline{S}}$-matrices for this group is a permutation matrix $\underline{\underline{D}}(g')$. Then $g'\varepsilon G$ and $g'(1)=k\neq 1$. If $g'$ is an involution, i.e., if $g'^2=e$ holds, then $\underline{\underline{M}}_k=\underline{\underline{D}}(g')$. If $g'$ is not an involution, then one has

$$\underline{\underline{M}}_k = \underline{\underline{D}}(g') + \underline{\underline{D}}(g'^{-1}) \;; \qquad \underline{\underline{S}}_k = \underline{\underline{D}}(g'^{-1}) - \underline{\underline{D}}(g'). \tag{10}$$

Proof. It follows immediately from the proof of Theorem 1, that every antisymmetric matrix $\underline{\underline{S}}$ which is invariant with respect to $G$ commutes with all $\underline{\underline{M}}_k$-matrices. Therefore, $\underline{\underline{D}}(g')$ commutes with all $\underline{\underline{M}}_k$. But then $g'\varepsilon G$ since $G$ is permissible. Since $g'$ also commutes with all other elements of $G$, it must map 1 onto a value $k\neq 1$. Now consider the matrix $\underline{\underline{A}}_k$ defined by

$$A_k(i,j) = \begin{cases} 1 \text{ if there is a } g\varepsilon G \text{ with } g(1)=i,\ g(k)=j, \\ \\ 0 \text{ otherwise.} \end{cases} \tag{11}$$

One has $\underline{\underline{M}}_k=\underline{\underline{A}}_k+\underline{\underline{A}}_k^T$ (transposed matrix: $A^T(i,j)=A(j,i)$) and $\underline{\underline{S}}_k=\underline{\underline{A}}_k-\underline{\underline{A}}_k^T$. But since $gg'=g'g$ for all $g\varepsilon G$, $\underline{\underline{A}}_k^T=\underline{\underline{D}}(g')$ and $\underline{\underline{A}}_k=\underline{\underline{D}}(g'^{-1})$, from which the statements of the lemma follow. ¶

The set of all elements $z\varepsilon G$ which commute with all $g\varepsilon G$, $zg=gz$, form a subgroup of $G$, called the center $Z(G)$ of $G$.

Lemma 2. If a permissible group $G$ has a center $Z(G)\neq\{e\}$, then $H_1$ keeps also all letters $z(1)$ with $z\varepsilon Z(G)$, $z\neq e$, fixed. Conversely, if $G$ is a permissible group with a P-algebra and if $H_1$ keeps the letters $k_1,..,k_t$ fixed $(k_i\neq 1)$, then $G$ has a center with $|Z(G)|=t+1$, so that for every $k_i$, there is a unique $z_i$ with $z_i(1)=k_i$.
Proof. Let $z\varepsilon Z(G)$; then $z\neq e$ cannot keep the letter 1 fixed, for if it did, it would belong not only to $H_1$, but also to all $H_k$, see Lemma 2.4. Therefore, set $z(1)=k\neq 1$; then one has $h(k)=hz(1)=zh(1)=k$ for all $h\varepsilon H_1$, so that $H_1$ leaves all letters fixed which occur as images of 1 by elements of the center of $G$. The number of fixed letters is then $|Z(G)|$.
Now let all elements of $H_1$ keep the letter $k$ fixed. If $g_k$ is an element of $G$ with $g_k(1)=k$, this implies $H_k=g_kH_1g_k^{-1}=H_1$. The permutation $\sigma_k$ given as

$$\sigma_k(j) = g_k^{-1}g_jg_k(1) \tag{12}$$

is now well-defined since replacement of $g_j$ by $g_jh$ yields

$$g_k^{-1}g_jhg_k(1) = g_k^{-1}g_jg_kh'(1) = \sigma_k(j). \tag{13}$$

The following formulae show that $g_k\sigma_k$ commutes with all $g\varepsilon G$:

$$g_k\sigma_kg_s(t) = g_k\sigma_kg_sg_t(1) = g_kg_k^{-1}g_sg_tg_k(1) = g_sg_tg_k(1), \tag{14a}$$

$$g_sg_k\sigma_k(t) = g_sg_k\sigma_kg_t(1) = g_sg_tg_k(1). \tag{14b}$$

Therefore, $g_k\sigma_k$ is a linear combination of $\underline{\underline{M}}$- and $\underline{\underline{S}}$-matrices. Now if $G$ has a P-algebra, then $g_k\sigma_k\varepsilon Z(G)$ by Lemma 1, so that $\sigma_k\varepsilon H_1$. It follows that $z_k=g_k\sigma_k$, since this maps 1 onto k. It easily follows that this is the unique element of $Z(G)$ which has this property. ¶

Lemma 3. The center of a completely permissible group consists of involutions only. If a permissible group has a P-algebra and a center which does not consist of involutions only, then there are no real irreducible representations of the second kind in $\theta_{per}$.

Proof. By Lemma 1, a permissible group $G$ with a P-algebra and a center element $z$ which is not an involut ion, admits an antisymmetric matrix $\underline{\underline{S}}_z=\underline{\underline{D}}(z)-\underline{\underline{D}}(z^{-1})$ invariant with respect to the group. Since a completely permissible group has a P-algebra and admits no such antisymmetric matrices, the first part of the lemma follows. To prove the second part, eq. (11) is generalized:

$$A_k(i,j) = \begin{cases} 1 \text{ if } g_i^{-1}g_j\varepsilon H_1g_kH_1, \\ 0 \text{ otherwise.} \end{cases} \tag{15}$$

Since one always has $\underline{\underline{M}}_k=\underline{\underline{A}}_k+\underline{\underline{A}}_k^T$ and $\underline{\underline{S}}_k=\underline{\underline{A}}_k-\underline{\underline{A}}_k^T$, the existence of a P-algebra implies

$$[\underline{\underline{M}}_k,\underline{\underline{M}}_\ell]=0,\ [\underline{\underline{M}}_k,\underline{\underline{S}}_\ell]=0,\ [\underline{\underline{M}}_\ell,\underline{\underline{S}}_k]=0, \tag{16}$$

where the commutator is $[\underline{\underline{A}},\underline{\underline{B}}]=\underline{\underline{AB}}-\underline{\underline{BA}}$. For $k=\ell$, this implies

$$\underline{\underline{A}}_k\underline{\underline{A}}_k^T= \underline{\underline{A}}_k^T\underline{\underline{A}}_k, \tag{17}$$

i.e., all $\underline{\underline{A}}_k$ are normal matrices. For $k\neq\ell$, one finds

$$[\underline{\underline{A}}_k,\underline{\underline{A}}_\ell]=-[\underline{\underline{A}}_k{}^T,\underline{\underline{A}}_\ell]=-[\underline{\underline{A}}_k,\underline{\underline{A}}_\ell{}^T]=[\underline{\underline{A}}_k{}^T,\underline{\underline{A}}_\ell{}^T]. \tag{18}$$

Now let $z\varepsilon\mathcal{Z}(\mathcal{G})$; then the product $\underline{\underline{A}}_k\underline{\underline{D}}(z)$ is also of the type of eq. (15), since one has

$$\{\underline{\underline{A}}_k\underline{\underline{D}}(z)\}(i,j) = A_k(i,z(j)) = \begin{cases} 1 \text{ if } & g_i{}^{-1}g_j\varepsilon\mathcal{H}_1 g_k z^{-1}\mathcal{H}_1, \\ 0 \text{ otherwise.} \end{cases} \tag{19}$$

Since the $\underline{\underline{D}}(z)$ matrices commute with all $\underline{\underline{A}}_k$, the corresponding symmetric matrix is

$$\underline{\underline{A}}_k\underline{\underline{D}}(z) + \underline{\underline{A}}_k{}^T\underline{\underline{D}}(z)^{-1}. \tag{20}$$

The commutator of such a symmetric matrix with any $\underline{\underline{A}}_\ell$ must vanish:

$$0=[\underline{\underline{A}}_k\underline{\underline{D}}(z)+\underline{\underline{A}}_k{}^T\underline{\underline{D}}(z)^{-1},\underline{\underline{A}}_\ell]=[\underline{\underline{A}}_k,\underline{\underline{A}}_\ell]\{\underline{\underline{D}}(z)-\underline{\underline{D}}(z^{-1})\}, \tag{21}$$

where eq. (18) has been used. Now if $z$ is not an involution (or the identity), this implies $[\underline{\underline{A}}_k,\underline{\underline{A}}_\ell]=0$ for all $k,\ell$, so that all antisymmetric matrices also commute. However, if $\theta_{per}$ contains an irreducible representation of the second kind, then there exist anticommuting antisymmetric invariant matrices, see the proof of Theorem 1. This contradiction proves the second part of the lemma. ¶

Finally, a simple lemma concerning nonpermissible groups is proved:

Lemma 4. Let $\mathcal{G}$ be a permissible group, $\mathcal{G}_1$ a transitive subgroup of $\mathcal{G}$ so that $\mathcal{G}_1{}^{(p)}=\mathcal{G}$. Then the number of antisymmetric (symmetric) matrices commuting with all $\underline{\underline{D}}(g_1)$, $g_1\varepsilon\mathcal{G}_1$ is larger or equal (equal) to the number of antisymmetric (symmetric) matrices commuting with all $\underline{\underline{D}}(g)$, $g\varepsilon\mathcal{G}$. If $\mathcal{G}$ has a P-algebra, then the $\underline{\underline{M}}_k$ matrices of its MI commute with all $\underline{\underline{S}}_k$ matrices of $\mathcal{G}_1$.
Proof. Since the MI's of $\mathcal{G}$ and $\mathcal{G}_1$ are equal, the sets of symmetric matrices mentioned in the lemma are equal too. It is, however, not impossible that $\mathcal{G}_1$ contains less elements $\sigma_k$ with $\sigma_k(1)=k,\sigma_k(k)=1$, than does $\mathcal{G}$, from which the statement concerning the antisymmetric matrices follows. The last part of the lemma follows as in the proof of Theorem 1. It is remarked, that this must hold for all $\mathcal{G}_1<\mathcal{G}$ such that $\mathcal{G}_1{}^{(p)}=\mathcal{G}$ holds, which may be a much stronger requirement than the corresponding one for $\mathcal{G}$ alone. ¶

REFERENCES.

(1). K. Wagner, Graphentheorie (Bibliographisches Institut, Mannheim, 1970) p. 9.
(2). Ref. (1), p. 118.
(3). R.B. Potts, Proc. Camb. Phil. Soc. 48 (1952) 106.
(4). B. Huppert, Endliche Gruppen I (Springer-Verlag, Berlin, Heidelberg, New York, 1963) p. 145 ff.
(5). Ref. (4), p. 11.
(6). See, e.g., J.S. Lomont, Applications of Finite Groups (Academic Press, New York, London, 1959) p. 46 ff., or Ref. (4), p. 456 ff.
(7). J.P. Serre, Lineare Darstellung endlicher Gruppen (Vieweg, Braunschweig, 1972) p. 68 ff.
(8). Ref. (4), p. 597.
(9). Ref. (6), p. 261.

GENERAL REFERENCES.

Much of the material in the first six chapters of these lectures first appeared in a series of papers :

H. Moraal, Physica 113A (1982) 44, 67; 117A (1983) 109.

Independently, a number of similar results were obtained by :

M. Marcu, A. Regev and V. Rittenberg, J. Math. Phys. 22 (1981) 2740.
M. Marcu and V. Rittenberg, J. Math. Phys. 22 (1981) 2753.
M. Marcu, Ph.D. Dissertation, Bonn University (1981, unpublished).

## 2. SUBGROUPS AND PRODUCTS OF PERMISSIBLE GROUPS.

### 2.1. Permissible subgroups.

If $\mathcal{G}$ is a permissible group on M letters, then a subgroup $\mathcal{R}$ of $\mathcal{G}$ must certainly be transitive on M letters if it is also to be permissible. The general situation is described by the following Lemma 1:

Lemma 1. Let $\mathcal{G}$ be permissible, $\mathcal{R}<\mathcal{G}$ a transitive proper subgroup. Then (a) the permissible group $\mathcal{R}^{(p)}$ corresponding to $\mathcal{R}$ is also a subgroup of $\mathcal{G}$, $\mathcal{R}^{(p)} \leqq \mathcal{G}$, and (b) the MI of $\mathcal{G}$ consists of graphs which are sums of one or more graphs from the MI of $\mathcal{R}^{(p)}$. Further, complete permissibility of $\mathcal{R}^{(p)}$ implies the same property of $\mathcal{G}$.

Proof. Statement (b) is trivial. Let g be a permutation from $\mathcal{S}(M)$ which keeps all graphs of the MI of $\mathcal{R}$ invariant, but which does not belong to $\mathcal{R}$. Then by (b), g keeps all graphs of the MI of $\mathcal{G}$ invariant, but since $\mathcal{G}$ is permissible, $g \varepsilon \mathcal{G}$ follows. This implies (a). The statement concerning complete permissibility is trivial. ¶

In group theory, the normal subgroups play a special role; these are defined by the requirement that $g^{-1}\mathcal{R}g=\mathcal{R}$ holds for all $g \varepsilon \mathcal{G}$. In order to study the situation of Lemma 1 for this special case, a number of extra lemmas is needed. These are also of independent interest, since they throw some light on the nature of permissible groups.

Lemma 2. Let $\mathcal{G}$ be permissible; the group $\mathcal{A}(\mathcal{G})$ is defined as the set of all permutations, which keep the MI of $\mathcal{G}$ invariant, i.e., which map any graph from the MI of $\mathcal{G}$ onto the same or another (isomorphic) graph from this MI. Then $\mathcal{A}(\mathcal{G})=\mathcal{N}_M(\mathcal{G})$, defined as the group of all $g \varepsilon \mathcal{S}(M)$ such that $g^{-1}\mathcal{G}g=\mathcal{G}$ holds. This latter group is called the normalizer of $\mathcal{G}$ in $\mathcal{S}(M)$. For a nonpermissible, transitive group $\mathcal{G}$ one has

$$\mathcal{N}_M(\mathcal{G}) \leqq \mathcal{A}(\mathcal{G}) = \mathcal{A}(\mathcal{G}^{(p)}) = \mathcal{N}_M(\mathcal{G}^{(p)}). \tag{1}$$

Proof. For the proof of this lemma, it is expedient to denote the set of undirected edges which make up the graph $G_k$ by the notation $\{g(1),g(k)\}$, where $G_k$ belongs to the MI of $\mathcal{G}$ and is obtained by operating on the edge $\{1,k\}$ with the group elements $g \varepsilon \mathcal{G}$. The fact that $\sigma \varepsilon \mathcal{A}(\mathcal{G})$ can then be written as

$$\sigma\{g(1),g(k)\} = \{g(1),g(k')\}, \tag{2}$$

where $k=k'$ is not excluded. Eq. (2) implies in particular that there is an element $g'\varepsilon G$ such that the (undirected) edges equality

$$\sigma\{1,k\} = g'\{1,k'\} \tag{3}$$

holds. This implies

$$\{g\sigma(1),g\sigma(k)\} = \{gg'(1),gg'(k')\} = \{g(1),g(k')\}. \tag{4}$$

Combination of eqs. (2) and (4) yields

$$\{g(1),g(k)\} = \{\sigma^{-1}g\sigma(1),\sigma^{-1}g\sigma(k)\} . \tag{5}$$

Since this holds for all $k$, the permutation $\sigma^{-1}g\sigma$ leaves all graphs of the MI of $G$ invariant. If $G$ is permissible, this implies $\sigma^{-1}g\sigma\varepsilon G$ for all $g$ or $\sigma^{-1}G\sigma=G$, so that $\sigma\varepsilon N_M(G)$. This shows that $A(G)\leqq N_M(G)$ holds if $G$ is permissible.

Now let $G$ be arbitrary transitive and $\sigma\varepsilon N_M(G)$. Then eq.(5) holds; since $G$ is transitive, so is $N_M(G)$ and by Lemma 1.2.4 one can assume $\sigma(1)=1$. Setting $\sigma(k)=k'$, this gives

$$\sigma G_k= \{\sigma g(1),\sigma g(k)\} = \{g\sigma(1),g\sigma(k)\} = \{g(1),g(k')\} =G_{k'}, \tag{6}$$

so that $\sigma\varepsilon A(G)$ follows. Therefore, $N_M(G)\leqq A(G)$ holds for an arbitrary transitive group $G$. Now if $G$ is permissible, this, together with the first part of the proof shows that $N_M(G)=A(G)$ for this case. Eq. (1) then follows immediately. ¶

As already noted in the proof above, $N_M(G)$ and $A(G)$ can, by Lemma 1.2.4, be replaced by their letter-1-fixing subgroups, to be denoted by $H_M(G)$ and $A_1(G)$, respectively. These groups are closely connected with certain other groups associated with $G$ which are defined as follows. Two groups $G_1$ and $G_2$ are called <u>isomorphic</u> (notation: $G_1\simeq G_2$) if there is a one-to-one onto mapping $\pi:G_1\to G_2$, which conserves the group product: $\pi(g_1g_2)=\pi(g_1)\pi(g_2)$. The <u>automorphism group</u> $\mathrm{AUT}(G)$ of a group $G$ consists of all isomorphisms of $G$ with itself, i.e., $\alpha\varepsilon\mathrm{AUT}(G)$ means that $\alpha$ is a one-to-one mapping of $G$ onto itself, which satisfies $\alpha(g_1g_2)=\alpha(g_1)\alpha(g_2)$. For transitive permutation groups, the subgroup $\mathrm{AUT}_1(G)$ of the automorphism group is defined by the extra requirement that these automorphisms keep the letter-1-fixing subgroup $H_1$ of $G$ invariant:

$$\alpha\varepsilon\mathrm{AUT}_1(G) \text{ iff (i) } \alpha\varepsilon\mathrm{AUT}(G) \text{ and (ii) } \alpha(h)\varepsilon H_1 \text{ for all } h\varepsilon H_1. \tag{7}$$

The next lemma shows an interesting connection between $\mathrm{AUT}_1(G)$ and $H_M(G)$:

Lemma 3. $\mathrm{AUT}_1(G)$ and $H_M(G)$ are isomorphic groups; for $G$ permissible, this implies $H_M(G)=A_1(G)\simeq\mathrm{AUT}_1(G)$.

Proof. Let $\alpha\varepsilon\mathrm{AUT}_1(G)$; if $g_k h$ is an element of the coset $g_k H_1$ of $G$ which maps $1$ onto the letter $k$, then one has $\alpha(g_k h)=\alpha(g_k)\alpha(h)$. Here $\alpha(h)\varepsilon H_1$ and $\alpha(g_k)$ will be an element of the coset $g_{k'}H_1$ which maps $1$ onto the letter $k'$. Since $\alpha$ is an automorphism, this implies $\alpha(g_k H_1)=g_{k'}H_1$, so that $\alpha$ maps cosets onto cosets. Further, $\alpha(H_1)=H_1$ by definition, so that each $\alpha\varepsilon\mathrm{AUT}_1(G)$ defines a permutation $\sigma$ by

$$\sigma(1)=1,\ \sigma(k)=k' \text{ if } \alpha(g_k H_1)=g_{k'}H_1. \tag{8}$$

To prove that this $\sigma$ is given by $\alpha$ unambiguously, assume that $\alpha'$ leads to the same $\sigma$. By eq. (8) this implies that $\alpha$ and $\alpha'$ are related by $\alpha'(g)=\alpha(g)h(g)$ with $h(g)\varepsilon H_1$. But since $\mathrm{AUT}_1(G)$ is a group, the product $\beta=\alpha^{-1}\alpha'$ is also an element of this group; this satisfies $\beta(g)=g\ h'(g)$ with $h'(g)=\alpha^{-1}(h(g))\varepsilon H_1$. But such a mapping cannot be an automorphism:

$$\beta(g_1g_2) = g_1g_2\ h'(g_1g_2) = \beta(g_1)\beta(g_2) = g_1\ h'(g_1)\ g_2\ h'(g_2) \tag{9}$$

implies that

$$g_2^{-1}h'(g_1)g_2 = h'(g_1g_2)\ h'(g_2)^{-1}\ \varepsilon H_1 \text{ for all } g_1,g_2\varepsilon G, \tag{10}$$

so that $h'(g_1)=e$ must hold and $\alpha=\alpha'$ follows. Eq. (8), therefore, associates a different $\sigma$ with every $\alpha\varepsilon\mathrm{AUT}_1(G)$.

Since $\sigma$ leaves the letter $1$ fixed, one has by eq. (8)

$$\sigma g\sigma^{-1}(1) = \sigma g(1) = \alpha(g)(1), \tag{11}$$

so that $\sigma g\sigma^{-1}$ differs from $\alpha(g)$ by a factor that is at most a letter-1-fixing element of $S(M)$:

$$\sigma g\sigma^{-1} = \alpha(g)\ h_\alpha(g),\quad h_\alpha(g)(1)=1 \text{ for all } g\varepsilon G. \tag{12}$$

One can now calculate $\sigma g_1g_2\sigma^{-1}$ in two ways again:

$$\sigma g_1g_2\sigma^{-1}= \sigma g_1\sigma^{-1}\sigma g_2\sigma^{-1}= \alpha(g_1)\ h_\alpha(g_1)\ \alpha(g_2)\ h_\alpha(g_2)\ , \tag{13a}$$

$$\sigma g_1 g_2 \sigma^{-1} = \alpha(g_1 g_2)\ h_\alpha(g_1 g_2) = \alpha(g_1)\ \alpha(g_2)\ h_\alpha(g_1 g_2). \tag{13b}$$

Equality of eqs. (13a) and (13b) can only be obtained for $h_\alpha(g)=e$, so that one has $\sigma g \sigma^{-1}=\alpha(g)$ and $\sigma \varepsilon H_M(G)$. On the other hand, it is clear that if $\sigma \varepsilon H_M(G)$ then $\alpha(g)$ defined by $\sigma g \sigma^{-1}$ is an automorphism of $\mathrm{AUT}_1(G)$. This establishes the isomorphism $\mathrm{AUT}_1(G) \simeq H_M(G)$. The second assertion of the lemma then follows from Lemma 2. ¶

The above Lemma 2 now allows for a formulation of the normal subgroup problem as

<u>Lemma 4</u>. Let $G$ be a permissible group, $R$ a proper, transitive, normal subgroup. Then $R^{(p)}$ is normal in $G$, $R$ is normal in $R^{(p)}$ and the MI of $G$ consists of graphs which are sums of one or more isomorphic graphs from the MI of $R^{(p)}$.

<u>Proof</u>. Since $R$ is a subgroup of $G$, $R^{(p)}$ is also a subgroup by Lemma 1. This implies that $R$ is normal in $R^{(p)}$ since $R$ is normal in all of $G$. This latter fact also ensures that $G$ is a subgroup of $N_M(G)$. By Lemma 2 one has

$$R^{(p)} \leqq G \leqq N_M(R) \leqq N_M(R^{(p)}) = A(R^{(p)}), \tag{14}$$

which implies that $R^{(p)}$ is normal in $G$. If $R^{(p)} \neq G$, the elements of $G$ not in $R^{(p)}$ belong to $A(R^{(p)})$, from which the statement concerning the MI's follows. ¶

The above Lemmas 1 and 4 can be useful to obtain new permissible groups from known ones if these new groups are maximal in the given group:

<u>Lemma 5</u>. Let $G$ be permissible, $R$ a maximal transitive subgroup, i.e., there is no subgroup $T$ with $R<T<G$; then either $R$ is permissible or $R^{(p)}=G$.

<u>Proof</u>. By Lemma 1, $R^{(p)}$ is also a subgroup and $R \leqq R^{(p)} \leqq G$ holds. The maximality of $R$ implies the lemma. ¶

<u>Lemma 6</u>. Let $G$ be permissible, $R$ a maximal normal transitive subgroup, i.e., there is no normal subgroup in between $R$ and $G$. Then either $R$ is permissible or $R^{(p)} = G$.

<u>Proof</u>. By Lemma 4, $R^{(p)}$ is normal in $G$, from which the result follows immediately. ¶

The above Lemmas 5 and 6 are most useful in the form of the next corollary:

Corollary 1. Let $G$ be permissible, $R$ a transitive subgroup of index a prime $p$, i.e., $|G|/|R|=p$. If the MI of $R$ is different from the MI of $G$, then $R$ is permissible.

Proof. A subgroup of index a prime $p$ is clearly maximal (for $p=2$ it is even maximal normal ([1])). The condition on the MI's excludes the possibility $R^{(p)}=G$ in Lemmas 5 and 6, so that $R$ is necessarily permissible. ¶

## 2.2. Wreath products.

In this section, a very useful device for the construction of new permissible groups, the wreath product of two groups, is described. For a more general definition, see ([2]). Here, all that is needed is the wreath product of two transitive permutation groups:

Definition 1. Let $G_1$ and $G_2$ be transitive permutation groups on $M_1$ and $M_2$ letters, respectively. Denoting the sets of letters by $S_1$ and $S_2$, the wreath product $G_1 \sim G_2$ is defined as a set of permutations on $S_3=S_1\times S_2$:

$$S_3=\{j_3=(j_2-1)M_1+j_1 \mid j_1\varepsilon S_1, j_2\varepsilon S_2\} \tag{1}$$

by the following prescription: an element $\tilde{g}\varepsilon G_1\sim G_2$ is determined by an arbitrary element $g^{(2)}\varepsilon G_2$ and by $M_2$ arbitrary elements $g^{(1)}{}_k\varepsilon G_1$ $(k=1,2,..,M_2)$:

$$\tilde{g}=\{g^{(2)};g^{(1)}{}_1,g^{(1)}{}_2,\dots,g^{(1)}{}_{M_2}\}; \tag{2}$$

the action of $\tilde{g}$ on an element $j_3\varepsilon S_3$ is given as

$$\tilde{g}(j_3)=j_3';\ j_2'=g^{(2)}(j_2);\ j_1'=g^{(1)}{}_{j_2'}(j_1); \tag{3}$$

here $j_3$ and $j_3'$ are related to $j_1,j_2$ and to $j_1',j_2'$ as in eq. (1).

From this definition immediately follow a number of properties of the wreath product which are collected in the next lemma:

Lemma 1. The wreath product of two transitive permutation groups is a transitive permutation group on $S_3$. The number of elements is given as

$$|G_1\sim G_2| = |G_1|^{M_2}\,|G_2|. \tag{4}$$

The wreath product is associative:

$$(G_1 \sim G_2) \sim G_3 = G_1 \sim (G_2 \sim G_3) = G_1 \sim G_2 \sim G_3 . \tag{5}$$

Proof. That the set of permutations defined above forms a group is clear from the product rule, which follws from the definition as

$$\tilde{h}\tilde{g} = \{h^{(2)}g^{(2)};\ h^{(1)}{}_1 g^{(1)}{}_{a(1)}, \ldots, h^{(1)}{}_{M_2} g^{(1)}{}_{a(M_2)}\}\ ,\ a=h^{(2)^{-1}} . \tag{6}$$

The transitivity of $G_1 \sim G_2$ is then obvious, as is eq. (4) for the number of elements of this group. The associativity, eq. (5), follows from a straightforward calculation. ¶

The usefulness of the wreath product is due to the simplicity of its MI, as shown in the proof of the next lemma:

Lemma 2. Let $G_i$ (i=1,2), be transitive permutation groups on $S_i$ with MI's consisting of $s(G_i)$ graphs $G_{ik}$, $k=1,..,s(G_i)$. Then the MI of $G_1 \sim G_2$ consists of

$$s(G_1 \sim G_2) = s(G_1) + s(G_2) \tag{7}$$

graphs as shown below. The corresponding permissible groups satisfy

$$(G_1 \sim G_2)^{(p)} = G_1{}^{(p)} \sim G_2{}^{(p)} , \tag{8}$$

so that, in particular, the wreath product of two permissible groups is again permissible. It is completely permissible iff both factors are completely permissible.

Proof. The structure of the MI of $G_1 \sim G_2$ is easily seen to be the following: (i) for every graph $G_{1k}$ from the MI of $G_1$, there is a graph in the MI of the wreath product which consists of $M_2$ disconnected copies of $G_{1k}$. The automorphism group of such a graph is $G(G_{1k}) \sim S(M_2)$. (ii) For every graph $G_{2k}$ from the MI of $G_2$, there is a graph in the MI of the wreath product which can be constructed as follows: replace every one of the $M_2$ vertices of $G_{2k}$ by a set of $M_1$ vertices and connect all pairs of different such sets by edges iff the original vertices are connected in $G_{2k}$. The automorphism group of a graph of this type is $S(M_1) \sim G(G_{2k})$. Now the intersection of all automorphism groups is clearly $G_1{}^{(p)} \sim G_2{}^{(p)}$. The statement concerning complete permissibility is obvious.¶

As examples, the MI's of the (completely permissible) groups $S(2) \sim S(4)$

and $\mathcal{S}(4)\sim\mathcal{S}(2)$ are shown in Fig. 1 below.

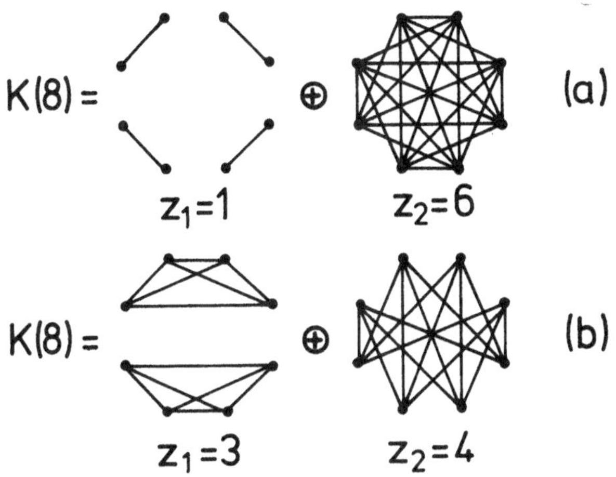


Fig. 1. The maximal interactions of (a) $\mathcal{S}(2)\sim\mathcal{S}(4)$ and of (b) $\mathcal{S}(4)\sim\mathcal{S}(2)$.

As an application of the wreath product concept, a lemma concerning the types of graphs, which can occur in the MI of a permissible group, is proved:

<u>Lemma 3</u>. Let $\mathcal{G}$ be permissible with MI

$$K(M) = G_1 + G_2 + \ldots + G_s, \quad s \geqq 2. \tag{9}$$

Then <u>either</u> (i) all $G_k$ satisfy $(\overline{G}_k)^{(2)} = G_k$ <u>or</u> (ii) one of the graphs (say $G_1$) satisfies $(\overline{G}_1)^{(\infty)} = G_1$, whereas the others still satisfy the condition (i). In case (ii), $\mathcal{G}$ is a wreath product, $\mathcal{G} = \mathcal{G}_1 \sim \mathcal{S}(M_2)$ with $\mathcal{G}_1$ permissible on $M_1$ letters, $M = M_1 M_2$.

<u>Proof</u>. Every regular graph $G_k$ satisfies the identity

$$G_k = (\overline{G}_k)^{(2)} + (\overline{G}_k)^{(3)} + \ldots + (\overline{G}_k)^{(\infty)}. \tag{10}$$

Since $\overline{G}_k$ has the same automorphism group as $G_k$, each term on the right-hand-side of eq. (10) is invariant with respect to the transformations of $\mathcal{G}(G_k)$ and then also with respect to those of $\mathcal{G}$. But since $G_k$ is from the MI of $\mathcal{G}$, the sum of eq. (10) can only have one term, so that every graph from the MI of a permissible group satisfies either $G_k = (\overline{G}_k)^{(2)}$ or $G_k = (\overline{G}_k)^{(\infty)}$. In the latter case, $G_k$ is the complement of $M_{k_2}$

complete $K(M_{k_1})$ graphs with $M=M_{k_1}M_{k_2}$, so that it has

$$z(G_k) = M-M_{k_1} \tag{11}$$

edges adjacent to each vertex. Now assume that there are $q$ such graphs (say the first $q$) in eq. (9). One then has

$$M-1 = \sum_{k=1}^{s} z(G_k) \geqq \sum_{k=1}^{q} (M-M_{k_1}) = qM - \sum_{k=1}^{q} M_{k_1}. \tag{12}$$

But since $M_{k_1}$ divides $M$, $M_{k_1} \leqq M/2$ certainly holds, so that eq. (12) gives the inequality

$$M-1 \geqq qM/2. \tag{13}$$

This shows that $q=0$ (case (i) of the lemma) or $q=1$ (case (ii)) are the only possibilities. In the latter case, $G_1$ is the complement of $M_2$ complete $K(M_1)$ graphs, so that the other graphs consist of $M_2$ graphs each on $M_1$ letters. Therefore, the group $\mathcal{G}$ is of the form $\mathcal{G}_1 \sim \mathcal{S}(M_2)$ with $\mathcal{G}_1$ permissible on $M_1$ letters by Lemma 2. ¶

Repeated application of Lemma 3 yields the next result, the proof of which is trivial:

Lemma 4. A permissible group is called of type A if it corresponds to case (i) of Lemma 3. An arbitrary permissible group on $M$ letters is a repeated wreath product $\mathcal{G}_1 \sim \mathcal{S}(M_2) \sim \ldots \sim \mathcal{S}(M_n)$ with $\mathcal{G}_1$ of type A on $M_1$ letters, $M=M_1M_2\ldots M_n$.

There is a slight restriction on the graphs of groups of type A, which is sometimes useful:

Lemma 5. If $G_k=(\overline{G}_k)^{(2)}$ holds, then $z_k$, the number of edges emanating from every vertex of $G_k$, satisfies $z_k \leqq M-1-\sqrt{(M-1)}$.

Proof. Since $\overline{G}_k$ must be connected, $z_k$ must be less than the number of edges of $\overline{G}_k$, which are not connected to a special vertex:

$$z_k \leqq (M-z_k-1)(M-z_k-2). \tag{14}$$

Solution of eq. (14) yields the lemma. ¶

An application of this lemma is made in Theorem 4.3.4.

## 2.3. Direct products.

A further useful product of groups is the direct product ([3]) of two transitive permutation groups, defined here in a somewhat special way:

Definition 1. Let $G_1$ and $G_2$ be transitive permutation groups on $S_1$ and $S_2$, respectively. The direct product $G_1 \otimes G_2$ on $S_3$ (defined as in eq. (2.2.1)) is a subgroup of elements $\bar{g}$ from $G_1 \sim G_2$:

$$\bar{g} = \{g^{(2)};\ g^{(1)}{}_k = g^{(1)} \text{ for } k=1,2,..,M_2\} \equiv \{g^{(2)};g^{(1)}\}. \tag{1}$$

It follows that $\bar{g}$ operates on an element $j_3 \varepsilon S_3$ as

$$\bar{g}(j_3) = j_3' \ ;\ \ j_2' = g^{(2)}(j_2) \ ;\ \ j_1' = g^{(1)}(j_1). \tag{2}$$

The product rule in $G_1 \otimes G_2$ is then simply

$$\bar{h}\,\bar{g} = \{h^{(2)}g^{(2)};\ h^{(1)}g^{(1)}\}. \tag{3}$$

Lemma 1. Let $G_1$ and $G_2$ be transitive permutation groups on $S_1$ and $S_2$, respectively. Then $G_1 \otimes G_2$ is transitive on $S_3$ and one has

$$G_1 \otimes G_2 \leqq (G_1 \otimes G_2)^{(p)} \leqq G_1{}^{(p)} \otimes G_2{}^{(p)}. \tag{4}$$

In particular, the direct product of two permissible groups is again permissible.

Proof. The subgroups $\{g^{(2)};e\}$ and $\{e;g^{(1)}\}$ of $G_1 \otimes G_2$ are isomorphic to $G_2$ and $G_1$, respectively. This makes sure that the MI of $G_1 \otimes G_2$ contains the following graphs: for any graph $G_{1k}$ from the MI of $G_1$, there is a graph consisting of $M_2$ disconnected copies of $G_{1k}$, whereas for any graph $G_{2k}$ from the MI of $G_2$, there is a graph consisting of $M_1$ disconnected copies of $G_{2k}$. Although the remaining graphs are not so easily characterized, this implies already, that the permissible group corresponding to $G_1 \otimes G_2$ must be a subgroup of the intersection of $G_1{}^{(p)} \sim S(M_2)$ and of $G_2{}^{(p)} \sim S(M_1)$ as shown in the proof of Lemma 2.2. But this intersection is the direct product of the permissible groups corresponding to $G_1$ and $G_2$, so that eq. (4) follows. ¶

As an example, Fig. 1 shows the MI of the (completely permissible, see below) group $S(2) \otimes S(4)$, compare also Fig. 2.1. It is not difficult to see that $S(2) \otimes S(4)$ is actually the group of the connected graph in this figure; this connected graph is the three-dimensional cube. It is also interesting to note that the other graphs in Fig. 1 are the distance graphs $G^{(2)}$ and $G^{(3)}$ of this cube $G$ as defined in the proof of Theorem 1.3.2.

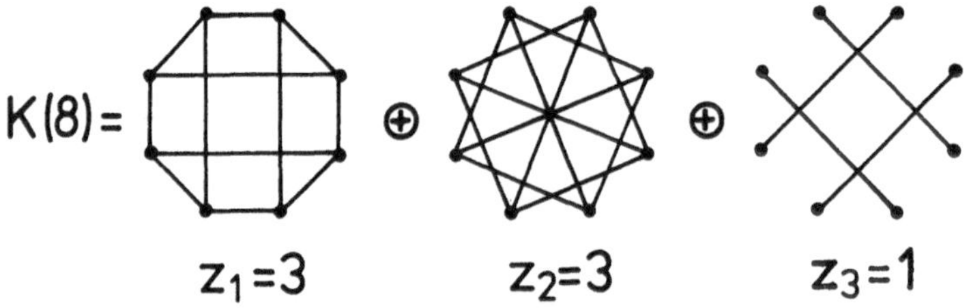


Fig. 1. The MI of the group $S(2)\otimes S(4)$ of the three-dimensional cube.

In the following, the number of graphs of $(G_1\otimes G_2)^{(p)}$ is considered in more detail. To this end, it is assumed that $G_1$ and $G_2$ have $s_1$ and $s_2$ graphs in their MI's and that they have $t_1$ and $t_2$ self-conjugate transitivity sets, respectively. Then $G_i$ (i=1,2) has, by Lemma 1.4.2, $2s_i-t_i+1$ different double cosets. Therefore, $G_1\otimes G_2$ has

$$(2s_1-t_1+1)(2s_2-t_2+1)$$

double cosets. Of these, $(t_1+1)(t_2+1)$ are self-conjugate (including $H_1$). Therefore, the number of graphs in the MI of $G_1\otimes G_2$ is given as

$$s(G_1\otimes G_2) = (t_1+1)(t_2+1) -1 + \tfrac{1}{2}\{(2s_1-t_1+1)(2s_2-t_2+1)-(t_1+1)(t_2+1)\} =$$

$$= s_1s_2+s_1+s_2+(s_1-t_1)(s_2-t_2) \quad . \tag{5}$$

This is independent of $t_1$ and $t_2$ only if $s_1=t_1$ and/or $s_2=t_2$ holds. This yields the following lemmas:

Lemma 2. The direct product of two transitive permutation groups $G_1$ and $G_2$ satisfies

$$(G_1\otimes G_2)^{(p)} = G_1^{(p)}\otimes G_2^{(p)} \tag{6}$$

if one or both of the groups has only self-conjugate transitivity sets. In particular, eq. (6) holds if one of the groups is completely permissible. Also, if $G_1$ is doubly transitive on $M_1$ letters,

$$(G_1\otimes G_2)^{(p)} = S(M_1)\otimes G_2^{(p)} \tag{7}$$

holds.

Proof. Eq. (5) implies

$$s(G_1^{(p)} \otimes G_2^{(p)}) = s_1 s_2 + s_1 + s_2 + (s_1 - t_1')(s_2 - t_2'), \tag{8}$$

with $t_i'$ the number of self-conjugate transitivity sets of $G_i^{(p)}$. Eqs. (5) and (8) certainly yield the same number if $t_i=0$ for one or both of $i=1,2$, since one has $t_i' \leqq t_i$, see the remark following Lemma 1.4.3. Eq. (7) follows directly from Lemma 1.3.3 and from the fact that a doubly transitive permutation group has only one self-conjugate transitivity set (except for the letter 1). ¶

Lemma 3. The direct product of two permissible groups is completely permissible iff both groups are completely permissible.

Proof. Complete permissibility of $G_1 \otimes G_2$ is equivalent to the vanishing of the term in braces in eq. (5). This can only be the case for $t_1=s_1$ and $t_2=s_2$. ¶

Since the only permissible groups available up to now are the Potts model groups $S(M)$, see Lemma 1.3.2, the constructions of this and the previous section yield the new (completely permissible) groups $S(M_1) \sim S(M_2)$, which have two-graph MI's, and $S(M_1) \otimes S(M_2)$, with three-graph MI's. (From these, many more can, of course, be constructed by repeated product formation.) The two simple examples of Fig. 1.2.1 are actually both of the direct product type, since one has $K(4) \simeq S(2) \otimes S(2)$ and $D(6) \simeq S(2) \otimes S(3)$.

## 2.4. Semidirect products.

In the present section, still another type of group product is defined, which is a very useful concept even though it cannot be used, at least in a general way, to construct new permissible groups. For an exception to this rule, see Section 3.2.

Definition 1. Let $G$ be a transitive permutation group on $M$ letters, $H_M(G)$ the letter-1-fixing subgroup of its normalizer in $S(M)$, see Lemma 1.2. Then if $L$ is a subgroup of this $H_M(G)$, the product $GL$ is a transitive permutation group with $|G||L|$ elements, since for every $k \in L$ one has $k^{-1}Gk=G$. This product $GL$ is called the semidirect product of $G$ and $L$.

By Lemma 1.3, $L$ is isomorphic to a subgroup of the automorphism group of $G$. This gives the connection with the standard definition [4].

It is, in general, not possible to decide on the permissibility of the semidirect product, not even if $G$ is (completely) permissible. It is, however, possible to say something about the MI of $GL$:

Lemma 1. Let $G$ be a permissible group; then the MI of the semidirect product $GL$ consists of graphs which are sums of isomorphic graphs from the MI of $G$.
Proof. By Lemma 1.2, the elements of $L$ belong also to $A(G)$, i.e., they leave the MI of $G$ invariant, so that they can only map isomorphic graphs onto each other. ¶

In the above Lemma 1, one only needs the equality $N_M(G)=A(G)$. Although this is certainly true for permissible groups, it turns out that it is true for a certain class of nonpermissible groups as well. These quasi-permissible groups have not yet been studied in detail. Here it is only remarked that $G$ is normal in its corresponding permissible group if it is quasi-permissible; this property by itself, however, does not imply quasi-permissibility yet.

## 2.5. Abelian groups, corresponding permissible groups and permissible subgroups of these.

A group $A$ is called Abelian if any pair of elements $a_1, a_2 \varepsilon A$ commute: $a_1a_2=a_2a_1$. As transitive permutation groups, the Abelian groups play a special role, since they are necessarily regular, i.e., they contain exactly as many elements $M$ as there are letters in the space $S$ in which their elements operate:

Lemma 1. An Abelian, transitive permutation group is regular.
Proof. Let $H_1$ be the letter-1-fixing subgroup of $A$ and let $g_k$ be an element mapping 1 onto $k$: $g_k(1)=k$. Then if $h \varepsilon H_1$, the commutativity implies

$$k = g_k(1) = g_kh(1) = hg_k(1) = h(k). \qquad (1)$$

By the transitivity of $A$, $h$ keeps all letters fixed, so that $h=e$ or $H_1=\{e\}$ follows. By Lemma 1.2.4, one then has $|A|=M$. ¶

The question of the permissibility of an Abelian regular group is resolved by

Theorem 1. The only permissible Abelian groups are the q-fold direct products $S(2)\otimes\ldots\otimes S(2)\equiv K(2^q)$. The MI of $K(2^q)$ consits of $2^q-1$ isomorphic graphs, each of which consists of $2^{q-1}$ isolated edges.

Proof. Since there is a one-to-one correspondence between the elements $g_k$ of $\mathfrak{A}$ and the letters $k$ by $g_k(1)=k$ (set $g_1=e$), the graphs of the MI of $\mathfrak{A}$ are given as the sets $G_k=\{g(1),gg_k(1)\}$. A permutation $\sigma$ can be defined unambiguously by the prescription

$$\sigma g(1) = g^{-1}(1). \tag{2}$$

Clearly, $\sigma^2=e$ holds, but $\sigma\neq e$ unless all elements of $\mathfrak{A}$ are involutions. The Abelian groups consisting of involutions only are exactly the $\mathfrak{K}(2^q)$ defined in the theorem; these groups are completely permissible, since they are direct products of completely permissible groups $\mathfrak{S}(2)$, see Lemma 3.3. Now for $\mathfrak{A}\neq\mathfrak{K}(2^q)$, $\sigma\neq e$ leaves all graphs of the MI invariant:

$$\sigma G_k=\sigma\{g(1),gg_k(1)\}=\{g^{-1}(1),g_k{}^{-1}g^{-1}(1)\}=\{g_k{}^{-1}g^{-1}(1),g^{-1}(1)\}=$$

$$=g_k{}^{-1}\{g^{-1}(1),g_kg^{-1}(1)\}=g_k{}^{-1}\{g^{-1}(1),g^{-1}g_k(1)\}=g_k{}^{-1}\{g(1),gg_k(1)\}=$$

$$=g_k{}^{-1}G_k=G_k. \tag{3}$$

Further, $\sigma$ cannot belong to $\mathfrak{A}$, since $\sigma(1)=1$. Therefore, $\mathfrak{A}\neq\mathfrak{K}(2^q)$ is not permissible. The statement concerning the MI of $\mathfrak{K}(2^q)$ is obvious. ¶

It is remarked, that a model with symmetry group $\mathfrak{K}(2^q)$ can be realized by positioning $q$ Ising spins at every site. For $q=2$, this model is known in the literature as the Ashkin-Teller model ([5]).

In order to derive the permissible groups corresponding to the non-permissible Abelian groups, attention is first focussed on the cyclic groups $\mathfrak{C}(M)$, defined as the group of powers of the cyclic permutation $g_M=(12...M)$. (Note that $\mathfrak{C}(2)=\mathfrak{S}(2)$).

Theorem 2. The MI of a cyclic group $\mathfrak{C}(M)$, $M\geq 3$, consists of the distance graphs $G^{(k)}$ derived from the circle $G$ on $M$ vertices. Therefore, the corresponding completely permissible group is the group $\mathfrak{G}(G)=\mathfrak{D}(M)$, the dihedral group generated by $g_M$ and by the $\sigma$ defined by eq. (2), $|\mathfrak{D}(M)|=2M$.

Proof. The graph $G$ is obtained by operating with the powers of $g_M$ on the edge $\{1,2\}$. Operating with these powers on the edge $\{1,k+1\}$, one obtains the distance graph $G^{(k)}$, defined as in the proof of Theorem 1.3.2. This theorem shows that the permissible group corresponding to $\mathfrak{C}(M)$ is the group $\mathfrak{D}(M)$, since this is the symmetry group of the circle on $M$ vertices. Since eq. (2) is easily seen to imply

$$\sigma g \sigma = g^{-1} \text{ for all } g \in \mathcal{C}(M), \tag{4}$$

$\mathcal{C}(M)$ is normal in $\mathcal{D}(M)$, so that $|\mathcal{D}(M)|=2M$. That $\mathcal{D}(M)$ is completely permissible follows from the fact that $g_M{}^{k-1}\sigma \in \mathcal{D}(M)$ and that this is such, that

$$g_M{}^{k-1}\sigma(1)=g_M{}^{k-1}(1)=k, \quad g_M{}^{k-1}\sigma(k)=g_M{}^{k-1}\sigma g_M{}^{k-1}(1)=g_M{}^{k-1}g_M{}^{1-k}(1)=1 \tag{5}$$

hold; therefore, $g_M{}^{k-1}\sigma$ can play the role of the $\sigma_k$ in criterion (b) of Corollary 1.4.1 for $k=2,3,..,M$. ¶

The case of a general nonpermissible Abelian group $\mathcal{A}$ can be reduced to the case of a cyclic group, since every finite Abelian group can be written as a direct product of a number of cyclic groups :

$$\mathcal{A} = \mathcal{C}(m_1) \otimes \mathcal{C}(m_2) \otimes \dots \otimes \mathcal{C}(m_n) \quad \text{with} \quad M=m_1 m_2 \dots m_n. \tag{6}$$

<u>Theorem 3</u>. If $\mathcal{A}$ is Abelian and nonpermissible, then $\mathcal{A}^{(p)}=\langle\mathcal{A},\sigma\rangle$, the group generated by $\mathcal{A}$ and $\sigma$ of eq. (2). $|\mathcal{A}^{(p)}|=2|\mathcal{A}|$ by eq. (4); $\mathcal{A}^{(p)}$ is completely permissible.

<u>Proof</u>. Let $\mathcal{A}$ be written as in eq. (6); let $\sigma_i=e$ for $m_i=2$, $\sigma_i=$ the $\sigma$ of eq. (2) or (4) for the cyclic group $\mathcal{C}(m_i)$ with $m_i \neq 2$. Clearly, the $\sigma$ of eq. (2) for the whole group is the product $\sigma_1\sigma_2\dots\sigma_n$. Now suppose the theorem proved for a product of $n-1$ factors; since it is true for $n=1$ (this is Theorem 2), the theorem is proved by induction if the case with $n$ factors can be derived from the case with $n-1$ factors. Write $\mathcal{D}(m_1,..,m_{n-1})$ for the permissible group in the case of $n-1$ factors and assume first $m_n \neq 2$. Then the group $\mathcal{D}(m_1,..,m_{n-1})\otimes\mathcal{D}(m_n)$ is permissible by Lemma 3.1, but its MI consists of less graphs than there are in the MI of $\mathcal{A}$, as follows from eq. (3.5) for this case. Therefore, the group defined as $\mathcal{D}(m_1,..,m_{n-1},m_n)$ is a permissible index 2 subgroup of this by Corollary 1.1, which proves the case of $n$ factors. For $m_n=2$, one has

$$\mathcal{D}(m_1,..,m_{n-1}) \otimes \mathcal{S}(2) = \mathcal{D}(m_1,..,m_{n-1},m_n),$$

which completes the proof. The complete permissibility of $\mathcal{A}^{(p)}$ can be proved in the same way as used in Theorem 2. ¶

<u>Corollary 1</u>. A permissible group which contains an Abelian regular subgroup $\mathcal{A}$ is completely permissible.

Proof. If $\mathcal{A}=\mathcal{K}(2^q)$, this is completely permissible; if $\mathcal{A}\neq\mathcal{K}(2^q)$, $\mathcal{A}^{(p)}$ is completely permissible. In both cases, Lemma 1.1 implies the corollary. ¶

Corollary 1.1 can be used to obtain permissible, regular and, in general, nonabelian subgroups of the $\mathcal{D}(m_1,..,m_n)$. The simplest case is described by the next theorem:

Theorem 4. For $M=2m$, the groups $\mathcal{D}(2m)$ contain permissible subgroups $\mathcal{F}(2m)$ of index 2. Abstractly, $\mathcal{F}(2m)$ is isomorphic to $\mathcal{D}(m)$.
Proof. The elements of the group $\mathcal{D}(2m)$ are explicitly given as

$$\mathcal{D}(2m) = \{e,g,g^2,..,g^{2m-1},\sigma,\sigma g,\sigma g^2,..,\sigma g^{2m-1}\}. \tag{7}$$

Of these, the set of elements

$$\mathcal{F}(2m) = \{e,g^2,..,g^{2m-2},\sigma g,\sigma g^3,..,\sigma g^{2m-1}\} \tag{8}$$

is easily seen to be a transitive subgroup of index 2, so that $\mathcal{F}(2m)$ is regular and permissible by Corollary 1.1. Also, this group is nonabelian by eq. (4) unless $m=2$, in which case the isomorphism (as permutation groups) $\mathcal{F}(4)\simeq\mathcal{K}(4)$ holds. Further, $\mathcal{F}(2m)$ is generated by the powers of $g^2$ and by $\sigma g$ with $(\sigma g)^2=\sigma g\sigma g=g^{-1}g=e$, so that, abstractly, $\mathcal{F}(2m)\simeq\mathcal{D}(m)$ holds. ¶

As examples of the developments of this section, Fig. 1.2.1 is referred to; this shows the MI of $\mathcal{K}(4)=\mathcal{S}(2)\otimes\mathcal{S}(2)$ in part (a) and the MI of $\mathcal{D}(6)$, which is also the MI of $\mathcal{C}(6)$, in part (b). With this, the MI of $\mathcal{F}(6)$, shown below in Fig. 1, should be contrasted.

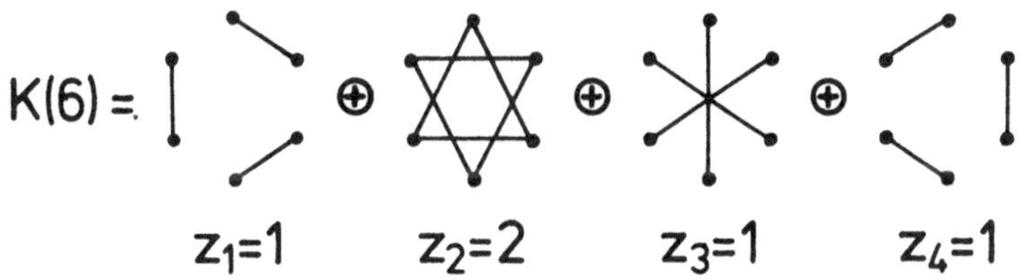


Fig. 1. The maximal interaction of the regular, permissible, nonabelian group $\mathcal{F}(6)$.

REFERENCES.

(1). B. Huppert, Endliche Gruppen I (Springer-Verlag, Berlin, Heidelberg, New York, 1963) p. 13.
(2). Ref. (1), p. 94 ff.
(3). Ref. (1), p. 45 ff.
(4). Ref. (1), p. 89.
(5). J. Ashkin and E. Teller, Phys. Rev. 64 (1943) 178.

## 3. PRIMITIVE AND IMPRIMITIVE PERMISSIBLE GROUPS.

### 3.1. Primitivity and imprimitivity.

A transitive permutation group $\mathcal{G}$ on $M$ letters is called imprimitive if the set $S$ of letters can be divided into $k$ disjoint imprimitivity sets $\Delta_i$ such that every $g \in \mathcal{G}$ either maps $\Delta_i$ onto itself or onto another $\Delta_{i'}$. If such a division of $S$ is not possible, $\mathcal{G}$ is called primitive. Since the elements of $\mathcal{G}$ are permutations, one easily sees that all imprimitivity sets must have the same number of elements, $|\Delta_i|=j$ for $i=1,..,k$ , so that $M=jk$ must hold. This already shows that every transitive permutation group on $p$ ($p$ a prime) letters must be primitive. The (im)primitivity of a group $\mathcal{G}$ can be read off directly from its MI:

Lemma 1. A transitive permutation group $\mathcal{G}$ is imprimitive iff its MI contains a disconnected graph. Therefore, $\mathcal{G}$ is primitive iff all graphs of its MI are connected.

Proof. Let $\mathcal{G}$ be imprimitive, $\Delta$ an imprimitivity set containing the vertex 1. Choose another vertex $k \in \Delta$ and consider the graph $\{g(1), g(k)\}$. By the definition of imprimitivity, each edge of this graph is contained within some imprimitivity set, so that the graph is disconnected. On the other hand, let $G$ be a disconnected graph from the MI of $\mathcal{G}$. Then the automorphism group $\mathcal{G}(G)$ is imprimitive; but since $\mathcal{G} \leq \mathcal{G}(G)$ holds, $\mathcal{G}$ must also be imprimitive. ¶

Lemma 2. Let $\mathcal{G}$ be a transitive, nonpermissible permutation group. Then $\mathcal{G}^{(p)}$ is (im)primitive iff $\mathcal{G}$ is (im)primitive.

Proof. Since $\mathcal{G}$ and $\mathcal{G}^{(p)}$ have the same MI by definition, this follows immediately from Lemma 1. ¶

As a first application of these lemmas, the nature of all permissible groups with two graphs in their MI's is clarified:

Theorem 1. A permissible group $\mathcal{G}$ on $M$ letters with two graphs in its MI is either primitive and the automorphism group of one of the graphs or it is a wreath product $\mathcal{S}(M_1) \sim \mathcal{S}(M_2)$ with $M=M_1M_2$. In any case, $\mathcal{G}$ has a P-algebra.

Proof. From Lemmas 2.2.3 and 2.2.4, there are two possibilities: (i) $\mathcal{G}$ is of type A and the graphs $G$ and $\overline{G}$ of the MI satisfy

$$G^{(2)} = \bar{G}, \quad \bar{G}^{(2)} = G, \tag{1}$$

so that $G$ and $\bar{G}$ are both connected. Then $\mathcal{G}$ is primitive by Lemma 1 and $\mathcal{G}=\mathcal{G}(G)=\mathcal{G}(\bar{G})$; this group has a P-algebra by Lemma 1.4.6. (ii) $\mathcal{G}$ is a wreath product $\mathcal{G}_1 \sim \mathcal{S}(M_2)$ with $\mathcal{G}_1$ permissible on $M_1$ letters, $M=M_1M_2$. But by eq. (2.2.7), the number of graphs in the MI of $\mathcal{G}$ is then $2= =s(\mathcal{G}_1)+1$, so that $\mathcal{G}_1=\mathcal{S}(M_1)$ follows. In this case, $\mathcal{G}$ is even completely permissible. ¶

Theorem 1 shows already that there is an intimate connection between imprimitive permissible groups and wreath products. The following construction associates a unique wreath product of permissible groups with any imprimitive permissible group:

Let $\mathcal{G}$ be an imprimitive permissible group on $M=M_1M_2$ letters such that there are $M_2$ imprimitivity sets of $M_1$ letters each. Then the graphs of the MI of $\mathcal{G}$ are of two types: (i) disconnected graphs consisting of $M_2$ copies of graphs completely embedded in imprimitivity sets and (ii) graphs which contain only edges between imprimitivity sets. The intersection of the automorphism groups of the graphs of type (i) will be a group $\mathcal{G}_1 \sim \mathcal{S}(M_2)$ with $\mathcal{G}_1$ permissible on $M_1$ letters. The graphs of type (ii) can be reduced to graphs on $M_2$ letters by shrinking each imprimitivity set to a single vertex and by connecting two such vertices iff the corresponding imprimitivity sets are connected. The intersection of the automorphism groups of the so-obtained graphs will be a permissible group $\mathcal{G}_2$ on $M_2$ letters. Now the intersection of the automorphism groups of the type (ii) graphs must needs be a subgroup of $\mathcal{S}(M_1) \sim \mathcal{G}_2$, so that $\mathcal{G}$ is a subgroup of the wreath product $\mathcal{G}_1 \sim \mathcal{G}_2$, where both groups are determined uniquely by $\mathcal{G}$. This construction will be referred to by saying that $\mathcal{G}$ <u>leads to</u> the wreath product $\mathcal{G}_1 \sim \mathcal{G}_2$. The numbers of graphs in the MI's of these three groups satisfy

$$s(\mathcal{G}) \geqq s(\mathcal{G}_1) + s(\mathcal{G}_2), \tag{2}$$

where equality implies $\mathcal{G}=\mathcal{G}_1 \sim \mathcal{G}_2$, see eq. (2.2.7).

The above construction could also have been used to prove Theorem 2; here it will be used to give a (not completely explicit) characterization of the permissible groups with three graphs in their MI's:

<u>Theorem 2</u>. Permissible groups on $M$ letters with three graphs in their MI's are of one of the following six types:
(a) primitive; (b) $\mathcal{G}_1 \sim \mathcal{S}(M_2)$, $\mathcal{G}_1$ primitive on $M_1$ letters with a two-graph MI, $M=M_1M_2$; (c) $\mathcal{S}(M_1) \sim \mathcal{G}_2$, $\mathcal{G}_2$ primitive on $M_2$ letters with a two-

graph MI, $M=M_1M_2$; (d) $\mathcal{S}(M_1)\sim\mathcal{S}(M_2)\sim\mathcal{S}(M_3)$, $M=M_1M_2M_3$; (e) $\mathcal{S}(M_1)\otimes\mathcal{S}(M_2)$, $M=M_1M_2$; (f) imprimitive groups which lead to the wreath product $\mathcal{S}(M_1)\sim\mathcal{S}(M_2)$, $M=M_1M_2$, the MI's of which consists of two connected graphs and one graph which consists of $M_2$ copies of $K(M_1)$.

Proof. If $\mathcal{G}$ is not primitive (case (a)), so that all three graphs in its MI are connected, then $\mathcal{G}$ leads to a wreath product $\mathcal{G}_1\sim\mathcal{G}_2$ and eq. (2) reads $3\geqq s(\mathcal{G}_1)+s(\mathcal{G}_2)$. This gives a number of cases: (i) $s(\mathcal{G}_1)=1$, $s(\mathcal{G}_2)=2$; this implies $\mathcal{G}=\mathcal{S}(M_1)\sim\mathcal{G}_2$ with $\mathcal{G}_2$ as described in Theorem 1, so that case (c) results for $\mathcal{G}_2$ primitive, case (d) for $\mathcal{G}_2$ imprimitive by the associativeness of the wreath product. (ii) $s(\mathcal{G}_1)=2$, $s(\mathcal{G}_2)=1$; now one has $\mathcal{G}=\mathcal{G}_1\sim\mathcal{S}(M_2)$ with $\mathcal{G}_1$ as in Theorem 1, leading to cases (b) ($\mathcal{G}_1$ primitive) and, again, (d) ($\mathcal{G}_1$ imprimitive). (iii) $s(\mathcal{G}_1)=s(\mathcal{G}_2)=1$. It is easy to see that the direct product $\mathcal{S}(M_1)\otimes\mathcal{S}(M_2)$ corresponds to this, yielding case (e). Groups corresponding to case (iii) which are not direct products are easily shown to have the properties listed for groups of type (f). As will be shown in Chapter 5, such groups actually do exist, although no explicit expressions in terms of simpler groups can be given.¶

The fact that already for MI's with three graphs there is a class of permissible groups, which cannot be given explicitly, shows, that the construction "$\mathcal{G}$ leads to $\mathcal{G}_1\sim\mathcal{G}_2$" is not very useful for a classification of permissible groups with even more graphs in their MI's.

The primitive permissible groups with two or three graphs in their MI's obtained in Theorems 1 and 2 are given by Lemmas 3.3 and 3.4 below if they are on $p$ ($p$ a prime) letters. If this is not the case, the graph-theoretical constructions of Section 5.1 give at least some of these groups.

## 3.2. Permissible groups on $p$ ($p$ a prime) letters.

All groups mentioned in the title of this section can be given explicitly. Before showing this, a few more group-theoretical definitions are necessary:

Definition 1. Let $\mathcal{G}$ be a group (abstractly); then the derived or commutator group $\mathcal{G}'$ is the subgroup generated by all commutators $g^{-1}h^{-1}gh$ with $g,h\in\mathcal{G}$. Higher derived groups are defined by $\mathcal{G}''=(\mathcal{G}')'$, etc.

Definition 2. A subgroup $\mathcal{R}\leqq\mathcal{G}$ is called characteristic if for all $\alpha\in$ $\mathrm{AUT}(\mathcal{G})$ the equality $\alpha(\mathcal{R})=\mathcal{R}$ holds.

A characteristic subgroup is a normal subgroup, since the inner automorphisms $\alpha_g$ defined by $\alpha_g(h)=g^{-1}hg$ belong to $\mathrm{AUT}(\mathcal{G})$. Clearly, all derived groups are characteristic subgroups and, hence, normal.

Definition 3. A group $G$ is solvable if the sequence of derived groups is finite, i.e., if there is a number $k$ such that the m-th derived group equals $\{e\}$ for $m>k$.

An Abelian group $A$ is certainly solvable, since $A'=\{e\}$ holds. A non-abelian group for which $G''=\{e\}$ is called metabelian. Further, it is clear that a subgroup of a solvable group is again solvable.

Definition 4. A group is called simple if it does not contain any non-trivial normal subgroups.

Trivial normal (and characteristic) subgroups of a group are the group itself and $\{e\}$.

Definition 5. A group is called characteristically simple if it contains no nontrivial characteristic subgroups.

The connection of these definitions with the subject of this section rests on the following lemmas:

Lemma 1. A solvable simple group is isomorphic to a cyclic group $C(p)$, $p$ a prime. A solvable, characteristically simple group is isomorphic to a q-fold direct product of such cyclic groups, $C(p)\otimes\ldots\otimes C(p)$.
Proof. Let $G$ be simple; since $G'$ is normal, this implies either $G=G'$ or $G'=\{e\}$. Since $G$ is solvable, the second alternative must hold and $G$ is Abelian. But every subgroup of an Abelian group is normal, so that $G$ must have no nontrivial subgroups. This implies $G\simeq C(p)$. It can be shown [1], that a characteristically simple group is the direct product of a number of isomorphic simple groups; this implies the second statement of the lemma. ¶

Lemma 2. A normal subgroup of a primitive permutation group is transitive.
Proof. Let $G$ be a primitive permutation group with normal subgroup $N$. Let the elements of $N$ map the letter $k$ on the set of letters $N(k)$:

$$N(k) = \{n(k)\,|\,n\varepsilon N\}. \tag{1}$$

Consider the action of an element $g\varepsilon G$ on $N(k)$:

$$g\,N(k) = \{gn(k)\,|\,n\varepsilon N\} = \{gng^{-1}g(k)\,|\,n\varepsilon N\} = \{ng(k)\,|\,n\varepsilon N\} = N(g(k)), \tag{2}$$

where the normality of $N$ has been used. Clearly, $g$ either maps $N(k)$ onto itself or onto a set $N(k')$, i.e., the $N(k)$ are imprimitivity sets of $G$. Since $G$ is primitive, this implies that $N$ is transitive.
¶

Theorem 1. A solvable, primitive permutation group $G$ contains an Abelian (and, hence, regular) characteristic subgroup $C^{(q)}(p)$, the q-fold product of cyclic groups of Lemma 1. $G$ is the semidirect product of $C^{(q)}(p)$ with a subgroup $L$ of its automorphism group $H_M(C^{(q)}(p))$. This latter group is isomorphic to the general linear group $GL(q,p)$, which is the group of all nonsingular transformations of the q-dimensional vector space $V(q,p)$ over the Galois field $GF(p)$.

Proof. Since $G$ is solvable, there is a number k, such that the k-th derived group of $G$ is Abelian. This Abelian group $A$ is characteristic and then certainly normal. Hence, it is transitive by Lemma 2 and regular by Lemma 2.5.1. If $R$ is characteristic in $A$, then it is also characteristic, and, hence, normal, in $G$; therefore, $R$ is also transitive. But this is impossible, since $A$ is regular, so that $A$ is characteristically simple and equal to $C^{(q)}(p)$ by Lemma 1 for some prime number p and some number of factors q. The semidirect product property is then obvious, as is the identification of $H_M(C^{(q)}(p))$ with $GL(q,p)$. ¶

The above Theorem 1 would be very useful, if all permissible, primitive groups were solvable. This is, however, not generally the case, unless q=1, i.e., unless $G$ is on p letters. This follows from the next theorem of Burnside, a proof of which can be found in [2].

Theorem 2. (Burnside). A permutation group on p letters, which is not doubly transitive, is solvable.

Theorem 3. A permissible group on p letters is either $S(p)$ or a semidirect product $C(p)\langle\alpha^k\rangle$, where $\alpha$ is a generator of the cyclic group $H_M(C(p))$ of order p-1 and k divides (p-1)/2, $k\neq 1$. Every graph of the MI of this group consists of (p-1)/2k graphs (which are circles) from the MI of $D(p)$; the group is completely permissible.

Proof. Since $G$ is on p letters, it is primitive. By Lemma 1.3.3, is not doubly transitive, unless it is $S(p)$. Therefore, $G\neq S(p)$ is solvable by Burnside's Theorem 2 above. Theorem 1 then implies that $G = C(p)L$ with $L \leqq H_M(C(p))$. Now the automorphism group of $C(p)$ is easily seen to be cyclic of order p-1 [3], so that $G$ has the form $C(p)\langle\alpha^k\rangle$ with $\alpha$ a generator of $H_M(C(p))$ and k a divisor of p-1. For k=1, this semidirect product is doubly transitive, so that $k\neq 1$ follows. Further, since $G$ is permissible but $C(p)$ is not, $G$ must contain $D(p)$, Theorem 2.5.2. This implies that $\sigma\varepsilon G$ with $\sigma(1)=1$, $\sigma g\sigma=g^{-1}$ for all $g\varepsilon C(p)$; one easily finds $\sigma=\alpha^{(p-1)/2}$, so that k must divide (p-1)/2. Since $\sigma$ is the only power of $\alpha$, which keeps all graphs of the MI of $D(p)$ invariant, $G$ is permissible for all such k. The rest of the theorem is then obvious. ¶

Corollary 1. If $p$ is a prime such that $(p-1)/2$ is also a prime, e.g., for $p=3,5,7,11,23...$, then the only permissible groups on $p$ letters are $\mathcal{D}(p)$ and $\mathcal{S}(p)$.
Proof. Immediate from Theorem 3. ¶

The first prime, for which Corollary 1 admits permissible groups unequal to $\mathcal{D}(p)$ or $\mathcal{S}(p)$, is 13. Here $(p-1)/2=6$, so that $k=2$ or $k=3$ are possible by Theorem 3. The corresponding permissible groups have 78 and 52 elements, respectively.

The primitive permissible groups with two- or three-graph MI's found in Theorems 1.1 and 1.2 follow directly from Theorem 3 if they are on $p$ letters; these results are stated as the next two lemmas:

Lemma 3. Let $\mathcal{G}$ be a permissible group on $p$ letters with a two-graph MI. Then $p=4k+1$ and $\mathcal{G}$ is given as the semidirect product

$$\mathcal{G} = \mathcal{C}(p)\langle\alpha^2\rangle, \tag{3}$$

with $\alpha$ a generator of $\mathcal{H}_M(\mathcal{C}(p))$, which is cyclic of order $4k$. Both graphs of the MI are isomorphic, each consisting of $k$ graphs from the MI of $\mathcal{D}(p)$.

Lemma 4. Let $\mathcal{G}$ be a permissible group on $p$ letters with a three-graph MI. Then $p=6k+1$ and $\mathcal{G}$ is given as the semidirect product

$$\mathcal{G} = \mathcal{C}(p)\langle\alpha^3\rangle, \tag{4}$$

with $\alpha$ a generator of $\mathcal{H}_M(\mathcal{C}(p))$, which is cyclic of order $6k$. Again, each of the three isomorphic graphs of the MI consists of $k$ graphs from the MI of $\mathcal{D}(p)$.

### 3.3. The p-wreath product.

The p-wreath product is a useful construction to obtain primitive groups on $M^q$ letters from those on $M$ letters. Before giving this construction, a simple lemma concerning primitive permutation groups is needed:

Lemma 1. Let $\mathcal{G}$ be a transitive permutation group on $M$ letters, $\mathcal{H}_1$ its letter-1-fixing subgroup. Then $\mathcal{G}$ is primitive iff $\mathcal{H}_1$ is maximal.
Proof. Assume first that there is a group $\mathcal{M}$ satisfying $\mathcal{H}_1<\mathcal{M}<\mathcal{G}$. By Lemma 1.2.4, $\mathcal{G}$ has the right coset decomposition ($g_k$ is an element mapping 1 onto k):

$$G = \bigcup_{k=1}^{M} g_k H_1 . \tag{1}$$

Similarly, $M$ has the right coset decomposition

$$M = \bigcup_{k\varepsilon\Delta} g_k H_1 , \tag{2}$$

with $\Delta<S$, so that $M$ cannot be transitive. It is easy to see that the transitivity sets of $M$ are imprimitivity sets of $G$, so that $G$ is imprimitive. Conversely, let $G$ be imprimitive and $\Delta$ the imprimitivity set containing the letter 1. Then the subgroup $M$ defined by

$$M = \{g \mid g\varepsilon G, g(\Delta)=\Delta\} \tag{3}$$

obviously satisfies $H_1<M<G$. ¶

The p-wreath product is an extension of the q-fold direct product $G\otimes G\otimes..\otimes G$. The elements of $S\times S\times..\times S$ (q factors) can be represented by q-dimensional column vectors and the elements of the q-fold direct product by diagonal $q\times q$ matrices:

$$\begin{pmatrix} g_1 & & & \\ & g_2 & & \\ & & \cdot & \\ & & & \cdot \\ & & & & g_q \end{pmatrix} \begin{pmatrix} i_1 \\ i_2 \\ \cdot \\ \cdot \\ i_q \end{pmatrix} = \begin{pmatrix} g_1(i_1) \\ g_2(i_2) \\ \cdot \\ \cdot \\ g_q(i_q) \end{pmatrix} \tag{4}$$

This follows directly from the definition of the direct product in Section 2.3. Now let $G'$ be a permutation group on q letters; its elements can be represented by $q\times q$ matrices as in Section 1.5:

$$D_{ij}(g') = \delta(i,g'(j)) \quad \text{for all} \quad g'\varepsilon G'. \tag{5}$$

The group generated by the diagonal matrices of eq. (4) and by the permutation matrices of eq. (5) is easily seen to be isomorphic to the wreath product $G\sim G'$ as defined in Section 2.2. Since this group now operates on $M^q$ letters, it is designated as $G\underset{p}{\sim}G'$ and called the p-wreath product of $G$ and $G'$ to distinguish it from the standard wreath product $G\sim G'$ on $Mq$ letters. Some properties of the p-wreath product are listed in the next theorem, which corrects and generalizes a result of Marcu ([4]).

Theorem 1. (i) The group $(G\underset{p}{\sim}G')^{(p)}$ is completely permissible if $G$ is.

If $G \underset{p}{\sim} G'$ is completely permissible, then also $G$. (ii) $G \underset{p}{\sim} G'$ is primitive iff (a) $G$ is primitive, (b) $G \neq C(p)$ and (c) $G'$ is transitive. (iii) The q-fold direct product is normal in $G \underset{p}{\sim} G'$.

Proof. If $G$ is completely permissible, then also the q-fold direct product of this group with itself, so that $(G \underset{p}{\sim} G')^{(p)}$ is completely permissible as an extension of the q-fold direct product. Conversely, if $G \underset{p}{\sim} G'$ is compltely permissible, there must be an element in this group which exchanges the vectors

$$\underline{1} = \begin{pmatrix} 1 \\ 1 \\ \vdots \\ 1 \end{pmatrix} \quad \text{and} \quad \underline{i} = \begin{pmatrix} i_1 \\ i_2 \\ \vdots \\ i_q \end{pmatrix} . \tag{6}$$

Since $G'$ leaves the first vector invariant, such an element must belong to $G \otimes \ldots \otimes G$. This implies that $G$ is completely permissible, so that part (i) of the theorem is proved.

(ii). The letter-1-fixing subgroup of $G \underset{p}{\sim} G'$ leaves the vector $\underline{1}$ invariant. It is, therefore, equal to $H_1 \underset{p}{\sim} G'$ with $H_1$ the letter-1-fixing subgroup of $G$. If $G \underset{p}{\sim} G'$ is primitive, $H_1 \underset{p}{\sim} G'$ is a maximal subgroup by Lemma 1. This certainly implies that $H_1$ is a maximal subgroup of $G$, so that this is primitive. Further, $G'$ must be transitive, for if it is not, then the group $G \otimes .. \otimes G \otimes H_1 \otimes .. \otimes H_1$, with the $G$'s corresponding to a transitivity set of $G'$ can be extended by this latter group and the resulting group is a proper subgroup of the p-wreath product and contains the letter-1-fixing subgroup of this. Finally, consider the subgroup of $G \underset{p}{\sim} G'$ generated by the multiples of the unit matrix $g\underline{\underline{I}}$, $g \in G$, and by $G'$. This subgroup contains $H_1 \underset{p}{\sim} G'$ if $H_1 = \{e\}$, which implies $G = C(p)$ by primitivity. Therefore, the primitivity of the p-wreath product implies $G \neq C(p)$. The converse statement can be proved as follows: if $G$ is primitive with $H_1 \neq \{e\}$ and if $G'$ is transitive, then the group generated by $H_1 \underset{p}{\sim} G'$ and any one element of the p-wreath product which does not keep $\underline{1}$ fixed is easily seen to be the whole of this product. (iii) is obvious. ¶

Remark. The complete permissibility of $(G \underset{p}{\sim} G')^{(p)}$ does, in general, not even imply the permissibility of $G$. As an example, consider $C(3) \underset{p}{\sim} S(2)$; this group is not permissible, since it does not contain $D(3,3)$ as a subgroup. However, $(C(3) \underset{p}{\sim} S(2))^{(p)} \simeq S(3) \otimes S(3)$, which is completely permissible.

Of most interest is, of course, the question as to the permissibility of the p-wreath product. Here one has the next three lemmas:

Lemma 2. For $M>2$, the group $S(M)\tilde{_p}S(q)$ is primitive and completely permissible, since it is the group of the (unique) connected graph in the MI of the q-fold direct product $S(M)\otimes\ldots\otimes S(M)$.
Proof. This follows immediately from the following symbolic description of the graphs of the MI of the q-fold direct product: each of these graphs may be represented by a q-dimensional vector, the components of which are either 1 or the transitivity set $K=\{2,..,M\}$. Of these, only $\underline{1}$ is not represented by a graph. Obviously, $S(M)\otimes\ldots\otimes S(M)$ leaves each of these vectors invariant. The only vector corresponding to a graph, which is left invariant by all permutations from $S(q)$, is the vector $\underline{K}$ with all entries equal to K. The corresponding graph is easily seen to be the only connected one from the MI of the q-fold direct product. Primitivity and complete permissibility now follow from Theorem 1 and from the permissibility of the group of a graph, Theorem 1.3.2. ¶
The MI of $S(M)\tilde{_p}S(q)$ consists of q graphs, the graph of Lemma 2 having $z_1=(M-1)^q$, the others, as sums of isomorphic graphs, having $z_n=\binom{q}{k-1}\times (M-1)^{q-k+1}$, $k=2,3,..,q$. This latter result is also valid for $M=2$: the group $S(2)\tilde{_p}S(q)$ has a q-graph MI with $z_k=\binom{q}{k-1}$, $k=1,2,..,q$. It is not difficult to see, that this is the MI of the automorphism group of the q-dimensional hypercube, so that $S(2)\tilde{_p}S(q)$ is also a (completely) permissible group by Theorem 1.3.2.

Lemma 3. The group $G\tilde{_p}S(q)$ is permissible for $G$ permissible.
Proof. This follows directly from Lemma 2, which implies

$$G\tilde{_p}S(q) \leqq (G\tilde{_p}S(q))^{(p)} \leqq S(M)\tilde{_p}S(q), \qquad (7)$$

so that an extra element of $(G\tilde{_p}S(q))^{(p)}$ would be a diagonal matrix; but then, this would also belong to $(G\otimes..\otimes G)^{(p)}=G\otimes..\otimes G$, where the permissibility of $G$ has been used. ¶

Lemma 4. For $G$ permissible, $(G\tilde{_p}G')^{(p)}\leqq G\tilde{_p}S(q)$ holds. $G\tilde{_p}C'$ is permissible for all $G'<S(q)$ if $s(G)\geqq q$ holds.
Proof. The first part of the lemma follows immediately from Lemma 3. The second part follows from the fact that, for $s(G)\geqq q$, there are vectors $\underline{i}$ such that each permutation $g'$ of $S(q)$ maps $\underline{i}$ onto a vector corresponding to a different graph from the MI of $G\otimes..\otimes G$. ¶

REFERENCES.

(1). B. Huppert, Endliche Gruppen I (Springer-Verlag, Berlin, Heidelberg, New York, 1963) p. 51.

([2]). Ref. ([1]), p. 609.
([3]). Ref. ([1]), p. 20.
([4]). M. Marcu, Ph. D. Dissertation, Bonn University (1981, unpublished).

## 4. REGULAR GROUPS.

### 4.1. Spin models defined on groups.

Let $\hat{R}$ be an abstract finite group with M elements. This group can be mapped isomorphically onto a regular group (as defined in Section 2.5), denoted by $R$, by assigning the letter 1 to the unit element $\hat{e}$ of $\hat{R}$, letters $2,3,\ldots,M$ to the other elements of $\hat{R}$ and by defining a permutation $g \varepsilon R$ corresponding to each $\hat{g} \varepsilon \hat{R}$ by

$$g(k) = m \quad \text{iff} \quad \hat{g}\hat{g}_k = \hat{g}_m. \tag{1}$$

This isomorphism is known as the left-regular representation of $\hat{R}$. Now let $E(i,j)$ be an energy function invariant with respect to $R$, i.e., this energy function satisfies the three requirements of Section 1.2. Then the isomorphism of eq. (1) shows that this energy function may also be defined on $\hat{R}$ by

$$E(\hat{g},\hat{h}) = E(i,j) \quad \text{iff} \quad g(1)=i \quad \text{and} \quad h(1)=j\ , \tag{2}$$

where $\hat{g},\hat{h} \varepsilon \hat{R}$ are the isomorphic images of $g,h \varepsilon R$. The three requirements of Section 1.2 for $E(i,j)$ can then be translated into requirements for $E(\hat{g},\hat{h})$. The result is

$$E(\hat{g},\hat{h}) = f(\hat{g}^{-1}\hat{h})\ , \quad \text{(symmetry of } E(i,j) \text{ with respect to } R),$$

$$f(\hat{g}) = f(\hat{g}^{-1}), \quad \text{(spin-exchange symmetry)}, \tag{3}$$

$$f(\hat{e}) = 0, \quad (\ E(i,i)=0 \text{ requirement}).$$

Conversely, an energy function defined on a group $\hat{R}$, which satisfies eqs. (3) leads to an energy function $E(i,j)$ satisfying the requirements of Section 1.2 with a regular symmetry group $R$.

### 4.2. Permissible groups corresponding to regular groups.

The permissible groups corresponding to all regular groups can be given explicitly; this is due to the simplicity of the MI for a regular group:

Lemma 1. Let $\mathcal{R}$ be a regular group with $M$ elements consisting of the identity $e$, $t$ involutions and $M-t-1$ other elements. Then the MI of $\mathcal{R}$ consists of $t$ graphs with $z=1$, each consisting of $M/2$ isolated edges, and of $(M-t-1)/2$ graphs with $z=2$, each consisting of a number (which divides $M$) of isomorphic closed circles.

Proof. The graphs of the MI are given by $(k\neq 1)$:

$$G_k = \{g(1), g(k)\} = \{g(1), gg_k(1)\}, \tag{1}$$

obtained from the edge $\{1,k\}$. This edge is left invariant by no element of the group if $g_k$ is not an involution, so that in this case a $z=2$ graph with $M$ edges results. If $g_k$ is an involution, then taking $g=g_k$, one sees that there is an order two subgroup $\{e,g_k\}$ which leaves $\{1,k\}$ invariant, implying that the graph has $M/2$ edges, so that it must have $z=1$. ¶

In order to guide the investigation, the situation for Abelian groups $\mathcal{A}$ (which are always regular) as studied in Section 2.5 is recalled: If such a group does not consist of involutions only, there is a nontrivial automorphism $\alpha$ with $\alpha(g)=g^{-1}$ and the corresponding permutation $\sigma$,

$$\sigma g_k(1) = g_k^{-1}(1), \quad \sigma g_k \sigma = g_k^{-1} = \alpha(g_k), \tag{2}$$

must be added to the group to obtain the corresponding permissible group, $\mathcal{A}^{(p)} = \langle \mathcal{A}, \sigma \rangle$. Now in case $\mathcal{R}$ is nonabelian, there is no automorphism $\alpha$ with $\alpha(g)=g^{-1}$ for all $g\in\mathcal{R}$. Nonetheless, the results for the Abelian case have inspired the following program:

(a) Firstly, all nonabelian groups $\mathcal{R}$ such that $\mathcal{R}^{(p)}$ contains a permutation $\sigma$ with

$$\sigma g_k(1) = g_k^{-1}(1) \tag{3}$$

are determined. For these, the corresponding permissible groups are easily found.

(b) Secondly, all nonabelian groups which have an automorphism $\alpha$ such that

$$\alpha(g) = g \text{ or } g^{-1} \tag{4}$$

holds, are studied. For these also, the corresponding permissible groups are obtained.

(c) Thirdly, it is shown that all other nonabelian groups are permissible.

Step (a). Let $\mathcal{G}$ be a nonabelian regular group and $\sigma$ as given by eq. (3). The action of $\sigma$ on a graph $G_k$ is given as

$$\sigma G_k = \{\sigma g(1), \sigma g g_k(1)\} = \{g^{-1}(1), g_k^{-1} g^{-1}(1)\} = g^{-1}\{1, g g_k^{-1} g^{-1}(1)\}. \quad (5)$$

Clearly, $G_k$ contains edges from graphs generated by $g g_k^{-1} g^{-1}$; therefore, $\sigma G_k = G_k$ will hold iff one has

$$g g_k^{-1} g^{-1} = g_k \text{ or } g_k^{-1}, \text{ for all } g, g_k \varepsilon \mathcal{R}. \quad (6)$$

All regular groups with the property of eq. (6) must now be determined. Firstly, it is noted that if $g_k$ is an involution, then eq. (6) implies that $g_k$ commutes with all $g$, so that all involutions of $\mathcal{R}$ are in the center $\mathcal{Z}(\mathcal{R})$, which is a normal subgroup of $\mathcal{R}$. Now consider a fixed element $g$ and define the subgroup $C(g)$ by

$$C(g) = \{c \mid c \varepsilon \mathcal{R}, \ cg = gc\}. \quad (7)$$

Then for $d \varepsilon D(g) = \mathcal{R} - C(g)$, one has $d^{-1} g d = g^{-1}$ necessarily by eq. (6). For $c$ and $d$ as above, the products $cd$ and $dc$ necessarily belong to $D(g)$. For $d_1, d_2 \varepsilon D(g)$, however, $d_1 d_2$ and $d_2 d_1$ must belong to $C(g)$ again. This implies that the mapping $\pi$ of $\mathcal{R}$ onto $\mathcal{S}(2) = \{e, \tau\}$, $\tau^2 = e$, defined by

$$\pi(g') = e \text{ for } g' \varepsilon C(g); \quad \pi(g') = \tau \text{ for } g' \varepsilon D(g) \quad (8)$$

conserves the group multiplication. Such a mapping is a homomorphism, and the set of elements mapped onto e, the kernel of the homomorphism, is a normal subgroup of $\mathcal{R}$. Eq. (8) now allows for two possibilities: (i) $C(g) = \mathcal{R}$, so that $g \varepsilon \mathcal{Z}(\mathcal{R})$ or (ii) $C(g)$ is an index two subgroup of $\mathcal{R}$. Since $\mathcal{R}$ is nonabelian, there are elements $g$ for which the second alternative holds. Now consider the commutator of two elements $g_1, g_2 \varepsilon \mathcal{R}$; one has

$$g_1^{-1} g_2^{-1} g_1 g_2 = \begin{cases} e & \text{if } g_1 \text{ and } g_2 \text{ commute,} \\ g_2^{\,2} = g_1^{\,-2} & \text{if not.} \end{cases} \quad (9)$$

Therefore, all elements $g \notin Z(R)$ have the same square and, since these elements are not involutions, it follows from $g^2 = g^{-2}$ that they are fourth-order elements: $g \notin Z(R)$ implies $g^4 = e$ and $g^2 = i$, the same involution for all such $g$, $i \in Z(R)$. Now suppose $z \in Z(R)$, $g$ a fourth-order element; then $gz$ cannot be in $Z(R)$, so that one has

$$i = (gz)^2 = g^2 z^2 = iz^2, \tag{10}$$

i.e., $z$ is an involution. If $g_1$ and $g_2$ do not commute, then they generate a subgroup $Q = \langle g_1, g_2 \rangle$ of $R$ with the defining relations

$$g_1{}^4 = g_2{}^4 = e, \quad g_1{}^2 = g_2{}^2 = i, \quad g_1 g_2 = g_2 g_1{}^{-1} = g_2 g_1{}^3. \tag{11}$$

Such a group is known as the <u>quaternion group of order 8</u> ([1]). Now if $g_3$ is a third element of $R$, one has

$$g_3{}^{-1} g_1 g_3 = g_1{}^{\pm 1} \quad \text{and} \quad g_3{}^{-1} g_2 g_3 = g_2{}^{\pm 1}. \tag{12}$$

For any of these four possibilities, there is already an element $g_3 \varepsilon Q$ which satisfies this equation. Therefore, $R$ can be written as a product

$$R = QE, \tag{13}$$

with $E$ the subgroup of those elements of $R$ which commute with all elements of $Q$. It follows from eq. (6), that the group generated by the powers of any element of $R$ must be a normal subgroup; therefore, if $f \varepsilon E$, $\langle g_1 f \rangle$ must be normal, but on the other hand one has

$$g_2{}^{-1} g_1 f g_2 = g_1{}^{-1} f. \tag{14}$$

For this to belong to $\langle g_1 f \rangle$, one must have $f^2 = e$ since $g_1{}^2 = i \neq e$. Hence, $E$ consists of involutions only, yielding $E = Z(R)$. Since eq. (6) shows that $Q$ is a normal subgroup, these results all togther imply that $R$ is a direct product of $Q$ with an Abelian group $K(2^q)$:

$$R = Q \otimes K(2^q), \quad Z(R) = \{e, i\} \otimes K(2^q). \tag{15}$$

Conversely, it is easily checked that eq. (6) holds if $R$ is of the form of eq. (15): Any $g$ from $R$ is <u>either</u> an involution and $C(g) = R$ holds <u>or</u> it is a product $g = qz$ with $q \varepsilon Q$, $q^2 = i$, $z \varepsilon Z(R)$; in this latter case, $C(g) = \langle q \rangle \otimes K(2^q)$, which is indeed an index two subgroup of $R$. Since the quaternion group contains exactly six different fourth-order elements, $C(g)$ is one of the three Abelian subgroups of $R$:

$$C_1=\langle g_1\rangle \otimes K(2^q), \quad C_2=\langle g_2\rangle \otimes K(2^q), \quad C_3=\langle g_1 g_2\rangle \otimes K(2^q), \tag{16}$$

with

$$C_1 \cap C_2 = C_1 \cap C_3 = C_2 \cap C_3 = C_1 \cap C_2 \cap C_3 = Z(Q) \otimes K(2^q) \simeq K(2^{q+1}). \tag{17}$$

It remains to determine the permissible groups corresponding to $Q \otimes K(2^q)$. By Lemma 2.3.2, this is given as

$$(Q \otimes K(2^q))^{(p)} = Q^{(p)} \otimes K(2^q), \tag{18}$$

since $K(2^q)$ is completely permissible. To find $Q^{(p)}$, its MI,shown in Fig. 1 below, is considered; This figure is constructed from the following realization of $Q$ as a regular permutation group:

$$Q=\langle g_1, g_2\rangle; \; g_1=(1234)(5678), \; g_2=(1537)(2846). \tag{19}$$

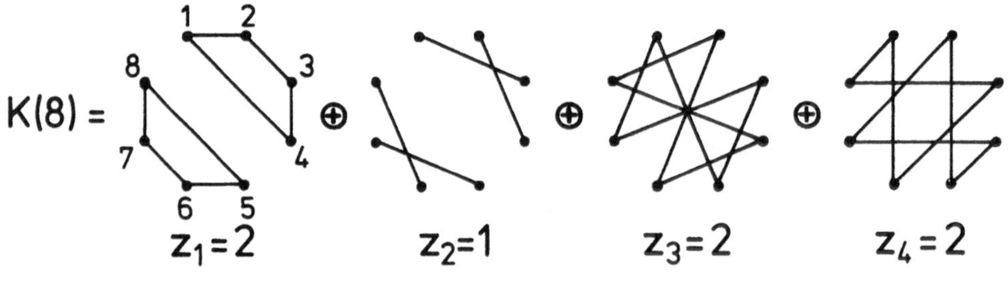


Fig. 1. The maximal interaction of the quaternion group as given by eq. (19).

It is immediately clear from this figure that $Q^{(p)}$ is much larger than $Q$. In fact, one has

$$Q^{(p)} = S(2) \sim K(4), \quad |Q^{(p)}| = 64 = 8|Q|, \tag{20}$$

so that eq. (18) implies for the general case:

$$(Q \otimes K(2^q))^{(p)} = (S(2) \sim K(4)) \otimes K(2^q). \tag{21}$$

This completes step (a).

Step (b). Let $R$ now be a regular group with an automorphism $\alpha$ as in eq. (4). The permutation $\tau$ defined by

$$\tau g(1) = g^{\pm 1}(1) \quad \text{iff} \quad \alpha(g) = g^{\pm 1} \tag{22}$$

is such that

$$\tau g \tau = \alpha(g) \tag{23}$$

holds; this permutation keeps all graphs of the MI of $R$ invariant:

$$\tau G_k = \tau\{g(1), gg_k(1)\} = \{\alpha(g)(1), \alpha(gg_k)(1)\} = \{\alpha(g)(1), \alpha(g)\alpha(g_k)(1)\} =$$

$$= \{g(1), g\alpha(g_k)(1)\} = \{g(1), gg_k^{\pm 1}(1)\} = G_k = \{g(1), gg_k(1)\}. \tag{24}$$

Now a subgroup $T$ is defined by

$$T = \{g \mid g \in R, \ \alpha(g) = g\}. \tag{25}$$

Let $t \in T$, $s \notin T$; then the product $ts$ either belongs to $T$ or it does not. In the first case, $\alpha(ts) = \alpha(t)\alpha(s) = ts^{-1}$ holds and $s$ is an involut ion, which must be in $T$. This contradiction shows, that only the second case applies, $\alpha(ts) = s^{-1}t^{-1} = \alpha(t)\alpha(s) = ts^{-1}$; this leads to the requirement

$$tst = s \quad \text{for all} \quad t \in T, \ s \notin T. \tag{26}$$

Now let $s_1$ and $s_2$ both not belong to $T$. Suppose first that the product $s_1 s_2$ does not belong to $T$ either; then eq. (26) implies (i) $ts_1 s_2 t = s_1 s_2$ and (ii) $ts_1 tts_2 t = s_1 s_2$ for all $t \in T$, so that $t$ is always an involution; but then $\alpha(g) = g^{-1}$ for all $g \in R$, so that $R$ cannot be nonabelian. Therefore, $s_1 s_2 \in T$ holds and one has $\alpha(s_1 s_2) = s_1 s_2 = \alpha(s_1)\alpha(s_2) = s_1^{-1} s_2^{-1}$, so that every $s \notin T$ satisfies $s^4 = e$, $s^2 = i$ with $i$ a fixed involution from $T$. The above analysis shows that the mapping $\pi$ with

$$\pi(t) = e \quad \text{for} \quad t \in T, \quad \pi(s) = \tau \quad \text{for} \quad s \notin T, \tag{27}$$

is again a homomorphism of $R$ onto $S(2) = \{e, \tau\}$. Since $R$ is nonabelian, it cannot equal $T$, so that this must be an index two subgroup of $R$. Therefore, each $t \in T$ can be written as a product $s_1 s_2$ with $s_1, s_2 \notin T$ and one of the two factors can be taken arbitrarily. In this way, it is easy to show that $\mathcal{T}$ is Abelian: let $t_1,t_2 \in \mathcal{T}$ and set $t_1=s_1 s$, $t_2=ss_2$; then one has

$$t_1 t_2=s_1 sss_2=s_1 is_2=s_1^{-1} s_2=s_1^{-1}(s_1 s) s_2 (s_1 s)=ss_2 s_1 s=t_2 t_1, \tag{28}$$

where the crucial fourth equality follows from eq. (26). The general group $\mathcal{R}$ is then specified completely by an Abelian group $\mathcal{T}$ with at least one involution $i$ (i.e., an Abelian group of even order) and by a fourth-order element $g_1$ with the defining relations:

$$g_1^4=e; \quad g_1^2=i; \quad g_1 t=t^{-1} g_1 \quad \text{for all} \quad t \in \mathcal{T}. \tag{29}$$

Conversely, a group $\mathcal{R}$ defined by eq. (29) has an automorphism $\alpha(g)$ with $\alpha(g)=g$ or $g^{-1}$ by defining $\alpha(t)=t$ for all $t \in \mathcal{T}$, $\alpha(g_1)=g_1^{-1}$ and requiring the automorphism property $\alpha(gh)=\alpha(g)\alpha(h)$.

The groups defined by eq. (29) must be nonabelian; this is the case if $\mathcal{T}$ does not consist of involutions only: $\mathcal{T} \neq \mathcal{K}(2^q)$. Further, it is easily seen that the group of eq. (29) is identical with $\mathcal{Q} \otimes \mathcal{K}(2^q)$ if $\mathcal{T}$ has the form $\mathcal{C}(4) \otimes \mathcal{K}(2^q)$. This will also be excluded in the following. For $\mathcal{T}$ not of these types, $\mathcal{R}$ will be denoted by $\mathcal{C}(4) * \mathcal{T}$, since $\mathcal{C}(4)$ is the group generated by $g_1$. It is easy to see that this group admits only one automorphism with the property of eq. (4). Since eq. (24) shows that the permutation $\tau$ derived from $\alpha$ by eq. (22) leaves all graphs of the MI of $\mathcal{C}(4) * \mathcal{T}$ invariant, this group is not permissible. As an example, the MI of $\mathcal{C}(4) * \mathcal{C}(6)$ generated by

$$g_1=(1278)(45\ 10\ 11)(12\ 963), \quad t=(13579\ 11)(2468\ 10\ 12) \tag{30}$$

is shown in Fig. 2 on the next page. The permutation $\tau$ is here

$$\tau=(28)(4\ 10)(6\ 12)$$

and this is the only element of the letter-1-fixing subgroup of $\mathcal{C}(4) * \mathcal{C}(6)$, as can be seen from Fig. 2. In general, the permissible group corresponding to $\mathcal{C}(4) * \mathcal{T}$ is given by $\langle \tau, \mathcal{C}(4) * \mathcal{T} \rangle$, but this will be shown as a by-product of step (c).

Step (c). It remains to prove, that all nonabelian regular groups, which are neither equal to $\mathcal{Q} \otimes \mathcal{K}(2^q)$ nor to $\mathcal{C}(4) * \mathcal{T}$, are permissible. To this end, consider the MI of $\mathcal{R}$ and assume that there is a permutation $\tau'$ with $\tau'(1)=1$, which leaves all graphs invariant, but which is neither given by eq. (3) nor derived from an automorphism as in eq.

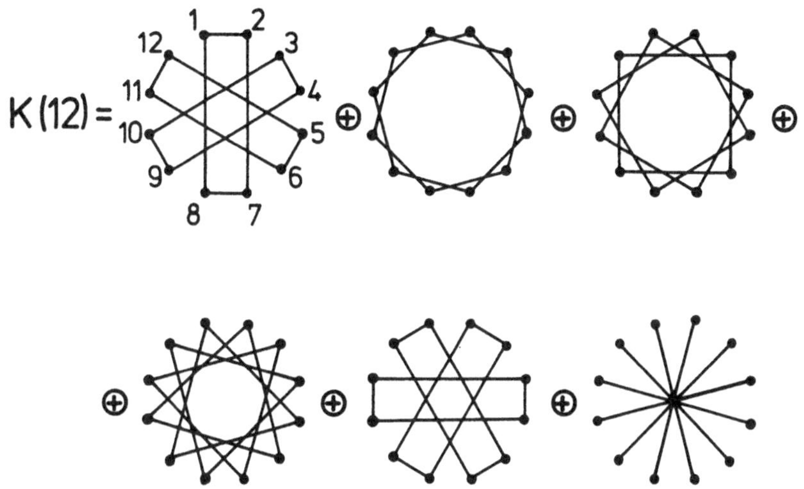


Fig. 2. The maximal interaction of $C(4)*C(6)$ as given by eq. (30).

(4). It immediately follows from Lemma 1, that $\tau'$ must be such that

$$\tau' g(1) = g(1) \quad \text{or} \quad \tau' g(1) = g^{-1}(1) \tag{31}$$

holds for all $g$. Then two disjoint sets $C$ and $D$ are defined by

$$C=\{g \mid g\varepsilon R,\ \tau' g(1)=g(1)\}; \quad D=\{g \mid g\varepsilon R,\ \tau' g(1)=g^{-1}(1)\}. \tag{32}$$

Now the invariance of the MI must be used in a more detailed fashion; let $G_s=\{g(1), gg_s(1)\}$ be a graph from this MI; then the condition $\tau' G_s = G_s$ yields four possible cases:
(i) $g\varepsilon C$ and $gg_s \varepsilon C$: this edge is kept invariant by $\tau'$, so that this case is trivial;
(ii) $g\varepsilon C$ and $gg_s\varepsilon D$: this edge is mapped onto $\{g(1), g_s^{-1}g^{-1}(1)\}$; for this edge to belong to $G_s$, one needs either

$$(\alpha)\ g^{-1}g_s^{-1}g^{-1}=g_s \quad \text{or} \quad (\beta)\ g^{-1}g_s^{-1}g^{-1}=g_s^{-1}.$$

Case ($\alpha$) implies $(gg_s)^2=e$, but then $gg_s\varepsilon C$ by definition, so that only case ($\beta$) remains; by varying $s$, this implies

$$cdc = d \quad \text{for all} \quad c\varepsilon C,\ d\varepsilon D. \tag{33}$$

(iii) $g\varepsilon D$, $gg_s\varepsilon C$: this edge has $\{g^{-1}(1), gg_s(1)\}$ as image under $\tau'$; for this to belong to $G_s$, $g^2 g_s = g_s$ or $g_s^{-1}$ must hold. The first case implies $g^2=e$, $g\varepsilon C$, which is impossible; the second case leads to eq. (33) again.

(iv) $g\varepsilon D$, $gg_s\varepsilon D$: the image of this edge under $\tau'$ is $\{g^{-1}(1), g_s^{-1}g^{-1}(1)\}$ so that there are again two cases;

(α) $gg_s^{-1}g^{-1}=g_s^{-1}$, and $g$ and $gg_s$ commute, or

(β) $gg_s^{-1}g^{-1}=g_s$, which implies $g^2=(gg_s)^2$.

By varying $s$, this yields: for $d_1, d_2\varepsilon D$ one or both of the conditions

$$d_1d_2=d_2d_1, \quad d_1{}^2d_2{}^2=e, \tag{34}$$

hold(s). From eq. (33), it follows that if $d_1, d_2\varepsilon D$, then $d_1d_2\varepsilon C$: if $d_1d_2$ would also belong to $D$, one has

$$cd_1c=d_1, \; cd_2c=d_2, \; cd_1d_2c=d_1d_2=cd_1c^2d_2c \quad \text{and} \quad c^2=e \quad \text{for all} \quad c\varepsilon C;$$

But then $\tau'$ equals the $\sigma$ of eq. (3) and this has been excluded. This then implies that the second of the conditions of eq. (34) always holds: since $d_1d_2\varepsilon C$, one has $d_1d_2d_1d_1d_2=d_1$ by eq. (33); if $d_1d_2=d_2d_1$, this implies $d_1{}^2d_2{}^2=e$. Finally, let $d\varepsilon D$, $d^{-1}\varepsilon C$; then eq. (33) implies $d^{-1}dd^{-1}=d=d^{-1}$ or $d\varepsilon C$, which is a contradiction. The above yields the following characterization of $C$ and $D$:

(α) For all $c\varepsilon C$, $d\varepsilon D$, one has $cdc=d$;

(β) For all $d_1, d_2\varepsilon D$, $d_1d_2\varepsilon C$;

(γ) All $d\varepsilon D$ satisfy $d^4=e$, $d^2=i$, $i$ a fixed involution from $C$.

Conversely, if $R$ is split up into two disjoint sets $C$ and $D$ so that conditions (α),(β),(γ) are fulfilled, then the permutation $\tau'$ defined by the inverse of eq. (32):

$$\tau'g(1)=g(1) \quad \text{for} \quad g\varepsilon C; \quad \tau'g(1)=g^{-1}(1) \quad \text{for} \quad g\varepsilon D; \tag{35}$$

leaves all graphs of the MI of $R$ invariant.

Now an element $c\varepsilon C$ is calles <u>proper</u> if $cd\varepsilon D$ for all $d\varepsilon D$, so that $c'\varepsilon C$ <u>improper</u> implies that there exists a $d\varepsilon D$ with $c'd\varepsilon C$. The following two statements will be proved: (i) $c'^2=i$ for $c'$ improper; (ii)

$cc'c=c'$ for $c'$ improper, $c$ proper.

(i) Let $c'$ be improper, $d\varepsilon D$ such that $c'd\varepsilon C$; eq. (33) implies

$$(c'd)d(c'd)=d \quad \text{or} \quad c'^{-2}=d^2=i, \tag{36}$$

so that $c'^2=i$ follows.

(ii) Let $c$ be proper, $c'$ improper, $d\varepsilon D$ such that $c'd\varepsilon C$; set $c=dd_1$, then $d_1\varepsilon D$ necessarily. Set $cc'c=x$; multiplying this with $c'$ on the left yields

$$c'cc'c = c'dd_1c'dd_1 = (c'd)d_1(c'd)d_1 = d_1^{\,2} = i = c'x. \tag{37}$$

Since $c'^2=i$ by (i), $x=c'$ follows.

From the above, the sets $C'$ and $D'$ defined by

$$C'=\{c\,|\,c\varepsilon C,\ c \text{ proper}\},\quad D'=\{c'\,|\,c'\varepsilon C,\ c \text{ improper}\}\cup D, \tag{38}$$

are such that the conditions ($\alpha$), ($\beta$) and ($\gamma$) above are satisfied, so that the permutation $\tau$ defined by

$$\tau g(1)=g(1) \quad \text{for} \quad g\varepsilon C'; \quad \tau g(1)=g^{-1}(1) \quad \text{for} \quad g\varepsilon D', \tag{39}$$

leaves all graphs of the MI of $R$ invariant. But now, the proper elements of $C$ form a group: the product $c_1c_2$ is such that $c_1c_2d\varepsilon D$ if $c_1d'$ and $c_2d''$ belong to $D$ for all $d',d''\varepsilon D$. Actually, this group is even Abelian: set $c_1=d_1d_2$, $c_2=d_2d_3$; then one has

$$c_1c_2=d_1d_2d_2d_3=d_1d_3^{\,-1} \quad \text{and} \quad c_2c_1=d_2d_3d_1d_2=c_2d_1c_2d_3^{\,-1}=d_1d_3^{\,-1}.$$

Clearly then, the $\tau$ defined by eq. (39) is an automorphism of $R$ as described in step (b), since $d_1'd_2'\varepsilon C'$ for all $d_1',d_2'\varepsilon D'$ follows again if $\tau$ is not equal to the $\sigma$ of eq. (3). However, $R$ is not supposed to be of this type, so that a contradiction ensues. Therefore, all non-abelian, regular groups, which are neither of the type $Q\otimes K(2^q)$ nor of the type $C(4)*T$, are permissible.

It remains to determine the permissible group corresponding to the $C(4)*T$ groups. By the above proof, a permutation $\tau'\neq\tau$ leaves the MI invariant, if $D'$, which consists of all elments of the form $gt$ with $g^2=i$ and $t\varepsilon T$, $T$ Abelian, $T=C'$, can be split into two classes such that for $gt_1$ in one class (improper elements of $C$) and $gt_2$ in the second class (proper elements of $D$), one always has

$$gt_1 gt_2 gt_1 = gt_2. \tag{40}$$

By the commutation relation $gt=t^{-1}g$, all $t\varepsilon T$, this implies $it_2{}^2=t_1{}^2$, or, with $t_1=qt_2$, $q^2=i$. Such an induced splitting of $C'$ in two classes is only possible if $T=C(4)\otimes K(2^q)$, which has been excuded, since then the group is the $Q\otimes K(2^q)$ of step (a). This then implies that the permissible group corresponding to $C(4)*T$ is the group $<\tau, C(4)*T>$ with twice as many elements. This finishes steps (b) and (c).

The above results may be summarized in terms of the following theorem:

Theorem 1. Nonabelian regular groups are of three distinct types:
Type (a): $R=Q\otimes K(2^q)$ with $Q$ the quaternion group; the corresponding completely permissible group is $(S(2)\sim K(4))\otimes K(2^q)$ with eight times as many elements.
Type (b): $C(4)*T$, the group defined by eq. (29) above; the corresponding permissible group has twice as many elements as shown above.
Type (c): all other regular groups; they are permissible.

Remark. The groups of types (a) and (b) are metabelian. For $R$ of type (a), the commutator group is $R'=\{e,i\}$, $i$ the involution of the quaternion group, whereas for $R$ of type (b), the commutator group is generated by the squares of the elements of the Abelian subgroup, $R'=<t^2|t\varepsilon T>$.

The question as to the existence of P-algebras for the permissible groups corresponding to regular groups is solved by the next theorem:

Theorem 2. For $R$ of type (a), the complete permissibility of $R^{(p)}$ implies the P-algebra property trivially. For $R$ of types (b) and (c), $R^{(p)}$ has no P-algebra.
Proof. For cases (b) and (c), suppose that $R^{(p)}$ has a P-algebra. Then, by Lemma 1.5.4, the matrices $\underline{\underline{M}}_k$ describing the MI must commute with all symmetric and antisymmetric matrices which commute with all $\underline{\underline{D}}(g)$, $g\varepsilon R$. If $g$ is an involution, the matrix

$$A_{ij}(g) = \begin{cases} 1 & \text{if } g_i{}^{-1}g_j=g, \\ 0 & \text{otherwise,} \end{cases} \tag{41}$$

is symmetric; if $g$ is not an involution, then $\underline{\underline{A}}(g)+\underline{\underline{A}}(g)^T$ is symmetric and $\underline{\underline{A}}(g)-\underline{\underline{A}}(g)^T$ is antisymmetric. Further, all matrices $\underline{\underline{A}}(g)$ commute with all $\underline{\underline{D}}(g)$. On the other hand, $\underline{\underline{A}}(g)$ may be interpreted as a matrix representing a permutation $\tilde{g}$:

$$\underline{\underline{A}}(g) = \underline{\underline{D}}(\tilde{g}), \quad \tilde{g}(j) = \tilde{g}g_j(1) \equiv g_j g^{-1}(1). \tag{42}$$

The matrices $\underline{\underline{M}}_k$ must, therefore, commute with all $\underline{D}(\tilde{g})$. But then, all $\tilde{g}$ must belong to $R^{(p)}$ by the definition of permissibility. For $R$ of types (b) and (c), this is clearly not the case, so that the corresponding permissible groups cannot have P-algebras. ¶

Remark. The permutations $\tilde{g}$ actually form the group $R^*$ defined in the next section. By Lemma 3.3 below, $R^*<R^{(p)}$ implies $H(R)=R^{(p)}$ with $H(R)$ the permissible class function group defined in Theorem 3.1 below. The theorem above then also follows from Corollary 3.1.

## 4.3. Permissible class function groups.

In eqs. (1.3), the energy function corresponding to a regular group $R$ has been expressed in terms of a function $f(\hat{g})$ defined on the isomorphic abstract group $\hat{R}$. In this section, the extra symmetries (if any) of the energy function are investigated for the case that $f(\hat{g})$ is a class function on $\hat{R}$, i.e., one has

$$f(\hat{g}) = f(\hat{h}^{-1}\hat{g}\hat{h}) \quad \text{for all} \quad \hat{g},\hat{h}\varepsilon\hat{R}. \tag{1}$$

Since eq. (1) is trivially fulfilled if $R$ is Abelian, one may assume that $\hat{R}$ is one of the types of nonabelian, regular groups described by Theorem 2.1. Eq. (1) can also be formulated by the requirement, that $f(\hat{g})$ be invariant with respect to all inner automorphisms $\alpha_{\hat{h}}$ of $\hat{R}$, $\alpha_{\hat{h}}(\hat{g})=\hat{h}^{-1}\hat{g}\hat{h}$. This formulation induces then a study of such functions invariant with respect to all automorphisms of $\hat{R}$, see Section 4.

For $f(\hat{g})$ a class function, the values that $E(i,j)$ can take may be indexed by $E_k$, where this value is taken on (by $f$) on a union of two equivalence classes:

$$\hat{R}_k = \{\hat{h}^{-1}\hat{g}\hat{h}\,|\,\hat{h}\varepsilon\hat{R}\} \cup \{\hat{h}^{-1}\hat{g}^{-1}\hat{h}\,|\,\hat{h}\varepsilon\hat{R}\}. \tag{2}$$

Here, $\hat{g}$ can be taken as that element of $\hat{R}$ to which the (unique) element $g_k$ of $R$ with $g_k(1)=k$ belongs, see eq. (1.1). The graph $H_k$ belonging to the value $E_k$ can be expressed as the union of a number of graphs from the MI of $R$:

$$H_k=\{i,j\,|\,\hat{g}_i^{-1}\hat{g}_j=\hat{g}_s,\ \hat{g}_s\varepsilon\hat{R}_k\}=\{g(1),gg_s(1)\,|\,g\varepsilon R,\ g_s\varepsilon R_k\}= \bigcup_{g_t\varepsilon R_k} G_t. \tag{3}$$

These graphs are certainly hyperregular and the intersection $H(R)$ of their automorphism groups is transitive since it contains $R$. Therefore, the group $H(R)$ is permissible by Theorem 1.3.1. $H(R)$ is called the permissible class function or PCF group corresponding to $R$ (or to $\hat{R}$ ). This group is completely decribed by the next theorem:

Theorem 1. $H(R) = \langle R,\sigma\rangle^{(p)}$, the completely permissible group corresponding to the group generated by $R$ and $\sigma$ defined as in eq. (2.3):

$$\sigma(s) = g_s^{-1}(1) \quad \text{for} \quad s\neq 1, \quad \sigma(1)=1. \tag{4}$$

The MI of $H(R)$ consists exactly of the graphs $H_k$.

Proof. Let $G_t$ be a graph from the MI of $R$. Then $\sigma G_t$ is given as

$$\sigma G_t=\{\sigma g(1),\sigma g g_t(1)\}=\{g^{-1}(1),g_t^{-1}g^{-1}(1)\}. \tag{5}$$

Hence, $\sigma G_t$ contains edges from each graph $G_{t'}$, with $g_{t'}=gg_t^{-1}g^{-1}$, so that (i) $\sigma$ keeps all graphs $H_k$ invariant and (ii) $\sigma$ keeps no subgraphs of these invariant. This implies the statements of the theorem, except for the complete permissibility, which follows from $g_s\sigma \varepsilon H(R)$ and $g_s\sigma(1)=s$, $g_s\sigma(s)=1$. ¶

The above Theorem 1 together with Theorem 2.1 yields a useful corollary:

Corollary 1. A nonabelian, regular group has $H(R)=R^{(p)}$ if $R$ is of type (a). For $R$ of types (b) or (c), $H(R)>R^{(p)}$ holds.

Proof. The permutation $\sigma$ defined by eq. (4) belongs to $R^{(p)}$ only if $R$ is of type (a). Since $H(R)$ is permissible and contains $R$, $H(R)\geq R^{(p)}$ holds generally, Lemma 2.1.1. These facts imply the corollary. ¶

There is an interesting connection between the (im)primitivity of $H(R)$ and the property of $H$ of being (non)simple. In order to prove this, a different description of $H(R)$ is needed:

Lemma 1. $H(R)$ is the permissible group corresponding to the semidirect product of $R$ with its group $\mathrm{INT}(R)$ of inner automorphisms, which consists of those elements $\sigma_i$ of $S(|R|)$ defined by

$$\sigma_i(k) = \sigma_i g_k(1) = g_i^{-1}g_k g_i(1) = \alpha_i(g_k)(1). \tag{6}$$

Proof. It is easy to see, that the graphs of the MI of $R\,\mathrm{INT}(R)$ are exactly those of $H(R)$, so that $H(R)=(R\,\mathrm{INT}(R))^{(p)}$. ¶

Lemma 2. $\langle R,\sigma\rangle$ contains $R\,\mathrm{INT}(R)$ as a (normal) subgroup of index two. Therefore, one has

$$|\langle R,\sigma\rangle| = 2|R|^2/|Z(R)| \,. \tag{7}$$

Proof. The inclusion $\langle R\,\mathrm{Int}(R),\sigma\rangle \geqq \langle R,\sigma\rangle$ is trivial. The formula

$$g_i^{-1}\sigma g_i^{-1}\sigma = \sigma_i \tag{8}$$

is easily checked, so that one also has $R\,\mathrm{Int}(R) \leqq \langle R,\sigma\rangle$. Since $\sigma$ commutes with all $\sigma_i$, this implies that $R\,\mathrm{Int}(R)$ has exactly half as many elements as $\langle R\,\mathrm{Int}(R),\sigma\rangle$. Two elements of $R$ lead to the same $\sigma_i$, if they differ by an element from the center $Z(R)$ of $R$. This implies $|\mathrm{Int}(R)|=|R|/|Z(R)|$ yielding eq. (7). ¶

Now it is possible to prove:

Theorem 2. $H(R)$ is primitive iff $R$ is simple.
Proof. By Lemmas 1 and 3.1.1, $H(R)$ is primitive iff $R\,\mathrm{Int}(R)$ is primitive. Suppose first, that $R$ is not simple. Then there is a proper normal subgroup $N$ of $R$. Then $N\,\mathrm{Int}(R)$ is a normal subgroup of $R\,\mathrm{Int}(R)$, since $N$ consists of full classes of conjugate elements by definition. If $R\,\mathrm{Int}(R)$ is primitive, then $N\,\mathrm{Int}(R)$ is transitive by Lemma 3.2.2. But this in obviously not true, since $N<R$ and $R$ is regular. Therefore, $R$ not simple implies that $H(R)$ is imprimitive. Now let $R$ be simple. Then $\mathrm{Int}(R)$ is a maximal subgroup of $R\,\mathrm{Int}(R)$: if $N\,\mathrm{Int}(R)$ is a larger subgroup, then $N$ would be normal in $R$. Lemma 3.3.1 now implies the theorem. ¶

For some purposes, a third description of $H(R)$ is useful. This is based on the fact that every element $g_i\sigma_k$ of $R\,\mathrm{Int}(R)$ can be written as $(g_ig_k^{-1})g_k\sigma_k$. Now the elements $g_k\sigma_k$ of this form a group, denoted by $R^*$, with the multiplication rule

$$g_j\sigma_j g_k\sigma_k = g_s\sigma_s \quad \text{iff} \quad g_s=g_kg_j. \tag{9}$$

The mapping $\pi:R\to R^*$ with $\pi(g_k)=g_k\sigma_k$ is one-to-one onto and is an anti-isomorphism:

$$\pi(g_jg_k) = \pi(g_k)\pi(g_j). \tag{10}$$

Lemma 3. One has $RR^*=R^*R=R\,\mathrm{Int}(R)$, so that $H(R)=(RR^*)^{(p)}$ holds. Every element of $R$ commutes with every element of $R^*$. The permutation $\sigma$ maps $R$ onto $R^*$ in a one-to-one fashion.
Proof. The equality $H(R)=(RR^*)^{(p)}$ is clear. The commutativity is easily checked:

$$g_jg_k\sigma_k(s)=g_jg_kg_k^{-1}g_sg_k(1)=g_jg_sg_k(1)=g_kg_k^{-1}g_jg_sg_k(1)=g_k\sigma_kg_j(s). \tag{11}$$

The action of $\sigma$ on elements of $R$ and of $R^*$ is easily calculated:

$$\sigma g_k^{-1}\sigma = g_k\sigma_k; \quad \sigma g_k\sigma_k\sigma = \sigma g_k\sigma\sigma_k = g_k^{-1}. \tag{12}$$

¶

The description of $H(R)$ afforded by Lemma 3 makes it possible to say something about the existence of regular Abelian subgroups of $H(R)$, which are important for duality transformations, see Section 7.3.

Theorem 3. Let $A$ and $B$ be two subgroups of $R$ with the properties (a) $R=AB$ and (b) $A\cap B\leq Z(R)$. Then $H(R)$ contains regular groups $A\pi(B)$ and $\pi(A)B$. If $A$ and $B$ are both Abelian, condition (b) follows from (a) and the regular groups $A\pi(B)$ and $\pi(A)B$ are Abelian. If, in addition, $A\cap B=\{e\}$ holds, then these groups are isomorphic to $A\otimes B$.
Proof. Since $R$ and $R^*$ commute elementwise by Lemma 3, $A\pi(B)$ and $\pi(A)B$ are, by condition (a), transitive subgroups of $H$. Since the mapping $\pi$ leaves all elements of the center $Z(R)$ fixed, these groups are regular by condition (b). The extra statements for $A,B$ Abelian are immediate. ¶

Corollary 2. For $R$ of types (a) or (b), $H(R)$ contains an Abelian, regular subgroup. For $R$ of type (c), no general statement can be made.
Proof. If $R$ is of type (a), it is a direct product $Q\otimes K(2^q)$. For $g_1$, $g_2$ the generators of the quaternion group, one has

$$Q\otimes K(2^q) = \langle g_1\rangle(\langle g_2\rangle\otimes K(2^q)), \tag{13}$$

so that the conditions of Theorem 3 are satisfied for $A=\langle g_1\rangle$, $B=\langle g_2\rangle\otimes K(2^q)$; both groups are Abelian, but since $A\cap B=\{e,i\}$, the Abelian regular subgroups $\pi(A)B$ and $A\pi(B)$ of $H(R)$ are not isomorphic to $A\otimes B$.
For $R$ of type (b), $R=C(4)*T$, one easily sees that $R=C(4)T$, so that again $\pi(A)B$ and $A\pi(B)$ are Abelian regular subgroups with $A=C(4)$, $B=T$, $A\cap B=\{e,i\}$. ¶

As examples of groups of type (c), consider the following. Let $A$ be an Abelian group, which is nonpermissible, so that $A^{(p)}=\langle A,\sigma\rangle$ as in Theorem 2.5.3. Let $A^{(1)}$ be an index two subgroup of $A$ and define $F_A(A^{(1)})$ by

$$F_A(A^{(1)}) = A^{(1)}\cup[(A-A^{(1)})\sigma]. \tag{14}$$

This is a generalization of the situation in Theorem 2.5.4. Now if this group is not accidentally Abelian, which is the case only for $A^{(1)}\simeq K(2^q)$, it is permissible by Corollary 2.1.1, and, therefore, of type (c) since it is regular. Now if $g \in A$, $g \notin A^{(1)}$, one has $A = A^{(1)} \cup gA^{(1)}$; therefore, eq. (14) implies $[(g\sigma)^2 = g\sigma g\sigma = gg^{-1} = e]$

$$F_A(A^{(1)}) = A^{(1)}\ \{e, g\sigma\}, \tag{15}$$

so that $H(F_A(A^{(1)}))$ contains an Abelian subgroup isomorphic to $A^{(1)} \otimes S(2)$. In particular, the $H(F(2m))$ corresponding to the groups $F(2m)$ of Theorem 2.5.4 contain Abelian subgroups isomorphic to $C(m) \otimes S(2)$.

For some of the groups of types (b) and (c), the corresponding PCF groups can be given explicitly in terms of direct and wreath products. This is shown in the next few lemmas:

Lemma 4. Let $R = C(4) * T$ be such that $|T| = 2k$, $k$ odd; then $H(R) \simeq T^{(p)} \sim S(2)$.
Proof. The equivalence classes of the elements of $C(4) * T$ are easily seen to be given by

$$\{e\};\ \{t\},\ t \in T,\ t^2 = e;\ \{t, t^{-1}\},\ t \in T,\ t^2 \neq e;\ \{gtt'^2 \mid t' \in T\},\ t \in T. \tag{16}$$

For the special case of the lemma, the MI of $H(R)$ consists of two types of graphs: for every graph from the MI of $T$, there is a graph consisting of two copies of this; there is one other graph only. This implies the lemma. ¶

Lemma 5. Let $R = C(4) * T$ be such that $T = K(2^q) \otimes \tilde{T}$ with $|\tilde{T}| = 2k$, $k$ odd, and $i = g^2 \in \tilde{T}$. Then $H(R) \simeq K(2^q) \otimes [\tilde{T}^{(p)} \sim S(2)]$.
Proof. Immediate from Theorem 1 and Lemma 4. ¶

Lemma 6. Let $R = F_A(A^{(1)})$ be such that $|A^{(1)}| = k$, $k$ odd; then $H(R) \simeq (A^{(1)})^{(p)} \sim S(2)$.
Proof. The equivalence classes of $F_A(A^{(1)})$ are, by eq. (14), given as

$$\{e\};\ \{a\},\ a \in A^{(1)},\ a^2 = e;\ \{a, a^{-1}\},\ a \in A^{(1)},\ a^2 \neq e;\ \{ba^2\sigma \mid a \in A^{(1)}\},\ b \in A - A^{(1)}. \tag{17}$$

For the special case of the lemma, there is again only one graph in the MI of $H(R)$ in addition to the ones consisting of two copies of graphs from the MI of $A^{(1)}$. ¶

Lemma 7. Let $R = F_A(A^{(1)})$ be such that $A^{(1)} = K(2^q) \otimes \tilde{A}^{(1)}$ with $|\tilde{A}^{(1)}| = k$, $k$ odd. Then $H(R) \simeq K(2^q) \otimes [(\tilde{A}^{(1)})^{(p)} \sim S(2)]$.
Proof. Immediate from Theorem 1 and Lemma 6. ¶

It is interesting to note, that in the cases of Lemmas 6 and 7 the group $H(R)$ actually equals $\langle R, \sigma \rangle$ as can be seen by comparing the number of elements of these PCF groups with eq. (7) of Lemma 2. In the cases of lemmas 4 and 5, $\langle R, \sigma \rangle$ is an index two subgroup of $H(R)$, however.

Examples of groups of type (c), which are not of the structure described by eq. (14), are afforded by the regular representations of the symmetric groups for $M \geqq 4$ : $S_r(M)$ is defined as a permutation group on its own $M!$ elements by eq. (1.1). Note that one has $S_r(2) \simeq S(2)$ and $S_r(3) \simeq F(6)$. Similarly, the alternating group $A_r(M)$ of all <u>even</u> permutations of $M$ letters, $M \geqq 3$, can be represented as a permutation group on its own $M!/2$ elements, since this is an index two subgroup of $S_r(M)$. Since $A_r(3) \simeq C(3)$, only the case $M \geqq 4$ need be considered. It is well-known ([2]), that $A_r(M)$ is simple for $M \geqq 5$; below, this result does not have to be used to find the PCF groups corresponding to $A_r(4)$ and $A_r(5)$:

<u>Theorem 4</u>. $H(A_r(4)) = S(4) \sim S(3)$; $H(A_r(5))$ is a primitive group with three graphs in its MI, so that $A_r(5)$ is simple.

<u>Proof</u>. (i) Consider first $A_r(4)$; this group has twelve elements, which are of three distinct types: ($\alpha$) the identity $e$, ($\beta$) permutations of the form $(ij)(mn)$; there are three of these, and they are all in the same class of conjugate elements; ($\gamma$) permutations of the form $(ijk)$; there are eight of these, which together make up two classes of conjugate elements consisting of each others inverses. Theorem 1 implies that the MI of $H(A_r(4))$ consists of two graphs, one with $z_1 = 3$, the other with $z_2 = 8$. If this group is primitive, both graphs satisfy $(\bar{G}_k)^{(2)} = G_k$ by Theorem 3.1.1. This implies $z_k \leqq 7$ by Lemma 2.2.5, so that $H(A_r(4))$ must be imprimitive; it is then a wreath product $S(M_1) \sim S(M_2)$ with $12 = M_1 M_2$ by Theorem 3.1.1. The correct values $M_1 = 4$ and $M_2 = 3$ easily follow from the $z_k$-values.

(ii) Consider now $A_r(5)$ with 60 elements, the even permutations of five letters. These are easily seen to belong to four (unions of inverse) conjugacy classes: ($\alpha$) $e$, the identity; ($\beta$) 15 elements of the form $(ij)(mn)$; ($\gamma$) 20 elements of the form $(ijk)$; ($\delta$) 24 elements of the form $(ijkmn)$. Therefore, the MI of $H(A_r(5))$ consists of three graphs $G_k$ with $z_k = 15$, 20, 24. By Theorem 3.1.2, each imprimitive group with three graphs in its MI is such, that at least one of the numbers $z_k + 1$ or $M - z_k$ divides $M$. For the case at hand, none of these divides 60, so that $H(A_r(5))$ must be primitive. Then $A_r(5)$ is simple by Theorem 2. ¶

## <u>4.4. Permissible characteristic interaction groups.</u>

As already announced at the beginning of the previous section, a different type of extension of a regular group can be obtained by requiring the function f(g) defined on $R$ (which gives the energy function), to be invariant with respect to all automorphisms of $R$. It is easy to show, that, in analogy with Lemma 3.1, these permissible groups are given by

$$J(R) = (R\ \mathrm{Aut}(R))^{(p)}, \qquad (1)$$

where $\mathrm{Aut}(R)$ is the full automorphism group $H_M(R)$ of Lemma 2.1.3, since $H_1=\{e\}$ for a regular group. The group $J(R)$ is called the <u>permissible</u> <u>characteristic</u> <u>interaction</u> group or <u>PCI</u> group corresponding to the regular group $R$, which now also can be Abelian. One obviously has $H(R) \leqq J(R)$, so that the PCI group is always completely permissible. The (im)primitivity of $J(R)$ is described by the next theorem, which is analogous to Theorem 3.2:

<u>Theorem 1</u>. $J(R)$ is primitive iff $R$ is characteristically simple. In particular, $J(R)=S(|R|)$ iff $R$ is characteristically simple and solvable, i.e., if it is a group $C^{(q)}(p)$ as defined in Theorem 3.2.1.
<u>Proof</u>. By eq. (1), $J(R)$ is primitive iff $R\ \mathrm{Aut}(R)$ is primitive and this is the case iff $\mathrm{Aut}(R)$ is a maximal subgroup of this. By Definition 3.2.5, this is the case iff $R$ is characteristically simple, i.e., iff $R$ is the direct product of a number of isomorphic simple groups. It follows, that $J(R)$ is primitive for $R=C^{(q)}(p)$, the direct product of q simple, solvable groups $C(p)$. But for these groups, $\mathrm{Aut}(R)$ is clearly transitive on the elements of $C^{(q)}(p)$, so that $J(R)=S(|R|)$ follows. Let, conversely, this latter equality hold; then $J(R)$ is primitive, so that $R$ is characteristically simple. All graphs of the MI of $R$ must be isomorphic, which can only be the case if all its elements satisfy $g^p=e$ for some prime p, so that $R=C^{(q)}(p)$ holds. ¶

For a number of groups, the groups $J(R)$ can be given explicitly again. This is the subject of the following lemmas.

<u>Lemma 1</u>. The MI of $R$ consists of a number of graphs, which can be grouped together if they are isomorphic. Let the sums of isomorphic graphs make up a set of graphs $W_k$. Then the intersection of the automorphism groups of these graphs is a permissible group $W(R) \geqq J(R)$.
<u>Proof</u>. Two graphs $G_1=\{g(1),gg_1(1)\}$ and $G_2=\{g(1),gg_2(1)\}$ are isomorphic iff $g_1$ and $g_2$ are of the same order, since then these graphs consist of $M/q$ (q is the order) circles of length q for $q \neq 2$ or of $M/2q$ isolated edges for $q=2$. Since an automorphism of a group conserves the order of each element, the MI of $J(R)$ consists of graphs, which are sums of isomorphic graphs from the MI of $R$. This implies $J(R) \leqq W(R)$. ¶

Lemma 2. For $R$ a cyclic group $C(M)$, $J(R)=W(R)$. In particular, for

$$C(M) = C(p_1^{t_1}) \otimes C(p_2^{t_2}) \otimes \ldots \otimes C(p_n^{t_n}), \quad p_i \neq p_j \quad \text{for} \quad i \neq j,$$

one has

$$J(C(M)) = J(C(p_1^{t_1})) \otimes J(C(p_2^{t_2})) \otimes \ldots \otimes J(C(p_n^{t_n})), \tag{2}$$

where the factors are given as

$$J(C(p^t)) = S(p) \sim S(p) \sim \ldots \sim S(p), \quad t \text{ factors.} \tag{3}$$

Proof. Let $\alpha$ be an automorphism of a cyclic group $C(M)=\langle g \rangle$, $g^M=e$. Then one must have $\alpha(g)=g^k$ for some $k$; since $g^k$ must have the same order as $g$, $k$ cannot have a divisor in common with $M$, notation: $(k,M)=1$. Conversely, if $\alpha(g)=g^k$ with $(k,M)=1$, then $\alpha$ is an automorphism of $C(M)$. The automorphism group $\mathrm{AUT}(C(M))$, therefore, has order $\phi(M)$ with $\phi$ the Euler function, giving the number of $k$-values with $(k,M)=1$. This automorphism group is transitive on the elements $g \in C(M)$ which have $g^M=e$, $g^k \neq e$ for $k<M$. Now let $C(s)$ be the (unique) subgroup of $C(M)$ generated by elements $g$ with $g^s=e$, $g^k \neq e$ for $k<s$. Clearly, $C(s)$ is characteristic in $C(M)$, so that the restriction of $\mathrm{AUT}(C(M))$ to $C(s)$ is a subgroup of $\mathrm{AUT}(C(s))$. But, on the other hand, this restriction contains exactly $\phi(s)$ elements, so that it equals $\mathrm{AUT}(C(s))$. Therefore, for all $s$ (including $M$), which divide $M$, $\mathrm{AUT}(C(M))$ is transitive on the sets $\{g \mid g^s=e, g^k \neq e \text{ for } k<s\}$. This implies $J(R)=W(R)$ by Lemma 1. The rest of the lemma follows easily by consideration of the divisors $s$: (i) $C(p^t)$ contains (apart from the identity) $p-1$ elements with $s=p$, $p^2-p$ elements with $s=p^2, \ldots, p^t-p^{t-1}$ elements with $s=p^t$. Therefore, the MI of $J(C(p^t))$ consists of $t$ graphs; eq. (3) follows by induction from the simple cases $t=1$ or $2$, which are evident. (ii) For $p_1 \neq p_2$, one has, by Lemma 2.3.1,

$$J[C(p_1^{t_1}) \otimes C(p_2^{t_2})] \leqq J(C(p_1^{t_1})) \otimes J(C(p_2^{t_2})), \tag{4}$$

since the automorphism group of the direct product is the direct product of the automorphism groups in this case. But the number of divisors $s$ of $p_1^{t_1} p_2^{t_2}$ is easily seen to be $t_1+t_2+t_1t_2$, so that the numbers of graphs of the MI's of the groups on both sides of eq. (4) are equal by the complete permissibility of the PCI groups. This implies equality in eq. (4); eq. (3) follows again by induction. ¶

To show that $J(R)<W(R)$ occurs, consider the Abelian group $C(p)\otimes C(p^2)$. It is easily seen that the elements of order $p$ from $C(p^2)$ must form a characteristic set of this group, so that $J(C(p)\otimes C(p^2))$ has a three-graph MI with $z_1=p^3-p^2$ ($s=p^2$), $z_2=p^2-p$ ($s=p$ not from $C(p^2)$) and $z_3=p-1$ ($s=p$ from $C(p^2)$). Theorem 3.1.2 shows, that one must have

$$J(C(p)\otimes C(p^2)) = S(p)\sim S(p)\sim S(p). \tag{5}$$

On the other hand, one obviously has

$$W(C(p)\otimes C(p^2)) = S(p^2)\sim S(p). \tag{6}$$

Lemma 3. $J(F(2m))=J(C(m))\sim S(2)$ for odd $m$. (7)
Proof. $(2m)$ consists of the powers $g^{2k}$ and of the elements $g^{2k+1}\sigma$, $g^{2m}=e$, $g\sigma=\sigma g^{-1}$. Since $m$ is odd, every automorphism must map these two types of elements onto ones of the same type. An automorphism $\alpha_1$ of $C(m)=\langle g^2\rangle$ can be extended to the whole group by setting $\alpha_1(g\sigma)=g\sigma$. Further, the mapping $\beta$ defined by $\beta(g^2)=g^2$ and $\beta(g\sigma)=g^3\sigma$ is also an automorphism. The automorphism group of $F(2m)$ is then transitive on the set $\{g^{2k+1}\sigma\}$ and reduces to $\mathrm{AUT}(C(m))$ on $\{g^{2k}\}$, so that

$$J(F(2m))\geqq J(C(m))\sim S(2) \tag{8}$$

certainly holds. But it is easy to see, that $W(F(2m))$ equals the right-hand-side of eq. (8), so that Lemma 1 implies eq. (7). ¶

Lemma 4. $J(C(4)*C(2m))=J(C(2m))\sim S(2)$ for $m$ odd. (9)
Proof. Completely analogous to the proof of Lemma 3. ¶

Lemma 5. $J(Q\otimes K(2^q))=S(2)\sim S(2^q)\sim S(4)$; (10)

for $q=0$, this means

$$J(Q)=S(2)\sim S(4). \tag{11}$$

Proof. The involutions of $Q\otimes K(2^q)$ fall into two classes, one consisting of the involution $i$ of $Q$, the other of all others. The automorphism group of $Q\otimes K(2^q)$ is, therefore, transitive on three sets: (i) the involution $i$; (ii) $2^{q+1}-2$ other involutions; (iii) $2^{q+3}-2^{q+1}$ elements of order four. Theorem 3.1.2 implies the lemma for $q\neq 0$, whereas the $q=0$ result comes from Theorem 3.1.1. ¶

Lemma 6. $J(A_r(M))=H(A_r(M))$ for $M=4$ or $5$. (12)
Proof. By the proof of Theorem 3.4, the different graphs of the MI of

of these PCF groups are generated by elements of different order in both cases. These, however, cannot be mixed by automorphisms. ¶

REFERENCES.

(1). B. Huppert, Endliche Gruppen I (Springer-Verlag, Berlin, Heidelberg, New York, 1963) p. 91.
(2). Ref. (1), p. 156.

## 5. GRAPH-THEORETICAL CONSTRUCTIONS OF PERMISSIBLE GROUPS.

### 5.1. Primitive permissible groups with two-graph MI's.

Theorem 1.3.2 showed, that the automorphism group $\mathcal{G}(G)$ of a graph $G$ is permissible if it is transitive; in this case, the graph $G$ is hyperregular. Lemma 1.3.1 showed, that $G$ will occur in the maximal interaction of $\mathcal{G}(G)$ if this group is also transitive on the edge set of $G$. In this latter case, the graph $G$ will be called superregular.

By Theorem 3.1.1, the two graphs of the MI of a primitive permissible group $\mathcal{G}$ with a two-graph MI are superregular graphs $G$ and $\overline{G}$. These graphs are such that $(G)^{(2)}=\overline{G}$ and $(\overline{G})^{(2)}=G$ hold, so that $\mathcal{G}(G)=\mathcal{G}(\overline{G})=\mathcal{G}$ and this group has a P-algebra. Therefore, both $G$ and $\overline{G}$ are of the "two-generation-type" shown schematically in Fig. 1 below. In this figure,

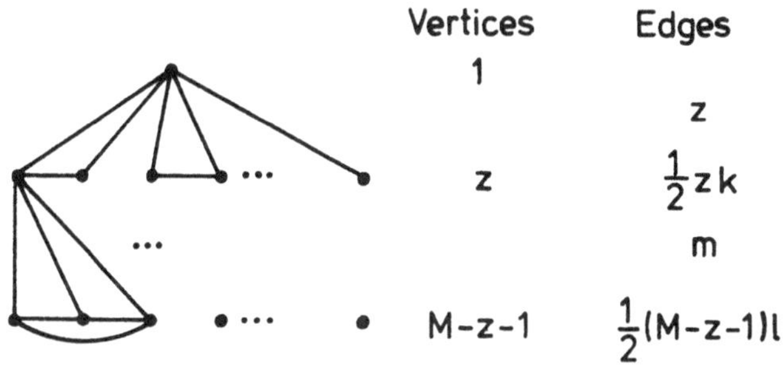


Fig. 1. Schematic representation of a graph with two generations.

$z$ denotes the number of edges adjacent to every vertex of $G$. Now since $G$ is superregular, the graphs formed by the vertices and edges within each generation have to be hyperregular. Therefore, each vertex in the first (second) generation must have the same number $k$ ($\ell$) of edges within each generation adjacent to it. This implies, that there are $zk/2$ edges within the first generation and $\ell(M-z-1)/2$ within the second one. Now the number $m$ of edges between the two generations can be calculated in two ways: (i) each vertex in the first generation "sticks out" $z-k-1$ edges toward the second one:

$$m = z(z-k-1);\tag{1}$$

(ii) each vertex in the second generation "sticks out" $z-\ell$ edges towards the first one:

$$m = (M-z-1)(z-\ell).\tag{2}$$

Equality of the two expressions for $m$ gives a relation between $k$ and $\ell$:

$$k = \frac{M-z-1}{z}\ell + 2z - M.\tag{3}$$

This equation immediately yields a useful result:

Lemma 1. For $M=p+1$, $p$ a prime, there is no permissible, primitive group on $M$ letters with a two-graph MI.

Proof. For $M=p+1$, $k$ as given by eq. (3) can be an integer only for $\ell=z$; but then $m=0$ by eq. (2) and $G$ is disconnected, so that $\mathcal{G}(G)$ is imprimitive. ¶

Since for the case $M=p$, the groups sought for are already given by Lemma 3.2.3, the values of $M\leq 20$, for which graphs have to be constructed, are $M=9$, 10, 15 and 16. For these numbers of letters, the constructions are given below.

(a). $M=9$. For this case, $z$ must be even; since $z=2$ is impossible (this yields a circle with imprimitive group $\mathcal{C}(9)$), only $z=4$ remains. Eq. (3) then shows that $k=\ell-1$. Since one must have $\ell<z$, the possible $(k,\ell)$ combinations are (0,1), (1,2) and (2,3). For (0,1) or (2,3), it is impossible to construct a graph, which "looks the same" from every vertex. The combination (1,2) yields the graph $G_9$ shown in Fig. 2.

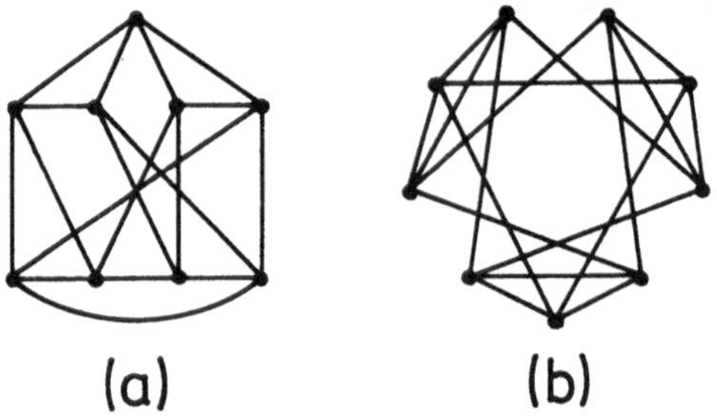

Fig. 2. (a) The graph $G_9$ in "generation representation", (b) the same graph in "circular" representation, showing its superregularity.

The group $\mathcal{G}(G_9)$ is then a group with the correct properties; in fact, $\mathcal{G}(G_9)=\mathcal{S}(3)\sim_p\mathcal{S}(3)$, which is a completely permissible primitive group with a two-graph MI by Lemma 3.3.2. The present construction shows, that there are no other groups on 9 letters with the same properties.

(b). M=10. Since z must have a factor in common with M-1=9 by eq. (3), and since $z\leqq(M-1)/2$ may always be assumed, z=3 follows. Eq. (3) then becomes $k=2\ell-4$, which, together with $\ell<z$, implies k=0, $\ell=2$. The corresponding superregular graph $G_{10}$ is shown in Fig. 3. It is known in graph theory as the Petersen graph. From Fig. 1, one immediately has $H_1(\mathcal{G}(G_{10}))\simeq\mathcal{D}(6)$ and, hence, $\mathcal{G}(G_{10})\simeq\mathcal{S}(5)$; the reason for this isomorphism is given in Section 3. $\mathcal{G}(G_{10})$ is completely permissible by Corollary 1.5.2.

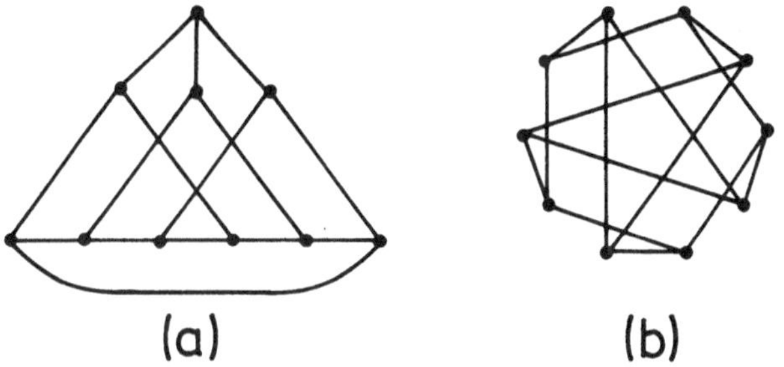


Fig. 3. The Petersen graph $G_{10}$ in "generation" (a) and "circular" (b) representations.

(c). M=15. Here z must be even again, so that z=4 or z=6 are possible. For z=4, eq. (3) yields $k=\frac{5}{2}\ell-7$, which does not give an integer for k for $\ell<z=4$. For z=6, eq. (3) gives $k=\frac{4}{3}\ell-3$, so that $\ell<z$ implies $\ell=3$, k=1. The corresponding graph $G_{15}$ has three disconnected edges in the first generation with symmetry group $\mathcal{S}(2)\sim\mathcal{S}(3)$, whereas the hyperregular graph in the second generation is a cube with group $\mathcal{S}(2)\otimes\mathcal{S}(4)$, see Fig. 2.3.1. Since the isomorphism $\mathcal{S}(2)\sim\mathcal{S}(3)\simeq\mathcal{S}(2)\otimes\mathcal{S}(4)$ holds (see Section 2), $H_1(\mathcal{G}(G_{15}))$ is isomorphic to these groups and $\mathcal{G}(G_{15})\simeq\mathcal{S}(6)$ follows, see again Section 3. This group is then completely permissible by Corollary 1.5.2 again.

(d). M=16. Since z must have a factor in common with M-1=15, the condition $z\leqq(M-1)/2$ gives 3, 5 and 6 as possible values for z. For z=3, eq. (3) becomes $k=4\ell-10$, which is not positive for $\ell<z$. For z=5, one gets $k=2\ell-6$, so that the $(k,\ell)$ pairs (0,3) and (2,4) are obtained. Of these, only the pair (0,3) give a superregular graph $G_{16,1}$. Since k=0, there is no graph in the first generation, but its second-generation graph is $G_{10}$, so that $H_1(\mathcal{G}(G_{16,1}))\simeq\mathcal{S}(5)\simeq\mathcal{G}(G_{10})$. This yields $\mathcal{G}(G_{16,1})$ = 1920. For z=6, eq. (3) reduces to $k=\frac{3}{2}\ell-4$, so that k=2, $\ell=4$

follow. The corresponding graph $G_{16,2}$ has two triangles in its first generation (group $\mathcal{S}(3)\sim\mathcal{S}(2)$) and $G_9$ in the second, so that $\mathcal{G}(G_{16,2}) = \mathcal{S}(4)\underset{p}{\sim}\mathcal{S}(2)$, which is completely permissible by Lemma 3.3.2.

The results of this section are summarized in Table 1 below.

Table 1. Primitive permissible groups and corresponding graphs with 2-graph MI's on $M \neq p$ letters for $M \leq 20$.

| M | Graph | z | First generation graph and its group | Second generation graph and its group | Isomorphic or equal |
|---|---|---|---|---|---|
| 9 | $G_9$ | 4 | two isolated edges, $\mathcal{S}(2)\sim\mathcal{S}(2)$. | circle on four vertices, $\mathcal{D}(4)$. | $=\mathcal{S}(3)\underset{p}{\sim}\mathcal{S}(2)$ |
| 10 | $G_{10}$ | 3 | empty, $\mathcal{S}(3)$. | circle on six vertices, $\mathcal{D}(6)$. | $\simeq\mathcal{S}(5)$ |
| 15 | $G_{15}$ | 6 | three isolated edges, $\mathcal{S}(2)\sim\mathcal{S}(3)$. | cube, $\mathcal{S}(2)\otimes\mathcal{S}(4)$. | $\simeq\mathcal{S}(6)$ |
| 16 | $G_{16,1}$ | 5 | empty, $\mathcal{S}(5)$. | $G_{10}$, $\mathcal{G}(G_{10})\simeq\mathcal{S}(5)$ | - |
| 16 | $G_{16,2}$ | 6 | two triangles, $\mathcal{S}(3)\sim\mathcal{S}(2)$. | $G_9$, $\mathcal{G}(G_9)$ | $=\mathcal{S}(4)\underset{p}{\sim}\mathcal{S}(2)$ |

## 5.2. Platonic graphs.

Let $F$ be a plane, i.e., a compact, two-dimensional differentiable manifold. A polyhedron $P$ divides the plane in $\alpha_2$ faces, which have $\alpha_1$ edges pairwise in common; these edges cut each other in $\alpha_o$ vertices. The formula of Euler asserts, that there is a relation between these three quantities:

$$\alpha_o - \alpha_1 + \alpha_2 = c(F), \tag{1}$$

where the Eulerian characteristic $c(F)$ depends on the type of plane only. For an orientable plane, this number is given as

$$c(F) = 2-2p, \tag{2}$$

where $p$, the genus of the plane, equals the number of independent "holes" of the plane, so that one has $p=0$ for the sphere, $p=1$ for the torus, etc. For non-orientable planes, the Eulerian characteristic is given by

$$c(F) = 1-q, \qquad (3)$$

with $q$ again the genus of the plane, so that one has $q=0$ for the projective plane, $q=1$ for the Klein bottle, etc. The dual polyhedron $P'$ is obtained from $P$ by placing a vertex on every face of $P$ and by connecting two such vertices if the corresponding faces are adjacent, i.e., if they have an edge in common. A polyhedron is called platonic if (i) every pair of faces from $P$ and $P'$ have at most one edge in common and (ii) the groups of transformations (homeomorphisms) of $F$ onto itself which leave $P$ and $P'$ invariant, are transitive on the vertices and on the edges of $P$ and of $P'$. Now clearly, every such polyhedron $P$ (or $P'$) defines a graph $G(P)$ consisting of the vertices and edges of $P$. Also, $G(P)$ is certainly superregular by the second platonic condition if it is simple. Note that the automorphism group $\mathcal{G}(G(P))$ may be larger than the group of transformations leaving $P$ invariant, since in defining the graph one is not interested in the plane in which it can be embedded as a polyhedron.

Let $z$ and $z'$ be the number of edges adjacent to every vertex of $P$ and $P'$, respectively. The regularity of $G(P)$ and $G(P')$ implies

$$\alpha_1 = \frac{1}{2}z\alpha_o = \frac{1}{2}z'\alpha_2. \qquad (4)$$

For $c(F)\neq 0$, eqs. (4) and (1) yield

$$\begin{aligned} \alpha_o &= c(F)\ 2z'\ [2(z+z')-zz']^{-1}, \\ \alpha_1 &= c(F)\ zz'\ [2(z+z')-zz']^{-1}, \qquad (5) \\ \alpha_2 &= c(F)\ 2z\ [2(z+z')-zz']^{-1}. \end{aligned}$$

For $c(F)=0$, one obtains the condition

$$zz' = 2(z+z'), \qquad (6)$$

together with the requirement that $\alpha_1$ is an integer multiple of $z/2$ and of $z'/2$. In the following, the platonic graphs of the sphere, of the projective plane and of the torus are studied; the Klein bottle has no platonic polyhedra.

(a). The sphere. Here one has $c(F)=2$ by eq. (2). The possible values of $z$, $z'$ and of the $\alpha$'s are easily obtained from eqs. (5) and are collected in the following table:

Table 1. Platonic polyhedra of the sphere.

| z | z' | $\alpha_0$ | $\alpha_1$ | $\alpha_2$ | Name | Duality |
|---|---|---|---|---|---|---|
| 3 | 3 | 4 | 6 | 4 | Tetrahedron | Self-dual |
| 3 | 4 | 8 | 12 | 6 | Cube | Dual pair |
| 4 | 3 | 6 | 12 | 8 | Octahedron | |
| 3 | 5 | 20 | 30 | 12 | Dodecahedron | Dual pair |
| 5 | 3 | 12 | 30 | 20 | Icosahedron | |

These were, of course, known to Plato already ([1]). The corresponding graphs are easily found: (i) The tetrahedron is simply the complete graph K(4) on four letters with group $\mathfrak{S}(4)$; (ii) The graph of the cube has already been considered in Fig. 2.3.1; (iii) The graph of the octahedron is the connected graph of the MI of $\mathfrak{S}(2) \sim \mathfrak{S}(3)$; the isomorphism $\mathfrak{S}(2) \otimes \mathfrak{S}(4) \simeq \mathfrak{S}(2) \sim \mathfrak{S}(3)$ noted in the previous section is due to the fact that cube and octahedron form a dual pair; (iv) The icosahedron I gives a superregular graph on 12 letters; the MI of its symmetry group $\mathfrak{G}(I)$ consists of three graphs as shown in Fig. 1 below; this is the first example of an imprimitive group of type (f) of Theorem 3.1.2; $\mathfrak{G}(I)$ is abstractly isomorphic to $\mathfrak{S}(2) \otimes \mathfrak{A}_r(5)$; (v) The dodecahedron yields a superregular graph D with 20 vertices; the corresponding group has a MI consisting of five graphs, two with z=3, two with z=6 and one with z=1; duality implies $\mathfrak{G}(I) \simeq \mathfrak{G}(D)$. For both of these groups, the MI's consist of the distance graphs of I and D, respectively, so that both have a P-algebra by Lemma 1.4.6; in fact, both are completely permissible, as can be checked by direct calculation.

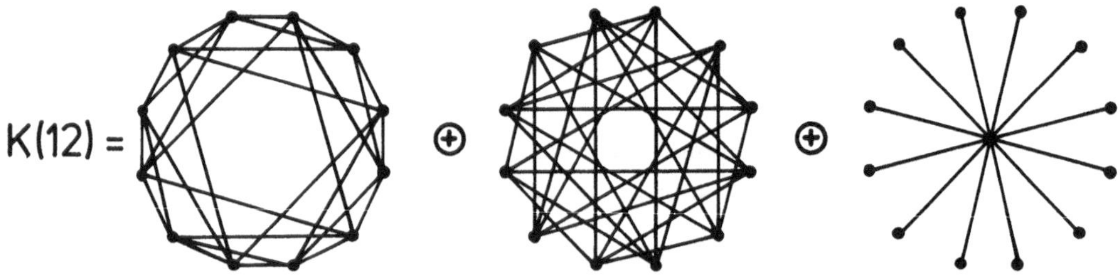


Fig. 1. The maximal interaction of the group $\mathfrak{G}(I)$ of the icosahedron.

(b). The projective plane. This has $c(F)=1$ by eq. (3). The possible values of $z$, $z'$ and of the $\alpha$'s follow again from eq. (5); two combinations are excluded by the first platonic condition, so that one is left with the three possibilities listed below.

Table 2. Platonic polyhedra of the projective plane.

| $z$ | $z'$ | $\alpha_o$ | $\alpha_1$ | $\alpha_2$ | Name | Duality |
|---|---|---|---|---|---|---|
| 3 | 4 | 4 | 6 | 3 | Trihedron | No dual |
| 3 | 5 | 10 | 15 | 6 | Hexahedron | Dual pair |
| 5 | 3 | 6 | 15 | 10 | Decahedron | |

The corresponding graphs are: (i) For the trihedron, the values of $\alpha_o$ and $\alpha_1$ imply that this is $K(4)$ again; (ii) The hexahedron can be depicted as in Fig. 2, where the projective plane is represented by a circle of which the diametrically opposed points have to be identified; this figure shows, that the corresponding graph is the Petersen graph of the previous section; (iii) For the decahedron, the values of $\alpha_o$ and $\alpha_1$ show that this is the complete graph $K(6)$.

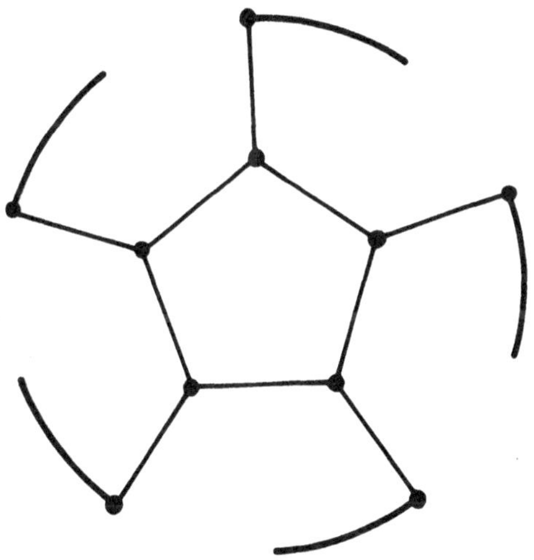

Fig. 2. The hexahedron in the projective plane is isomorphic to the Petersen graph.

(c). The torus. [2]. Here $c(F)=0$ by eq. (2), so that eq. (6) gives for $z,z'$ the possible pairs of values $(3,6)$, $(6,3)$ and $(4,4)$. These three possibilities correspond to finite hexagonal, triangular and quadratic lattices with periodic boundary conditions. This yields, in principle, infinitely many permissible groups as groups of superregular graphs with $z=3,6$ or 4. The groups with $z=4$ or 6 obviously contain regular, Abelian subgroups corresponding to translations, so that these are completely permissible by Corollary 2.5.1.

(i) Quadratic lattices, $z=4$ graphs. A finite quadratic lattice is obtained from the infinite one by drawing a line from the origin to an arbitrary lattice point $(b,c)$, by completion of this line to give a square and by the identification of the opposite sides of this square. Such a finite lattice has $b^2+c^2$ vertices, $2(b^2+c^2)$ edges and comprises $b^2+c^2$ elementary squares. The corresponding graph is simple if the number of edges does not exceed the number of edges of $K(b^2+c^2)$; this gives $b^2+c^2 \geqq 5$. Some examples are given in Fig. 3. The correspond-

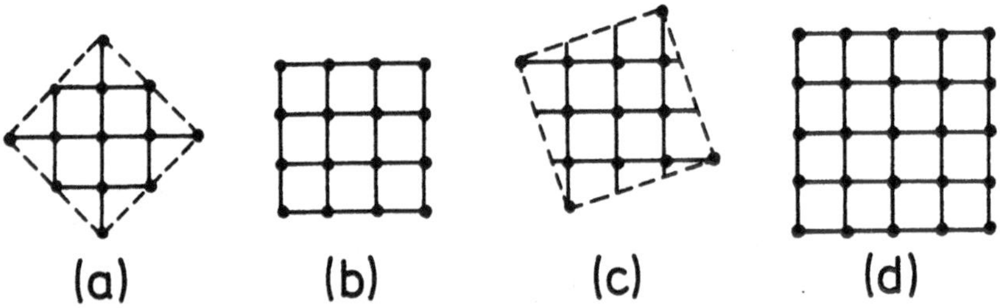


Fig. 3. Some quadratic polyhedra of the torus with nontrivial symmetry groups.

ing groups for the special cases with $b^2+c^2 \leqq 16$ are actually not new: (1) $b=1$, $c=2$ yields a graph on 5 letters with $z=4$: this is $K(5)$, so that the group is $S(5)$; (2) $b=2$, $c=2$, see Fig. 3(a); the group is $S(4) \sim S(2)$; (3) $b=0$, $c=3$, see Fig. 3(b); this is actually $G_9$, so that the group is $S(3) \underset{p}{\sim} S(2)$; (4) $b=1$, $c=3$, see Fig. 3(c); the group is $S(2) \otimes S(5)$; (5) $b=2$, $c=3$; this is on 13 letters, see the discussion following Corollary 3.2.1; (6) $b=0$, $c=4$, see Fig. 3(d); the group is $S(2) \underset{p}{\sim} S(4)$, the group of the four-dimensional hypercube, see the discussion following Lemma 3.3.2. Higher values of $b$ and/or $c$ have not been studied.

(ii) Triangular lattices, $z=6$ graphs. The finite triangular lattices are obtained from the infinite one by considering a $60^0$ rhombus constructed from a line extending from the origin to a point $(b,c)$ in a coordinate system, the axes of which include a $60^0$ angle. Again, opposite sides of this rhombus have to be identified to obtain the toroidal boundary conditions. The number of vertices is $\alpha_o=(b^2+bc+c^2)$ and one has $\alpha_1=3\alpha_o$, $\alpha_2=2\alpha_o$. The corresponding graphs are simple for $\alpha_o \geqq 7$. Some examples are shown in Fig. 4. The groups corresponding to the graphs

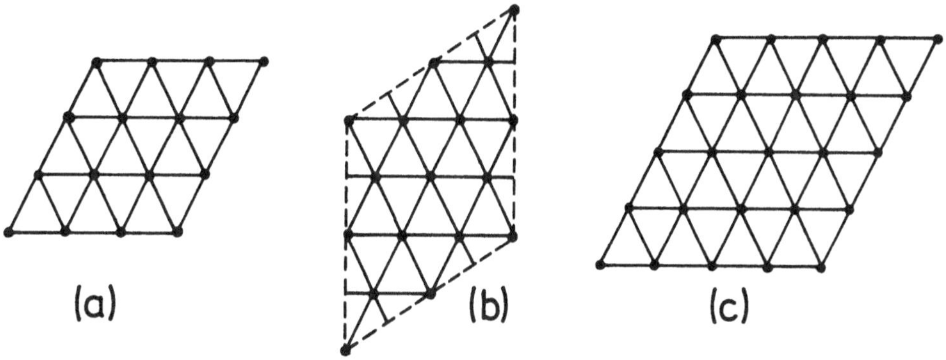


Fig. 4. Some triangular polyhedra of the torus.

on not more than 16 letters are: (1) b=2, c=1; this must be K(7) with group $\mathfrak{S}(7)$; (2) b=3, c=0, see Fig. 4(a); the group is $\mathfrak{S}(3)\sim\mathfrak{S}(3)$; (3) b=2, c=2, see Fig. 4(b); the group is $\mathfrak{S}(3)\otimes\mathfrak{S}(4)$; (4) b=3, c=1; this is on 13 letters again; (5) b=4, c=0, see Fig. 4(c); this yields a group of order 192 with a three-graph MI of type (f) of Theorem 3.1.2: one graph consists of four K(4)-graphs, whereas the other two are isomorphic graphs with z=6.

(iii) Hexagonal lattices, z=3 graphs. These are the duals of the z=6 graphs of the preceding subsection. Therefore, they contain $\alpha_o = 2(b^2+bc+c^2)$ vertices and $\alpha_1 = 3\alpha_o/2$ edges. The corresponding graphs are simple for $\alpha_o \geqq 4$. Some examples are given in Fig. 5. The graphs on not more than 16 letters are: (1) b=1, c=1; the group is $\mathfrak{S}(3)\sim\mathfrak{S}(2)$; (2) b=2, c=0, see Fig. 5(a); this is simply the cube with group $\mathfrak{S}(2)\otimes\mathfrak{S}(4)$; (3) b=2, c=1, see Fig. 5(b); this group of order 336 on 14 letters is again of type (f) of Theorem 3.1.2: The MI consists of three graphs, one of which consists of two K(7)-graphs, the other two being connected with z=3 and z=4, respectively.

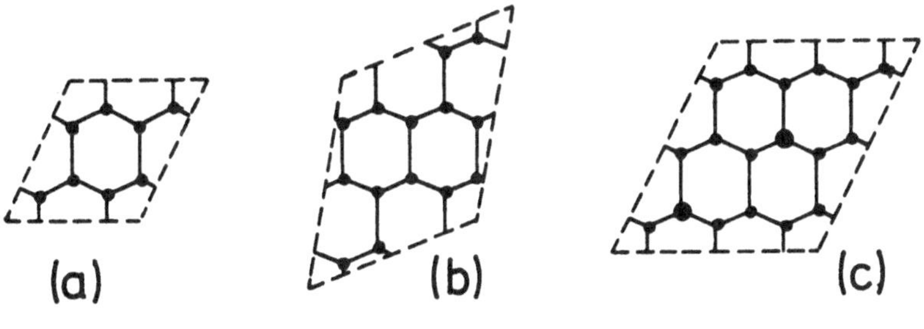


Fig. 5. Some hexagonal polyhedra of the torus.

As already remarked at the beginning of this section, the groups of finite quadratic and triangular lattices are completely permissible, since they contain regular, Abelian subgroups. The groups of the finite hexagonal lattices, which have $2(b^2+bc+c^2)$ vertices, also contain a regular subgroup $R_h$, but this consists of an index two Abelian subgroup $A$ and a reflection $\tau$ with respect to a vertex of the dual lattice:

$$R_h = A \cup A\tau, \quad \tau a = a^{-1}\tau \quad \text{for all} \quad a \in A. \tag{7}$$

Somewhat surprisingly, perhaps, the examples quoted above <u>are</u> completely permissible; the first two still contain a regular Abelian subgroup due to the extra symmetries which have their origin in the small size of the lattice, but in the third example, Fig. 5(b), the complete permissibility is not due to this cause: the group of order 336 does not contain $C(14)$, which is the only Abelian group on 14 letters. The smallest lattice, for which the group is not completely permissible, is the one shown in Fig. 5(c) on 18 letters, b=3, c=0. This group has a five-graph MI with z-values 3, 6, 3, 2, 2; the group does not have an associated P-algebra. This is due to the fact, that there is no element in the group which exchanges the two marked vertices in Fig. 5(c), even though the graph derived from an edge joining these vertices consists simply of six triangles.

## <u>5.3. Covering graphs.</u>

Let $G$ be a connected graph with $M$ vertices and $E$ edges; then the <u>covering graph</u> $G^c$ of $G$ is obtained by taking the <u>edges</u> of $G$ as the new vertices of $G^c$ and by connecting two of these new vertices by an edge iff the corresponding edges of $G$ have a common vertex. For an example, see Fig. 1 below. It is clear, that the covering graph of a regular graph with $z$ edges adjacent to every vertex is again regular; one has $M^c = zM/2$ and $z^c = 2(z-1)$. Here, the case that $G^c$ is hyperreg-

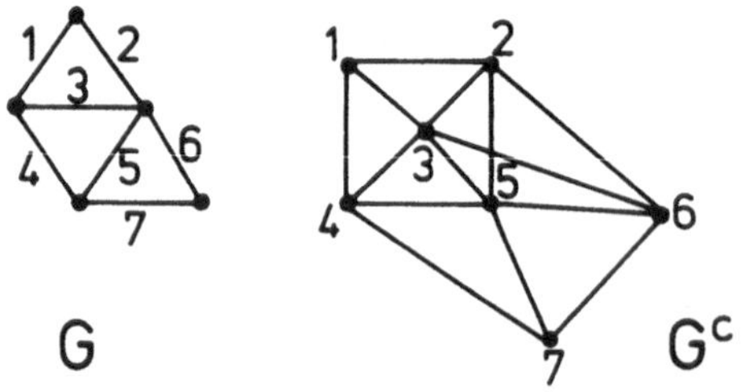


Fig. 1. A graph $G$ (left), and its covering graph $G^c$.

ular, i.e., $\mathcal{G}(G^c)$ must be transitive on the vertices of $G^c$. But since these are the edges of G originally, this will be the case if G is superregular:

Lemma 1. The covering graph of a graph with $z=1$ containing k edges has k vertices and no edges; the covering graph of a graph with $z=2$ is identical with the original graph; for $z\geq 3$, the covering graph of a superregular graph is hyperregular with $\mathcal{G}(G^c)\simeq\mathcal{G}(G)$ if the construction of $G^c$ is invertible unambiguously. The case $\mathcal{G}(G^c)>\mathcal{G}(G)$ is realized for the case $G=K(4)$; here, $G^c$ is the octahedron graph, so that $|\mathcal{G}(G^c)|=2|\mathcal{G}(G)|$ holds.

Proof. The statements concerning graphs with $z=1$ or $2$ follow from the definition of the covering graph. Further, $\mathcal{G}(G^c)\geqq\mathcal{G}(G)$ is clear if G is superregular with $z\geqq 3$. But, in general, the construction of $G^c$ is easily reversed to find G, since every vertex of G is replaced by a complete $K(z-1)$-graph and the vertices, where such complete graphs touch, are the original edges of G. This invertibility of the construction obviously implies the isomorphism $\mathcal{G}(G^c)\simeq\mathcal{G}(G)$. The case $G=K(4)$ is shown in Fig. 2; clearly, $G^c$ is the octahedron, so that $\mathcal{G}(G)=\mathcal{S}(4)$ with 24 elements and $\mathcal{G}(G^c)=\mathcal{S}(2)\sim\mathcal{S}(3)$ with 48 elements. This is here due to

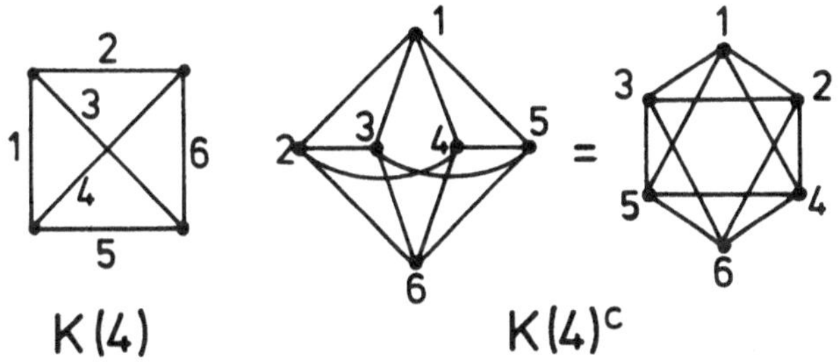


Fig. 2. The complete graph $K(4)$ and its covering graph, the octahedron.

the fact, that there are two choices for the set of four triangles with which to reconstruct $K(4)$. ¶

The covering graphs of some classes of superregular graphs give rise to infinite series of primitive permissible groups with two-graph MI's. These are described in the next two lemmas:

Lemma 2. The covering graph of the complete graph $K(M)$ is superregular as is its complement; for $M\geqq 5$, both are connected, so that these two form the MI of a primitive, completely permissible group on $\frac{1}{2}M(M-1)$ letters which is isomorphic to $\mathcal{S}(M)$.

Proof. By direct construction, it is seen that $[K(M)]^c\equiv G_1[\frac{1}{2}M(M-1)]$ has the following structure: it is a superregular graph with two generations, the graph in the first generation being the complement of the connected graph from the MI of $\mathfrak{S}(2)\otimes\mathfrak{S}(M-2)$; the second-generation graph is $G_1[\frac{1}{2}(M-2)(M-3)]$. The complement of this first graph, $G_2[\frac{1}{2}M(M-1)]$, is built up similarly: it has the connected $\mathfrak{S}(2)\otimes\mathfrak{S}(M-2)$-graph in its second generation, $G_2[\frac{1}{2}(M-2)(M-3)]$ in its first. Therefore, one has, in the notation of Section 1, the values of $k$, $\ell$ and $m$ listed below:

Table 1. Characteristic values of $G_{1,2}[\frac{1}{2}M(M-1)]$.

| Graph | z | k | ℓ | m |
|---|---|---|---|---|
| $G_1[\frac{1}{2}M(M-1)]$ | $2(M-2)$ | $M-2$ | $2(M-4)$ | $2(M-2)(M-3)$ |
| $G_2[\frac{1}{2}M(M-1)]$ | $\frac{1}{2}(M-2)(M-3)$ | $\frac{1}{2}(M-4)(M-5)$ | $M-3$ | $(M-2)(M-3)(M-4)$ |

From these values of $k$ and $\ell$, it follows that eq. (1.3) is satisfied. Further, $m\neq 0$ for $M\geqq 5$, so that both graphs are then connected. The complete permissibility follows from Corollary 1.5.2. ¶

Obviously, the graphs $G_{10}$ and $G_{15}$ of Table 1.1 are $G_2[10]$ and $G_2[15]$ in the present notation.

Lemma 3. The covering graph of the connected graph of the MI of $\mathfrak{S}(M)\sim\mathfrak{S}(2)$ is a superregular graph $G_1[M^2]$ with $M^2$ vertices. For $M\geqq 3$, this graph and its complement together define the MI of the completely permissible, primitive group $\mathfrak{S}(M)\underset{p}{\sim}\mathfrak{S}(2)$.

Proof. Direct construction gives the following structure for $G_1[M^2]$ and for its complement, $G_2[M^2]$: $G_1[M^2]$ is a two-generation graph; the graph inside its first generation is the disconnected graph from the MI of $\mathfrak{S}(M-1)\sim\mathfrak{S}(2)$, the graph inside its second generation is $G_1[(M-1)^2]$; similarly, $G_2[M^2]$ has $G_2[(M-1)^2]$ in the first generation and the connected graph with group $\mathfrak{S}(M-1)\sim\mathfrak{S}(2)$ in the second. The characteristic values of these graphs are listed in Table 2. Eq. (1.3) is fulfilled and

Table 2. Characteristic values of $G_{1,2}[M^2]$.

| Graph | z | k | ℓ | m |
|---|---|---|---|---|
| $G_1[M^2]$ | $2(M-1)$ | $M-2$ | $2(M-2)$ | $2(M-1)^2$ |
| $G_2[M^2]$ | $(M-1)^2$ | $(M-2)^2$ | $M-1$ | $2(M-1)^2(M-2)$ |

the graphs are superregular; they are connected for $M \geqq 3$. A comparison with Lemma 3.3.2 shows, that the groups of these graphs are equal to $S(M) \underset{p}{\sim} S(2)$. ¶

The graphs $G_9$ and $G_{16,2}$ of Table 1.1 are, in the present notation, $G_1[9]$ and $G_1[16]$. The graph $G_{16,1}$ of Table 1.1 has $z=5$, so that it cannot be the covering graph of a superregular graph; its complement has $z=10$, which, if it were a covering graph, would imply that there exists a graph with $z=6$ on $N$ letters with $16=zN/2=3N$. This is impossible, so that $G_{16,1}$ is not in any way related to a covering graph, even though it has $G_{10}=G_2[10]$ in its second generation.

Some further examples of covering graphs are: (i) the covering graph of the cube, shown in Fig. 3(a) and (ii) the covering graph of the Petersen graph shown in Fig. 3(b). Both of these are superregular with automorphism groups isomorphic to those of the original graphs. The covering graph of the octahedron, finally, yields a non-superregular graph, the symmetry group of which is the same as the group of the graph of Fig. 3(a).

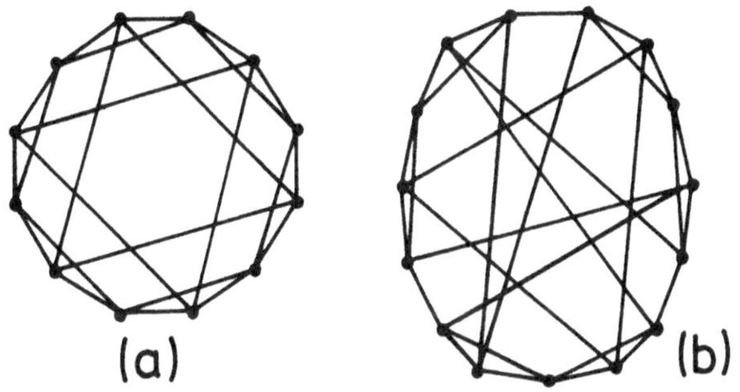


Fig. 3. The covering graphs of (a) the cube and (b) the Petersen graph.

REFERENCES.

([1]). Plato, Timaeus 54, 55.
([2]). H.S.M. Coxeter and W.O.J. Moser, Generators and Relations for Discrete Groups (Springer-Verlag, Berlin, Heidelberg, New York, 1957). H.S.M. Coxeter, Unvergängliche Geometrie (Birkhäuser Verlag, Basel, Stuttgart, 1963).

## 6. TABLES OF PERMISSIBLE GROUPS.

### 6.1. The permissible groups on $M\leq 10$ letters.

In this section, it is shown how all permissible groups on $M\leq 10$ letters can be obtained from the results of the first five chapters of these lectures.

M=2. Since the group $\mathcal{S}(2)$ has no nontrivial subgroups, this is the only permissible group on 2 letters, corresponding to the Ising model.

M=3. The group $\mathcal{S}(3)$ has only $\mathcal{C}(3)$ as nontrivial subgroup; by Theorem 2.5.1, $\mathcal{S}(3)$ is then the only permissible group on 3 letters. This corresponds to the 3-state Potts model.

M=4. a). The group $\mathcal{S}(4)$ is permissible.

b). By Lemma 5.1.1, there are no primitive groups with two-graph MI's for M=4; the only permissible group with a two-graph MI is, therefore, $\mathcal{D}(4)=\mathcal{S}(2)\sim\mathcal{S}(2)$ by Theorem 3.1.1.

c). If a permissible group on 4 letters has 3 graphs in its MI, these must all have z=1, so that $\mathcal{K}(4)=\mathcal{S}(2)\otimes\mathcal{S}(2)$ is the only group of this type on 4 letters by Theorem 3.1.2. This is the Ashkin-Teller model.

The subgroup structure for M=4 is simply $\mathcal{S}(4)>\mathcal{S}(2)\sim\mathcal{S}(2)>\mathcal{S}(2)\otimes\mathcal{S}(2)$. Therefore, all groups are completely permissible, since they all contain a regular, Abelian group, Corollary 2.5.1.

M=5. Since 5 is a prime, all permissible groups on 5 letters are given by Theorem 3.2.3: only $\mathcal{S}(5)$ and $\mathcal{D}(5)$ are found.

M=6. a). $\mathcal{S}(6)$ has one graph in its MI.

b). By Lemma 5.1.1 and Theorem 3.1.1, a permissible group on 6 letters with a two-graph MI is a wreath product: $\mathcal{S}(2)\sim\mathcal{S}(3)$ or $\mathcal{S}(3)\sim\mathcal{S}(2)$.

c). A group on 6 letters with a 3-graph MI is, by Theorem 3.1.2, equal to $\mathcal{S}(2)\otimes\mathcal{S}(3)=\mathcal{D}(6)$, the permissible group corresponding to the Abelian group $\mathcal{C}(6)$. This is seen as follows: (i) the group cannot be primitive, since this implies $z\geq 2$ for all three graphs and $M\geq 7$; (ii) the group cannot be of types (b) or (c), since there are no groups with 2-graph MI's on 2 or 3 letters; (iii) it cannot be of type (d), since 6 cannot be factorized into three factors larger than one; (iv) it cannot be of type (f), since this implies that there are two connected graphs, both of which must then be circles.

d). If a group on 6 letters has more than three graphs in its MI, these must include either a circle or two triangles, since the group must contain an element of order 3. The case of a circle is not possible, since the group would needs contain $D(6)$. The case of two triangles leads to a 4-graph MI, the other 3 graphs all having $z=1$. This determines the group uniquely as $F(6)$, see Fig. 2.5.1.

The subgroup structure for $M=6$ is indicated in Fig. 1, whereas some properties of these groups are listed in Table 1; here s is the number of graphs in the MI of the group, cp abbreviates complete permissibility and ras indicates regular, Abelian subgroups, if any.

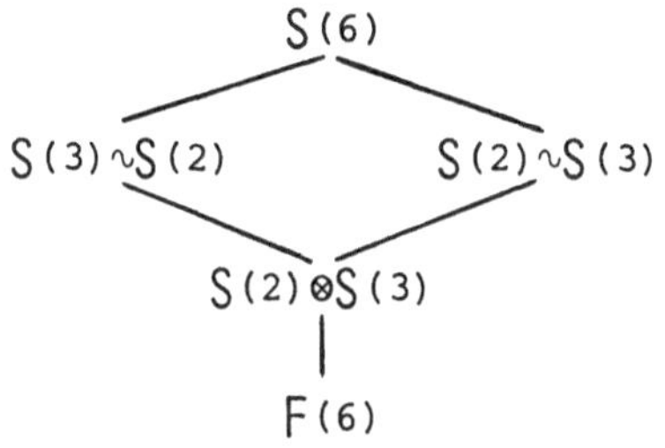


Fig. 1. The subgroup structure of the permissible groups on 6 letters.

Table 1. The permissible groups on 6 letters.

| Group | Order | s | cp | ras |
|---|---|---|---|---|
| $S(6)$ | 720 | 1 | yes | $C(6)$ |
| $S(3)\sim S(2)$ | 72 | 2 | yes | $C(6)$ |
| $S(2)\sim S(3)$ | 48 | 2 | yes | $C(6)$ |
| $S(2)\otimes S(3)$ | 12 | 3 | yes | $C(6)$ |
| $F(6)$ | 6 | 4 | no | - |

$M=7$. Theorem 3.2.3 yields $S(7)$ and $D(7)$ only.

$M=8$. a). $S(8)$ is permissible.

b). By Lemma 5.1.1 and Theorem 3.1.1, 2-graph MI's are generated only by the wreath products $S(2)\sim S(4)$ and $S(4)\sim S(2)$.

c). Theorem 3.1.2 yields two groups with 3-graph MI's, $S(2)\otimes S(4)$ and $S(2)\sim S(2)\sim S(2)$. This follows from similar arguments as in the case $M=6$.

d). Since $S(2)\otimes S(2)$ has a 3-graph MI, the groups $S(2)\sim(S(2)\otimes S(2))$ and $(S(2)\otimes S(2))\sim S(2)$ are permissible with 4-graph MI's.

e). Regular, Abelian groups on 8 letters are: (i) $K(8)$ with a

7-graph MI; this is permissible, since it equals $S(2)\otimes S(2)\otimes S(2)$, (ii) $S(2)\otimes C(4)$, which has $S(2)\otimes(S(2)\sim S(2))$ as corresponding permissible group with a 5-graph MI, and (iii) $C(8)$, which has $D(8)$ as permissible extension.with a 4-graph MI. These results follow from Theorems 2.5.1 and 2.5.3.

f). Regular, nonabelian groups with 8 elements are: (i) $Q$, the quaternion group, whose corresponding permissible group $S(2)\sim(S(2)\otimes S(2))$ has already been encountered, see point d), and (ii) $F(8)$, which is permissible; both results follow from Theorem 4.2.1.

A careful study of the possible MI's with four or more graphs for permissible groups on 8 letters shows, that the above list is exhaustive. The subgroup structure of these groups is shown in Fig. 2, whereas Table 2 lists them, together with some of their important properties.

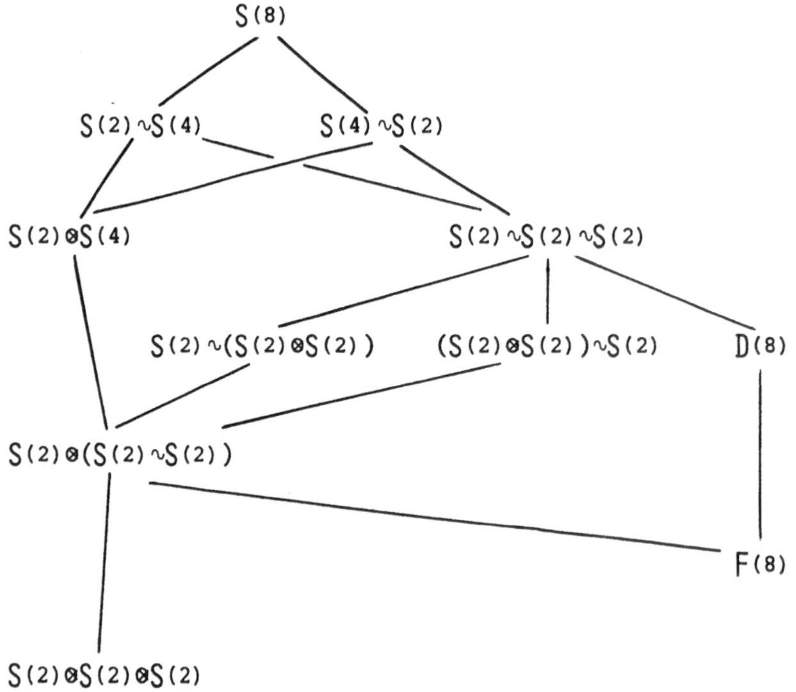


Fig. 2. The subgroup structure for the permissible groups on 8 letters.

M=9. a). $S(9)$ is permissible.

b). As primitive group with a 2-graph MI, $G(G_9)$ has been constructed in Section 5.1; as imprimitive group with a 2-graph MI, only $S(3)\sim S(3)$ is obtained from Theorem 3.1.1.

c). Theorem 3.1.2 yields only $S(3)\otimes S(3)$ as a group with a 3-graph

Table 2. The permissible groups on 8 letters.

| Group | Order | z-values | cp | ras |
|---|---|---|---|---|
| S(8) | 40320 | 7 | yes | C(8), S(2)⊗C(4), K(8) |
| S(4)~S(2) | 1152 | 3,4 | yes | " , " , " |
| S(2)~S(4) | 384 | 1,6 | yes | " , " , " |
| S(2)~S(2)~S(2) | 128 | 1,2,4 | yes | " , " , " |
| S(2)⊗S(4) | 48 | 1,3,3 | yes | S(2)⊗C(4), K(8) |
| S(2)~(S(2)⊗S(2)) | 64 | 1,2,2,2 | yes | " , " |
| (S(2)⊗S(2))~S(2) | 32 | 1,1,1,4 | yes | " , " |
| D(8) | 16 | 1,2,2,2 | yes | C(8) |
| S(2)⊗(S(2)~S(2)) | 16 | 1,1,1,2,2 | yes | S(2)⊗C(4), K(8) |
| F(8) | 8 | 1,1,1,1,1,2 | no | -- |
| K(8) | 8 | 1,1,1,1,1,1,1 | yes | K(8) |

MI. This follows again by arguments similar to the ones used for M=6.

d). If the group has a 4-graph MI, all graphs must have z=2; Now either there is a full circle in the MI and the group is D(9) corresponding to C(9), or all four graphs consist of three triangles and the group is D(3,3) corresponding to C(3)⊗C(3).
For the subgroup structure and for a list of these groups, see Fig. 3 and Table 3, respectively.

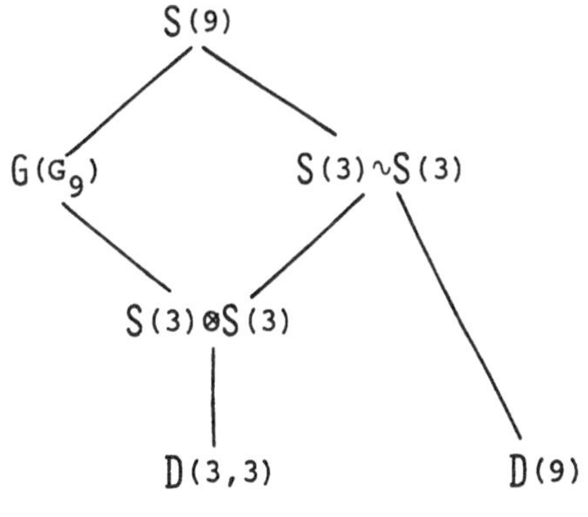


Fig. 3. The subgroup structure for the permissible groups on 9 letters.

Table 3. The permissible groups on 9 letters.

| Group | Order | s | z-values | cp | ras |
|---|---|---|---|---|---|
| S(9) | 9! | 1 | 8 | yes | C(9), C(3)⊗C(3) |
| S(3)∼S(3) | 1296 | 2 | 2,6 | yes | " , " |
| G($G_9$) | 72 | 2 | 4,4 | yes | C(3)⊗C(3) |
| S(3)⊗S(3) | 36 | 3 | 2,2,4 | yes | " |
| D(3,3) | 18 | 4 | 2,2,2,2 | yes | " |
| D(9) | 18 | 4 | 2,2,2,2 | yes | C(9) |

M=10. a). S(10) is permissible.

b). G($G_{10}$) is primitive with a 2-graph MI, constructed in Sect. 5.1; imprimitive groups with 2-graph MI's are S(2)∼S(5) and S(5)∼S(2).

c). Theorem 3.2.2 can be seen to yield S(2)∼D(5), D(5)∼S(2) and S(2)⊗S(5).

d). G($G_{10}$) can be seen to contain a nonabelian transitive subgroup R(10) of order 20, which is isomprphic to C(5)$H_M$(C(5)). This group is permissible with a 4-graph MI.

e). The only regular, Abelian group with 10 elements is C(10)= S(2)⊗C(5) with corresponding permissible group D(10)=S(2)⊗D(5).

f). The only regular, nonabelian group with 10 elements is the permissible group F(10).

Again, a careful study of the possible MI's with four or more graphs shows, that the above listed groups exhaust all possibilities. For the subgroup structure, see Fig. 4, for a list of these groups, Table 4. Note that G($G_{10}$) is the first example of a completely permissible group not containing a regular, Abelian subgroup. It is also interesting to note, that the group R(10) may be permissible or not, depending on the number of letters on which it is represented as a permutation group; one has

R(10) is permissible as a regular group on 20 letters;
R(10) is permissible as a permutation group on 10 letters;
R(10) is not permissible as a permutation group on 5 letters,
since C(5)$H_M$(C(5)) is doubly transitive, but not equal S(5).

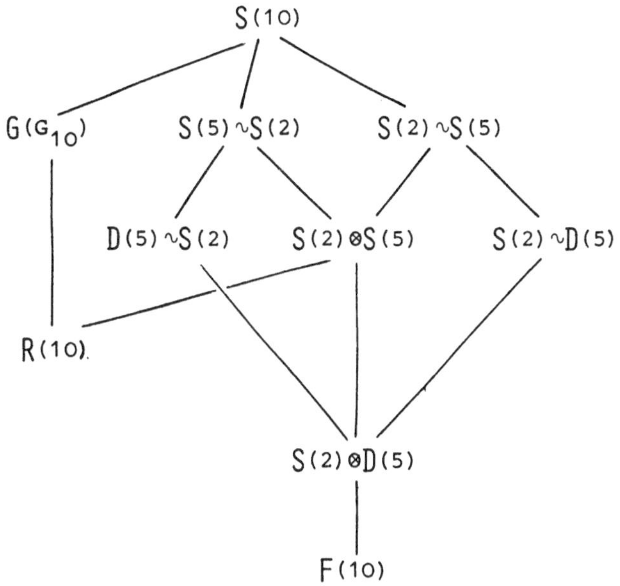


Fig. 4. The subgroup structure of the permissible groups on 10 letters.

Table 4. The permissible groups on 10 letters.

| Group | Order | s | z-values | cp | ras |
|---|---|---|---|---|---|
| S(10) | 10! | 1 | 9 | yes | C(10) |
| S(5)∼S(2) | 28800 | 2 | 4,5 | yes | " |
| S(2)∼S(5) | 3840 | 2 | 1,8 | yes | " |
| G($G_{10}$) | 120 | 2 | 3,6 | yes | -- |
| S(2)∼D(5) | 320 | 3 | 1,4,4 | yes | C(10) |
| D(5)∼S(2) | 200 | 3 | 2,2,5 | yes | " |
| S(2)⊗S(5) | 240 | 3 | 1,4,4 | yes | " |
| R(10) | 20 | 4 | 1,2,2,4 | no | -- |
| S(2)⊗D(5) | 20 | 5 | 1,2,2,2,2 | yes | C(10) |
| F(10) | 10 | 7 | 1,1,1,1,1,2,2 | no | -- |

## 6.2. The regular groups with M≤12 elements.

In the table below, all groups with not more than 12 elements are listed, together with their corresponding permissible, PCF and PCI groups. In the column "type", the letter A stands for Abelian, whereas N(a), N(b and N(c) refer to the types of nonabelian groups of Theorem 4.2.1. All results follow from results obtained in Sections 4.3 and 4.4, except for those marked by "§", which have been obtained by a separate calculation.

Table 1. Regular, corresponding permissible, PCF and PCI groups for M≤15.

| M | Group | Type | Perm. group | PCF group | PCI group |
|---|---|---|---|---|---|
| 2 | S(2) | A | S(2) | S(2) | S(2) |
| 3 | C(3) | A | S(3) | S(3) | S(3) |
| 4 | C(4) | A | S(2)∼S(2) | S(2)∼S(2) | S(2)∼S(2) |
| | K(4) | A | K(4) | K(4) | S(4) |
| 5 | C(5) | A | D(5) | D(5) | S(5) |
| 6 | C(6) | A | D(6) | D(6) | D(6) |
| | F(6) | N(c) | F(6) | S(3)∼S(2) | S(3)∼S(2) |
| 7 | C(7) | A | D(7) | D(7) | S(7) |
| 8 | C(8) | A | D(8) | D(8) | S(2)∼S(2)∼S(2) |
| | S(2)⊗C(4) | A | S(2)⊗D(4) | S(2)⊗D(4) | S(2)∼S(2)∼S(2) |
| | K(8) | A | K(8) | K(8) | S(8) |
| | Q | N(a) | S(2)∼(S(2)⊗S(2)) | S(2)∼(S(2)⊗S(2)) | S(2)∼S(4) |
| | F(8) | N(c) | F(8) | S(2)∼(S(2)⊗S(2)) | S(2)∼S(2)∼S(2) § |
| 9 | C(9) | A | D(9) | D(9) | S(3)∼S(3) |
| | C(3)⊗C(3) | A | D(3,3) | D(3,3) | S(9) |
| 10 | C(10) | A | D(10) | D(10) | S(2)⊗S(5) |
| | F(10) | N(c) | F(10) | D(5)∼S(2) | S(5)∼S(2) |
| 11 | C(11) | A | D(11) | D(11) | S(11) |
| 12 | C(12) | A | D(12) | D(12) | (S(2)∼S(2))⊗S(3) |
| | S(2)⊗C(6) | A | S(2)⊗D(6) | S(2)⊗D(6) | S(4)⊗S(3) § |
| | C(4)*C(6) | N(b) | $\langle \tau, C(4)*C(6) \rangle$ | D(6)∼S(2) | D(6)∼S(2) |
| | F(12) | N(c) | F(12) | S(2)⊗(S(3)∼S(2)) | D(6)∼S(2) § |
| | $A_r(4)$ | N(c) | $A_r(4)$ | S(4)∼S(3) | S(4)∼S(3) |

The reason to stop here is, that there are no "interesting" groups with 13, 14 or 15 elements, whereas the fifteen (!) different groups with 16 elements belong, for a large part, to classes of groups, for which no results have been derived in Chapter 4.

## 6.3. Special permissible groups.

In this section, the special permissible groups found by non-general methods in the preceding chapters, are listed in order of increasing number of letters.

M=12. a). The icosahedron $I$ (see Fig. 5.2.1), obtained as a platonic graph of the sphere, has the smallest permissible group of type (f) of Theorem 3.1.2, i.e., it has a 3-graph MI consisting of one disconnected graph (here with z=1) and of two connected graphs (here these have z-values 5 and 5). $\mathcal{G}(I)$ is of order 120 and is isomorphic to $\mathcal{S}(2)\otimes A_r(5)$.

b). The covering graph of the cube (Fig. 5.3.3(a)) yields a permissible group of order 48 (isomorphic to $\mathcal{S}(2)\otimes\mathcal{S}(4)$) with a 4-graph MI; the z-values are 1,2,4,4.

M=14. The platonic hexagonal graph of the torus with 14 vertices, see Fig. 5.2.5(b), yields again a group of type (f) of Theorem 3.1.2: it is of order 336, has one disconnected (two K(7)-graphs, z=6) and two connected graphs (z-values 3 and 4) in its MI.

M=15. The covering graph of the Petersen graph (Fig. 5.3.3(b)) has a symmetry group of order 120 (isomorphic to $\mathcal{S}(5)$) with a 3-graph MI of type (f) again: five triangles form a z=2 graph, the connected graphs have z=4 and z=8.

M=16. The platonic triangular graph of the torus (Fig. 5.2.4(c)) with 16 vertices has an automorphism group of order 192 of type (f): four K(4)-graphs form a z=3 graph, the connected ones both have z=6.

M=18. The platonic graph of the torus, which is hexagonal on 18 vertices, see Fig. 5.2.5(c), has the smallest group of this type which is not completely permissible. Its order is 216 and it has a 5-graph MI with z-values 2,3,3,3,6.

M=20. The dodecahedron $D$ is the dual of the icosahedron $I$; therefore, $\mathcal{G}(D)$ is isomorphic to $\mathcal{G}(I)$, but is on 20 letters; the z-values are 1,3,3,6,6, see Table 5.2.1.

M=60. $H(A_r(5))$ is the smallest primitive PCF group (which corresponds to a nonsolvable simple group). This group has 7200 elements and a three-graph MI with z-values of 15, 20 and 24. All three graphs are connected, see Theorem 4.3.4.

PART B : DUALITY.

# 7. DUALITY TRANSFORMATIONS AND DUAL MODELS.

## 7.1. Duality transformations and inequalities in two dimensions.

In this section, a spin system on an arbitrary graph is considered. The permissible group $\mathcal{G}$ of the interaction is taken such, that it contains an Abelian, regular subgroup $\mathcal{A}$:

$$\mathcal{A} = \mathcal{C}(\ell_1)\otimes\mathcal{C}(\ell_2)\otimes\ldots\otimes\mathcal{C}(\ell_n). \tag{1}$$

Then every letter $i$ of $S$ $(|S|=M=\ell_1\ell_2..\ell_n)$ can be represented by an n-dimensional vector $\underline{i}$:

$$i \leftrightarrow \underline{i} = \begin{pmatrix} i_1 \\ i_2 \\ \vdots \\ i_n \end{pmatrix}, \quad i_k=0,1,\ldots,\ell_k-1, \tag{2}$$

by means of the formula

$$i = 1+i_1+i_2\ell_1+i_3\ell_1\ell_2+\ldots+i_n\ell_1..\ell_n. \tag{3}$$

The vectors $\underline{i}$ form an Abelian group under addition if this addition is done componentwise modulo $\ell_k$. This additive group is isomorphic to $\mathcal{A}$ and will also be denoted by $\mathcal{A}$. The Boltzmann factor $\Omega(i,j)=\Omega(\underline{i},\underline{j})$ of eq. (1.1.5) is invariant with respect to the permutations of $\mathcal{G}$; in particular, the invariance with respect to an element $k\varepsilon\mathcal{A}$ is expressed as:

$$\Omega(\underline{i}+\underline{k},\underline{j}+\underline{k}) = \Omega(\underline{i},\underline{j}). \tag{4}$$

From Corollary 2.5.1, the group $\mathcal{G}$ is completely permissible, so that the matrices $\underline{\underline{\Omega}}$ all have the same set of eigenvectors for all values of the energy parameters. Due to eq. (4), these eigenvectors can be given explicitly:

$$(\mu_{\underline{t}})_{\underline{i}} = \frac{1}{\sqrt{M}} \prod_{k=1}^{n} \exp(\frac{2\pi i}{\ell_k} t_k i_k) \tag{5}$$

The corresponding (real) eigenvalues follow as:

$$\lambda_{\underline{t}} = \sum_{\underline{w}\in A} \Omega(\underline{w}) \prod_{k=1}^{n} \exp(-\frac{2\pi i}{\ell_k} w_k t_k). \tag{6}$$

Here $\Omega(\underline{w}) \equiv \Omega(\underline{w},\underline{0}) = \Omega(\underline{w}+\underline{i},\underline{i})$ by eq. (4). Since $\Omega(\underline{w}) = \Omega(-\underline{w})$ for $G$ permissible, $\lambda_{\underline{t}} = \lambda_{-\underline{t}}$ follows. The eigenvalue corresponding to the unit element (or $\underline{0}$) of $A$ plays a special role; it will be denoted by $\lambda_o$:

$$\lambda_o = \sum_{\underline{w}\in A} \Omega(\underline{w}). \tag{7}$$

The partition function for a spin model of the above type on a graph G=(V,E,I), which may have multiple edges in this chapter, is

$$Z(G) = \sum_{\substack{\underline{i}_v \in A \\ v\in V}} \prod_{e\in E} \Omega(\underline{i}_{v_1(e)}, \underline{i}_{v_2(e)}). \tag{8}$$

This can now be rewritten in two different ways. To this end, the edges of E are <u>oriented</u>, i.e., the incidence function I, which gives the unordered pair of vertices at the ends of an edge, is replaced by a function I', which gives one of the two ordered pairs corresponding to this. This orientation is done in a completely arbitrary fashion. For the first way to rewrite eq. (8), an edge variable $\underline{k}_e$ is attached to each edge by

$$\underline{k}_e = \begin{cases} \underline{i}_{v_1(e)} - \underline{i}_{v_2(e)} & \text{if } e \text{ is oriented from } v_1(e) \text{ towards } v_2(e), \\ \underline{i}_{v_2(e)} - \underline{i}_{v_1(e)} & \text{if } e \text{ is oriented from } v_2(e) \text{ towards } v_1(e). \end{cases} \tag{9}$$

Since $\Omega(\underline{w}) = \Omega(-\underline{w})$, eq. (8) can be rewritten as

$$Z(G) = M \sum_{\underline{k}_e \in A}{}' \prod_{e\in E} \Omega(\underline{k}_e), \tag{10}$$

where the prime indicates that the summation over the variables $\underline{k}_e$ is restricted. It is not difficult to see, that a particular configuration $\{\underline{k}_e\}$ is allowed iff, for every closed cicuit C of the graph G, the sum of the variables $\underline{k}_e$ on positively oriented (with respect to the direction in which C is traversed) edges equals the sum of the $\underline{k}_e$ on negatively oriented edges. Obviously, this condition does not depend on the direction in which C is traversed. Now for every finite graph G, all closed circuits can be made up of a number of elementary circuits $C_e$, see Fig. 1. Let a circuit C be made up of the elementary circuits

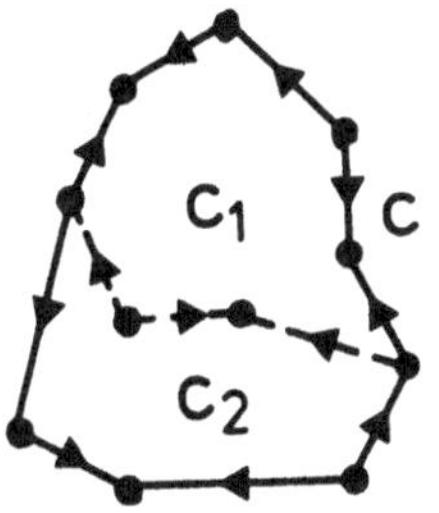


Fig. 1. A closed circuit C made up of two elementary circuits $C_1$ and $C_2$ in a directed graph.

$C_{e1}, C_{e2}, \ldots, C_{em}$. Then the condition for an allowable configuration for circuit C is, that

$$\delta_A[C] \doteq \delta_A\Big[ \sum_{\substack{e' \varepsilon C \\ e' \text{ oriented} \\ \text{positively}}} \underline{k}(e') - \sum_{\substack{e'' \varepsilon C \\ e'' \text{ oriented} \\ \text{negatively}}} \underline{k}(e'') \Big] = 1, \tag{11}$$

where

$$\delta_A[\underline{k}] = \begin{cases} 1 & \text{for } \underline{k}=\underline{0}, \\ 0 & \text{for all other } \underline{k} \varepsilon A. \end{cases} \tag{12}$$

It is now easy to see, that eq. (11) holds iff the corresponding equations for the elementary circuits $C_{e1}, \ldots, C_{em}$ hold, since one has

$$\delta_A[C] = \prod_{t=1}^{m} \delta_A[C_{et}]. \tag{13}$$

Therefore, eq. (10) can be written as an unrestricted sum upon insertion of $\delta$-functions for a minimal set of independent elementary circuits; denoting this set of elementary circuits by $\mathcal{C}$, one has

$$Z = M \sum_{\substack{\underline{k}_e \varepsilon A \\ e \varepsilon E}} \prod_{e \varepsilon E} \Omega(\underline{k}_e) \prod_{C \varepsilon \mathcal{C}} \delta_A[C]. \tag{14}$$

The second way to rewrite eq. (8) makes use of the fact, that eqs. (5) and (6) imply

$$\Omega(\underline{i},\underline{j}) = \sum_{\underline{t}\in A} \lambda_{\underline{t}} (\mu_{\underline{t}})_{\underline{i}} (\mu_{\underline{t}})^{*}_{\underline{j}}, \qquad (15)$$

where the star indicates complex conjugation. Insertion of eq. (15) into eq. (8) in such a way that the arrow of an edge always points towards the vertex which does not have the complex conjugate in eq. (15), also yields an expression for Z in terms of edge variables $\underline{k}_e$:

$$Z = \sum_{\underline{k}_e\in A} \prod_{e\in E} \lambda_{\underline{k}_e} \prod_{v\in V} \{ \sum_{\underline{i}_v\in A} \prod_{\substack{\text{edges } e' \\ \text{issuing} \\ \text{from } v}} (\mu_{\underline{k}_{e'}})^{*}_{\underline{i}_v} \prod_{\substack{\text{edges } e'' \\ \text{pointing} \\ \text{towards } v}} (\mu_{\underline{k}_{e''}})_{\underline{i}_v} \}. \qquad (16)$$

Use of the explicit form of eq. (5) for the eigenvectors yields

$$Z = M^{|V|-|E|} \lambda_o^{|E|} \sum_{\underline{k}_e\in A} \prod_{e\in E} \tilde{\Omega}(\underline{k}_e) \prod_{v\in V} \delta_A[ \sum_{\substack{\text{edges } e' \\ \text{issuing} \\ \text{from } v}} \underline{k}_{e'} - \sum_{\substack{\text{edges } e'' \\ \text{pointing} \\ \text{towards } v}} \underline{k}_{e''} ], \qquad (17)$$

where $\tilde{\Omega}(\underline{k})$ is the <u>dual</u> Boltzmann factor given by

$$\tilde{\Omega}(\underline{k}) = \lambda_{\underline{k}}/\lambda_o. \qquad (18)$$

If all eigenvalues $\lambda_{\underline{k}}$ are nonnegative (the set of values of the energy parameters for which this is the case is called the <u>ferromagnetic region</u>), eq. (17) looks very much like eq. (14) with a different (dual) interaction $\tilde{\Omega}(\underline{k})$. To make this more precise, the graph G is embedded in an orientable plane of genus $\chi$. This is possible for every finite graph G if $\chi$ is sufficiently large ([1]); of course, $\chi$ is taken as small as possible. Then the graph G has a dual $\tilde{G}$ obtained by putting a vertex of $\tilde{G}$ in each face of the polyhedron P(G) and by connecting two such vertices iff the corresponding faces of G have an edge in common, see Section 5.2. The edges of $\tilde{G}$ are now oriented by a rotation of the orientation of the corresponding edges of G over $\pi/2$ counter-clockwise, see Fig. 2. Further, the edge $\tilde{e}$ inherits the variable $\underline{k}_e$ associated with the edge e which crosses it. Eq. (17) then reads, with dual graph variables $\underline{k}_{\tilde{e}}$:

$$Z = M^{|V|-|E|} \lambda_o^{|E|} \sum_{\underline{k}_{\tilde{e}}\in A} \prod_{\tilde{e}\in\tilde{E}} \tilde{\Omega}(k_{\tilde{e}}) \prod_{\text{faces } F(\tilde{G})} \delta_A[F]. \qquad (19)$$

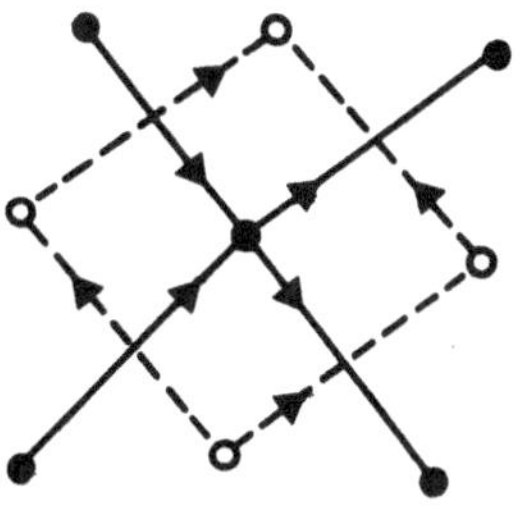

Fig. 2. The orientations of the graphs G (solid edges and vertex) and of its dual (broken edges and open vertices).

Since a plane of genus $\chi$ can be considered as a sphere with $\chi$ "holes", a graph $\tilde{G}$ embedded in a plane of minimal genus has two sets of independent elementary circuits: (i) the faces F of $\tilde{G}$ and (ii) $\chi$ elementary circuits, each circling a "hole". Remembering that $\tilde{\Omega}(\underline{k})$ is positive in the ferromagnetic region, one has here, from a comparison of eqs. (14) and (19):

$$Z(\Omega,G) = \alpha_1 \, M^{|V|-|E|-1} \, \lambda_o^{|E|} \, Z(\tilde{\Omega},\tilde{G}), \quad \alpha_1 \geq 1. \tag{20}$$

Similarly, the same procedure starting from the dual graph gives

$$Z(\tilde{\Omega},\tilde{G}) = \alpha_2 \, M^{|F|-|E|-1} \, (M/\lambda_o)^{|E|} \, Z(\Omega,G), \quad \alpha_2 \geq 1, \tag{21}$$

where use has been made of the equation

$$\tilde{\lambda}_o = \sum_{\underline{w}\in A} \tilde{\Omega}(\underline{w}) = \mathrm{Tr}\, \underline{\underline{\Omega}}/\lambda_o = M/\lambda_o. \tag{22}$$

Eqs. (20) and (21) are <u>duality inequalities</u> valid for any graph G. The product $\alpha_1\alpha_2$ is easily calculated as given by

$$\alpha_1\alpha_2 \, M^{|V|-|E|+|F|-2} = 1. \tag{23}$$

Use of the Euler relation, eq. (5.2.1), reduces this to

$$\alpha_1\alpha_2 = M^{2\chi}. \tag{24}$$

It might be thought, that $\alpha_1=\alpha_2=M$ holds, since this would take care of the missing $\delta$-functions of eq. (19) "on the average". A simple example, however, already shows that this is not the case: take for G the graph K(5); this can be embedded into the torus, so that $\chi=1$, see Section 5.2, quadratic polyhedra of the torus. For the Ising model (M=2), one has

$$\Omega(i,j) = \begin{pmatrix} 1 & \omega \\ \omega & 1 \end{pmatrix}, \quad \tilde{\Omega}(i,j) = \begin{pmatrix} 1 & \frac{1-\omega}{1+\omega} \\ \frac{1-\omega}{1+\omega} & 1 \end{pmatrix}, \quad \lambda_o=1+\omega. \tag{25}$$

The quantity $\alpha_1$ of eq. (20) is plotted for the Ising model on K(5) in Fig. 3 as a function of $\omega$; it is clear, that $\alpha_1=2$ is not a good approximation for most values of $\omega$.

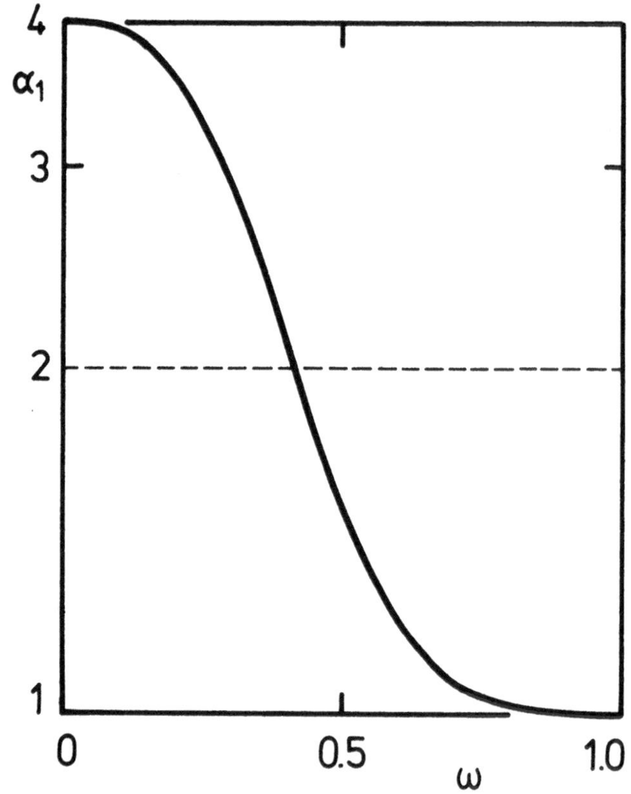


Fig. 3. The ratio $\alpha_1$ of eq. (20) as a function of $\omega$ for the Ising model on the graph K(5), which has $\chi=1$.

Duality transformations in two dimensions are obtained only for $\chi=0$, i.e., for G a planar graph:

$$Z(\Omega,G) = M^{|V|-|E|-1}\,\lambda_o^{|E|}\,Z(\tilde{\Omega},\tilde{G}), \quad [G \text{ planar}]. \tag{26}$$

This equation relates the partition function of a spin model with permissible group $\mathcal{G}$ containing an Abelian, regular subgroup $\mathcal{A}$ to a spin model with a permissible group $\tilde{\mathcal{G}}$, which also contains $\mathcal{A}$, as is clear from eq. (18). Some properties and examples of these dual permissible groups are given in the next section.

## 7.2. Dual permissible groups.

A glance at Tables 6.2.3 or 6.2.5 shows, that many permissible groups contain more than one regular, Abelian subgroup. Therefore, the first question one has to address is, whether the dual groups $\tilde{\mathcal{G}}_1$ and $\tilde{\mathcal{G}}_2$ , constructed using two different regular, Abelian subgroups of a permissible group, are identical or not. This question is answered by the next lemma:

Lemma 1. Let $\tilde{\Omega}(i,j)$ be the Boltzmann factors for a spin model with permissible symmetry group $\tilde{\mathcal{G}}$. Let $\mathcal{A}_1$ and $\mathcal{A}_2$ be nonisomorphic, regular and Abelian subgroups of $\tilde{\mathcal{G}}$ and let the dual Boltzmann factors be $\Omega_1(i,j)$ and $\Omega_2(i,j)$, if these are calculated using $\mathcal{A}_1$ and $\mathcal{A}_2$, respectively. Then there is a permutation $s \varepsilon \mathcal{S}(M)$ such that

$$\Omega_1(i,j) = \sum_{k,m=1}^{M} D_{ik}(s^{-1})\ \Omega_2(k,m)\ D_{mj}(s) \tag{1}$$

holds. Therefore, the groups $\mathcal{G}_1$ and $\mathcal{G}_2$ are isomorphic as permutation groups.

Proof. The vectors of eq. (1.2) corresponding to a letter $k$ will be denoted by $\underline{k} \varepsilon \mathcal{A}_1$ and $\hat{\underline{k}} \varepsilon \mathcal{A}_2$. Since the eigenvalues of $\underline{\underline{\tilde{\Omega}}}$ do, of course, not depend on $\mathcal{A}_1$ or $\mathcal{A}_2$, the matrices $\underline{\underline{\Omega}}_1$ and $\underline{\underline{\Omega}}_2$ must have the same eigenvalues $\lambda_k$; these matrices can, therefore, be written as

$$\Omega_1(i,j) = \sum_{k=1}^{M} (\mu_k)_i^* \lambda_k (\mu_k)_j, \tag{2}$$

$$\Omega_2(i,j) = \sum_{k=1}^{M} (\hat{\mu}_k)_i^* \lambda_k (\hat{\mu}_k)_j, \tag{3}$$

with $(\mu_k)_i$ and $(\hat{\mu}_k)_i$ given by

$$(\mu_k)_i = (\mu_{\underline{k}})_{\underline{i}}\ ; \quad (\hat{\mu}_k)_i = (\mu_{\hat{\underline{k}}})_{\hat{\underline{i}}}. \tag{4}$$

Since the $\underline{\underline{\Omega}}$-matrices are real symmetric, the eigenvectors of both form orthonormal bases in M-dimensional (complex) vector space, so that they must be related by a unitary transformation $\underline{\underline{A}}$ $(\underline{\underline{A}}^{\dagger}\equiv\underline{\underline{A}}^{T*}=\underline{\underline{A}}^{-1})$:

$$\underline{\mu}_k = \sum_{m=1}^{M} A_{km}\, \underline{\mu}_m \,. \tag{5}$$

The symmetry $(\mu_k)_i=(\mu_i)_k$ now implies

$$\underline{\underline{\Omega}}_1 = \underline{\underline{B}}^{\dagger}\, \underline{\underline{\Omega}}_2\, \underline{\underline{B}}, \tag{6}$$

with $\underline{\underline{B}}=\underline{\underline{A}}^T$ also unitary.
Consider now the planar graph consisting of two vertices and $q$ edges joining these; since $G_1$ and $G_2$ give the same partition function for any planar graph, this yields

$$\sum_{i,j} \{\Omega_1(i,j)\}^q = \sum_{i,j} \{\Omega_2(i,j)\}^q \tag{7}$$

for all $q$. Use of eq. (6) shows, that this can only be the case if $\underline{\underline{B}}$ is a permutation matrix, $\underline{\underline{B}}=\underline{\underline{D}}(s)$. This implies $G_1=s^{-1}G_2s$. ¶

In view of this lemma, one can from now on talk about the dual group $\tilde{G}$ of a permissible group $G$.

There is no simple, explicit way to express $\tilde{G}$ in terms of $G$. Some information concerning the automorphisms of the common Abelian subgroup (or subgroups) $A$ of $\tilde{G}$ and $G$ can be obtained by considering the characters of $A$:
The quantity

$$\chi_{\underline{t}}(\underline{w}) = \exp\Big(\sum_{k=1}^{n} \frac{2\pi i}{\ell_k}\, w_k t_k\Big), \tag{8}$$

which is the eigenvector of eq. (1.5) up to the normalization, is a character of the group $A$, i.e., a one-dimensional, irreducible representation, since it satisfies

$$\chi_{\underline{t}}(\underline{w}_1+\underline{w}_2) = \chi_{\underline{t}}(\underline{w}_1)\,\chi_{\underline{t}}(\underline{w}_2)\,. \tag{9}$$

These characters form a group isomorphic to $A$:

$$\chi_{\underline{t}_1}(\underline{w})\,\chi_{\underline{t}_2}(\underline{w}) = \chi_{\underline{t}_1+\underline{t}_2}(\underline{w}), \tag{10}$$

since the fundamental symmetry

$$\chi_{\underline{t}}(\underline{w}) = \chi_{\underline{w}}(\underline{t}) \tag{11}$$

follows directly from the definition, eq. (8). Further, eq. (8) yields all different characters of $A$. The symmetry of eq. (11) now defines a "duality" in the automorphism group $\mathrm{AUT}(A)$; to see this, consider some $\alpha \varepsilon \mathrm{AUT}(A)$. Since this is an automorphism, one has

$$\chi_{\underline{t}}(\alpha(\underline{w}_1+\underline{w}_2)) = \chi_{\underline{t}}(\alpha(\underline{w}_1))\ \chi_{\underline{t}}(\alpha(\underline{w}_2)) , \tag{12}$$

so that $\chi_{\underline{t}}(\alpha(\underline{w}))$ is a character, which must be of the form $\chi_{\alpha^T(\underline{t})}(\underline{w})$. Now the mapping $\alpha^T$ is also an automorphism:

$$\chi_{\alpha^T(\underline{t}_1+\underline{t}_2)}(\underline{w}) \equiv \chi_{\underline{t}_1+\underline{t}_2}(\alpha(\underline{w})) = \chi_{\underline{t}_1}(\alpha(\underline{w}))\chi_{\underline{t}_2}(\alpha(\underline{w})) =$$

$$= \chi_{\alpha^T(\underline{t}_1)}(\underline{w})\ \chi_{\alpha^T(\underline{t}_2)}(\underline{w}) . \tag{13}$$

The mapping $\phi:\alpha\to\alpha^T$ is an anti-automorphism of $\mathrm{AUT}(A)$:

$$\chi_{(\alpha_1\alpha_2)^T(\underline{t})}(\underline{w}) = \chi_{\underline{t}}(\alpha_1\alpha_2(\underline{w})) = \chi_{\alpha_1{}^T(\underline{t})}(\alpha_2(\underline{w})) = \chi_{\alpha_2{}^T\alpha_1{}^T(\underline{t})}(\underline{w}) \tag{14}$$

or

$$(\alpha_1\alpha_2)^T = \alpha_2{}^T\ \alpha_1{}^T. \tag{15}$$

$\alpha^T$ will be called the transpose of $\alpha$ from now on. This name will be justified shortly; here, it is shown that $\phi^2=e$, as should be for a transpose:

$$\chi_{\alpha(\underline{t})}(\underline{w}) = \chi_{\underline{w}}(\alpha(\underline{t})) = \chi_{\alpha^T(\underline{w})}(\underline{t}) = \chi_{\underline{t}}(\alpha^T(\underline{w})) = \chi_{(\alpha^T)^T(\underline{t})}(\underline{w}) , \tag{16}$$

so that

$$(\alpha^T)^T = \alpha \tag{17}$$

holds for all $\alpha\varepsilon\ \mathrm{AUT}(A)$. In the next lemma, $\mathrm{AUT}(A)$ is identified with $H_M(A)$ as explained in Section 2.1:

Lemma 2. Let $G$ be a permissible group with regular, Abelian subgroup

$A$. Then, if $\alpha \in G \cap AUT(A)$, then $\alpha^T \in G \cap AUT(A)$.

Proof. The interaction matrices $\Omega(\underline{i},\underline{j}) = \Omega(\underline{i}-\underline{j})$ and $\tilde{\Omega}(\underline{i},\underline{j}) = \tilde{\Omega}(\underline{i}-\underline{j})$ are each others Fourier transforms by eqs. (1.6) and (1.18):

$$\Omega(\underline{w}) = \sum_{\underline{t} \in A} \tilde{\Omega}(\underline{t}) \; \chi_{\underline{t}}(\underline{w}) \; / \; \sum_{\underline{t} \in A} \tilde{\Omega}(\underline{t}), \tag{18}$$

$$\tilde{\Omega}(\underline{t}) = \sum_{\underline{w} \in A} \Omega(\underline{w}) \; \chi_{\underline{w}}(\underline{t}) \; / \; \sum_{\underline{w} \in A} \Omega(\underline{w}). \tag{19}$$

If $\alpha \in G \cap AUT(A)$, then also $\alpha^{-1}$ has this property, so that eq. (19) yields, remembering eq. (1.7):

$$\lambda_0 \tilde{\Omega}(\alpha^T(\underline{t})) = \sum_{\underline{w} \in A} \Omega(\underline{w}) \; \chi_{\underline{w}}(\alpha^T(\underline{t})) = \sum_{\underline{w} \in A} \Omega(\underline{w}) \; \chi_{\alpha(\underline{w})}(\underline{t}) =$$

$$= \sum_{\underline{w} \in A} \Omega(\alpha^{-1}(\underline{w})) \; \chi_{\underline{w}}(\underline{t}) = \sum_{\underline{w} \in A} \Omega(\underline{w}) \; \chi_{\underline{w}}(\underline{t}) = \lambda_0 \; \tilde{\Omega}(\underline{t}). \tag{20}$$

¶

For some special Abelian groups, the transposes can be given explicitly:

(a). Let $A = C(M)$; the characters are simply

$$\chi_t(w) = \exp\left(\frac{2\pi i}{M} \, tw\right), \tag{21}$$

with $t$ and $w$ in the range $0,1,..,M-1$. An automorphism $\alpha$ of $C(M)$ is a mapping

$$\alpha(w) = nw \bmod M, \quad (n,M)=1, \tag{22}$$

so that one has

$$\chi_t(\alpha(w)) = \chi_t(nw) = \chi_{nt}(w) = \chi_{\alpha(t)}(w). \tag{23}$$

It follows, that $\alpha = \alpha^T$ holds for all automorphisms of $C(M)$.

(b). Let $A$ be a direct product of cyclic groups $C(p^f)$, $p$ a prime:

$$A = C(p^f) \otimes C(p^f) \otimes \ldots \otimes C(p^f), \quad q \text{ direct factors.} \tag{24}$$

Then the characters are

$$\chi_{\underline{t}}(\underline{w}) = \exp\left(\frac{2\pi i}{p^f} \, \underline{t} \cdot \underline{w}\right), \tag{25}$$

since the vectors $\underline{t}$ are q-dimensional and each entry takes values from $\{0,1,..,p^f\}$. An automorphism $\alpha$ is, in this case, a nonsingular $q\times q$ matrix $\underline{\underline{G}}$ with entries from the Galois field $GF(p^f)$, i.e., addition and multiplication are done modulo $p^f$ and

$$\det \underline{\underline{G}} \neq 0 \mod p^f \tag{26}$$

holds. It follows, that the transposed automorphism $\alpha^T$ is represented by the transposed matrix $\underline{\underline{G}}^T$. This explains the general name "transpose" for the mapping $\phi:\alpha\to\alpha^T$.

Another general feature of the group $\tilde{G}$ can be obtained from Theorem 1.5.1; since $G$ contains an Abelian, regular subgroup, it is completely permissible. As shown in the proof of Theorem 1.5.1, the permutation matrices of $G$ are completely reduced in the basis of the eigenvectors of $\underline{\underline{\Omega}}$ and a given (absolutely) irreducible representation $\theta_i$ occurs either once or not at all ( $c_i=1$ or 0). The number of different eigenvalues of a general $\underline{\underline{\Omega}}$ is, therefore, equal to the number of irreducible representations $\theta_i$ with $c_i=1$ occurring in $\theta_{per}$, which number equals $s+1$ by Corollary 1.5.1. The degeneracy of the eigenvalue corresponding to $\theta_i$ equals $n_i$, the dimension of $\theta_i$. Translated in terms of the MI of $\tilde{G}$, this yields

Lemma 3. $\tilde{G}$ is a completely permissible group with exactly as many graphs in its MI as $G$. The z-values of these graphs are the dimensions of the nontrivial irreducible representations $\theta_i$ occurring in $\theta_{per}$.

Proof. Immediate from the above and from the fact that the trivial representation occurs exactly once in $\theta_{per}$, see the remark following the proof of Theorem 1.5.1. ¶

Corollary 1. Let $A$ be Abelian and regular, $G=A^{(p)}$; then $\tilde{G}=G$.

Proof. Since $\tilde{G}$ has as many graphs in its MI as $A^{(p)}$ and contains $A$, $\tilde{G}=G$ follows; note that this is also in accordance with Lemma 2, since the automorphism $\sigma$ with $\sigma g\sigma=g^{-1}$ for all $g\in A$ satisfies $\sigma^T=\sigma$ and $G=\langle\sigma,A\rangle$. ¶

Corollary 2. If $G=S(M)$, then $\tilde{G}=G$.

Proof. Since $S(M)$ certainly contains all regular, Abelian groups with $M$ elements, $\tilde{G}$ exists. Now $S(M)$ has only one graph in its MI, so that $\tilde{G}$ has the same property, implying $\tilde{G}=G$. ¶

Corollary 3. Let $G_1$ and $G_2$ be permissible groups with regular, Abelian subgroups $A_1$ and $A_2$, respectively. Then one has

$$\widetilde{G_1\otimes G_2} = \tilde{G}_1\otimes\tilde{G}_2 \quad \text{and} \quad \widetilde{G_1\sim G_2} = \tilde{G}_2\sim\tilde{G}_1 .$$

Proof. The equality for the direct product follows immediately from the fact, that the absolutely irreducible representations of $G_1 \otimes G_2$ are the direct products $\theta_i^{(1)} \otimes \theta_j^{(2)}$ of absolutely irreducible representations $\theta_i^{(k)}$ of $G_k$. The second assertion follows, since $G_1 \sim G_2$ contains subgroups isomorphic to $G_1 \otimes G_1 \otimes \ldots \otimes G_1$ ($M_2$ factors) and to $G_2$, so that its irreducible representations have dimensions $M_2 \dim\theta_i^{(1)}$ or $\dim\theta_i^{(2)}$; these correspond exactly to the z-values of $\tilde{G}_2 \sim \tilde{G}_1$. ¶

The above three corollaries suffice to find all dual groups corresponding to the permissible groups (insofar these have Abelian subgroups) of Section 6.1, except for $G(G_9)$, for which Lemma 2 has to be used. The results are, that all are self-dual, except for the asymmetric wreath products.

## 7.3. The question of duality for nonabelian, regular groups.

In Section 1, the duality transformation has been derived for permissible groups containing regular, Abelian subgroups. The question immediately imposing itself is, whether such a requirement is necessary. To answer this question, the procedure followed in Section 1 is here reconsidered. First of all, the partition function, eq. (1.8), has been rewritten in terms of edge variables, which are differences of vertex variable, eqs. (1.9) and (1.14). Clearly, this is possible as soon as $G$ contains a regular subgroup $R$, which does not have to be Abelian; eq. (1.14) then reads

$$Z = M \sum_{r_e \in R} \prod_{e \in E} \Omega(r_e) \prod_{C \in C} \delta_R[C], \tag{1}$$

with

$$\delta_R[C] \equiv \delta_R[\prod_{e' \in C} r_e^{\alpha_e}], \tag{2}$$

where $\alpha_e = +1$ or $-1$ according to whether the direction in which $C$ is traversed coincides or is opposite to the direction of the edge $e$, and

$$\delta_R[r] = \begin{cases} 1 & \text{for } r=e, \\ 0 & \text{for all other } r \in R. \end{cases} \tag{3}$$

Finally, $\Omega(r)$ is defined by (it is assumed, that $r_i(1)=i$ holds):

$$\Omega(i,j) = \Omega(r_i,r_j) = \Omega(e,r_i^{-1}r_j) \equiv \Omega(r), \quad r=r_i^{-1}r_j. \tag{4}$$

The second way to rewrite the partition function is based on eq. (1.15); if $G$ contains a regular, but nonabelian subgroup $R$, there is no obvious way for labelling the eigenvectors and eigenvalues of $\underline{\underline{\Omega}}$ by elements $r$ of this group. Since, however, $|S|=M=|R|$, such a labelling is possible; assuming this done in some way, eq. (1.15) has an analogue as

$$\Omega(r_i,r_j) = \sum_{r\epsilon R} \lambda_r \, (\mu_r)_{r_i} \, (\mu_r)_{r_j}, \tag{5}$$

so that the equation corresponding to eq. (1.16) reads

$$Z = \sum_{r_e \epsilon R} \prod_{e\epsilon E} \lambda_{r_e} \; \{ \sum_{r_v \epsilon R} \prod_{\substack{\text{edges } e' \\ \text{issuing} \\ \text{from } v}} (\mu_{r_{e'}})^*_{r_v} \prod_{\substack{\text{edges } e'' \\ \text{pointing} \\ \text{towards } v}} (\mu_{r_{e''}})_{r_v} \}. \tag{6}$$

The crucial point is, therefore, the identification of the expression in braces in eq. (6) with $\delta_R[C]$ for the dual graph. But this clearly implies that $R$ must be Abelian, since the factors in eq. (6) can be rearranged arbitrarily (being numbers), whereas this is not possible in eq. (2) if $R$ is nonabelian.

A different approach to duality for nonabelian, regular groups is afforded by the permissible class function groups studied in Section 4.3: if a spin model is defined on a regular group by a class function, then the symmetry group of this spin model is the PCF group $H(R)$. Now this PCF group may contain a regular, Abelian subgroup, so that the dual group, denoted by $H_d(R)$ is well-defined. In fact, a glance at Section 6.2 shows, that for $M\leq 12$ all PCF groups contain Abelian subgroups. It is to be noted, that $H_d(R)$ does not necessarily contain $R$ again: for the quaternion group $Q$, for example, one has

$$H(Q) = S(2) \sim (S(2) \otimes S(2)) = Q^{(p)}$$

from Section 6.2. Corollary 3 then implies

$$H_d(R) = (S(2) \otimes S(2)) \sim S(2),$$

which is smaller than $Q^{(p)}$, so that it cannot contain $Q$. Secondly, even if $R<H_d(R)$ holds, then the dual spin model is, in general, not based on a class fuction, since this would imply $H(R)\leq H_d(R)$, which in

turn implies $H(R)=H_d(R)$ by Lemma 2.3. In Section 6.2, there are no self-dual PCF groups, however, since these all are (or contain) asymmetric wreath products.

In Corollary 4.3.2, a number of groups, for which duality transformations of the corresponding PCF group are possible, has already been exhibited. With more refined mathematical methods, the following lemma can be proved:

Lemma 1. $H(R)$ contains a regular, Abelian subgroup for $|R|=M$ of the following types: (i) $M=p^3$ or $p^4$, $p$ a prime; (ii) $M=p_1p_2$, $p_1^2p_2$ or $p_1^2p_2^2$, $p_1$ and $p_2$ different primes; (iii) $M=p_1p_2\dots p_n$, all $p_i$ different primes.
Proof. (i) Let $R$ be a group of prime power order $p^n$ and $A$ a maximal, Abelian, normal subgroup of order $p^a$. Then $2n\leq a(a+1)$ holds, see ([2]). For $n=3$, $a\geq 2$ follows; since $R$ is nonabelian, $a=2$ must hold. Let $r\in R$, $r\notin A$ and set $\langle r\rangle=B$, then one has $R=AB$ with $A$ and $B$ Abelian. Then $H(R)$ contains a regular, Abelian subgroup by Theorem 4.3.3.
(ii) The result $2n\leq a(a+1)$ implies, that a group of order $p$ or $p^2$ is Abelian. Therefore, the $p_1$- and $p_2$-Sylow subgroups of $R$ are Abelian for $M=p_1p_2$, $p_1^2p_2$ or $p_1^2p_2^2$. (Every finite group of order

$$p_1^{e_1}p_2^{e_2}\dots p_n^{e_n}$$

contains subgroups of order $p_i^{e_i}$, the $p_i$-Sylow subgroups, see ([3]).) Taking $A$= a $p_1$-Sylow subgroup and $B$= a $p_2$-Sylow subgroup yields the desired result by Theorem 4.3.3 again.
(iii) For $M=p_1p_2\dots p_n$, all Sylow subgroups of $R$ are cyclic groups $C(p_i)$; such a group is called metacyclic, and it is known ([4]), that such a group is generated by two elements $a$ and $b$ with

$$a^m = b^n = e, \quad b^{-1}ab = a^r, \quad r^n \equiv 1 \bmod m, \quad (m,n(r-1))=1, \quad M=mn. \qquad (7)$$

This implies $R=\langle a\rangle\langle b\rangle$, so that again Theorem 4.3.3 asserts the lemma. ¶

The smallest value of $M$, for which a group with $M$ elements possibly has no dual group $H_d(R)$ is 24 by Lemma 1 above, so that duality transformations of the type discussed above exist for all 25 nonabelian groups with $M<24$. The 12 nonisomorphic, nonabelian groups with 24 elements have not been exhaustively studied. The next number of elements not covered by Lemma 1, is $M=32$, for which there are as many as 44 nonisomorphic, nonabelian groups! (Numbers of groups taken from ([5]).)

It is at the moment a completely open question, whether or not there exist nonabelian groups, which cannot be written as a product of two Abelian subgroups, and which nevertheless are such, that their corresponding PCF groups contain regular, Abelian subgroups.

## 7.4. Duality transformations in dimensions higher than two; k-gauge models.

In this section, the graph G is taken such, that it can be embedded in d-dimensional Euclidean space $R^d$ in a way analogous to the embedding of a planar graph in $R^2$: the graph is space-filling in the sense that $R^d$ is partitioned into $N_d$ d-dimensional cells, the boundaries of which are (d-1)-dimensional cells, etc., until one reaches 1-dimensional cells, which are the edges of G, and the 0-dimensional cells, which are its vertices. The class of graphs, that can be embedded in $R^d$ in this way, is not known explicitly, except for the case d=2, for which the planar graphs are characterized by the well-known theorems of Kuratowski (6) and of MacLane (7). Of these two, the Kuratowski criterion is the most easy to visualize: a graph G can be embedded into $R^2$ in the way described above iff it cannot be derived from K(5) or from the graph of Fig. 1 by addition of vertices and/or edges. The nonexist-

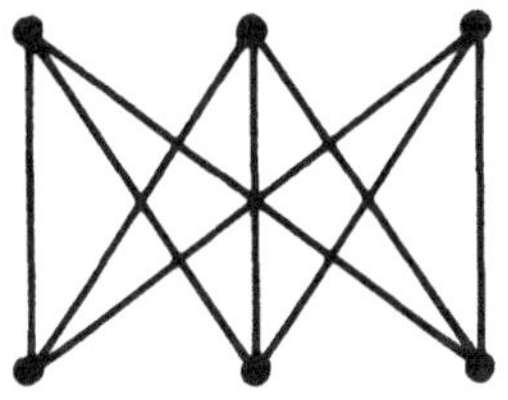

Fig. 1. The bipartite graph of the Kuratowski criterion for planar graphs.

ence of a similar theorem for $R^d$ should also not be confused with the well-known fact, that every finite graph G can be embedded in $R^3$ in such a way, that all edges are straight-line segments and no two edges cross, except at the vertices of G (8): such an embedding will, in general, not be space-filling in the sense used above. Examples of graphs which are emdeddable in d dimensions are all finite pieces of regular lattices in d dimensions.

In order to proceed, the dual graph $\tilde{G}$ has to be defined; this is done in a way similar to the d=2 case: the vertices of $\tilde{G}$ are placed inside each of the d-dimensional cells of G filling $R^d$; two such vertices are connected by an edge of $\tilde{G}$ iff the corresponding d-cells have a (d-1)-cell in common. It follows, that if the embedding of G contains $N_k$ k-cells, then $\tilde{G}$ will contain $\tilde{N}_k=N_{d-k}$ k-cells. The numbers $N_k$ (and, also, $\tilde{N}_k$) satisfy the generalized Euler relation:

$$N_o-N_1+N_2-N_3+\ldots+(-1)^d N_d = 1+(-1)^d. \tag{1}$$

In order to find the dual of a spin model with permissible group $\mathcal{G}$ (containing a regular Abelian subgroup $\mathcal{A}$), one starts from the general eq. (1.17), which now takes the form:

$$Z(\Omega,G) = M^{N_o-N_1} \lambda_o^{N_1} \sum_{\underline{k}_e} \prod_{e\in E} \tilde{\Omega}(\underline{k}_e) \prod_{v\in V} \delta_A[\sum_{e'\leftarrow v} \underline{k}_{e'} - \sum_{e''\to v} \underline{k}_{e''}]. \tag{2}$$

To each vertex now corresponds a d-cell of the dual graph $\tilde{G}$, which is bounded by as many (d-1)-cells as there are edges adjacent to this vertex. As in the case d=2, these (d-1)-cells now inherit the edge variables $\underline{k}_e$ as well as the directions of the edges to which they correspond. Now the (d-2)-cells bordering these (d-1)-cells are also given (arbitrary) directions and variables $\underline{\ell}$ are placed on them. It is now easy to see, that the $\delta$-function in eq. (2) is identically satisfied for all configurations of the $\underline{\ell}$-variables if one sets

$$\underline{k}_e = \sum_{\substack{\text{(d-2)-cells i}\\ \text{bordering the}\\ \text{(d-1)-cell}\\ \text{with } \underline{k}_e}} \alpha_i \underline{\ell}_i, \tag{3}$$

where

$$\alpha_i = \begin{cases} +1 & \text{if the direction of the i-th (d-2)-cell coincides with the direction of the (d-1)-cell it borders,} \\ -1 & \text{otherwise.} \end{cases} \tag{4}$$

Eq. (2) then takes the form

$$Z(\Omega,G) = M^{N_o-N_1} \lambda_o^{N_1} M^{-\tilde{T}(d-2)} \sum_{\substack{\underline{\ell}\text{-variables}\\ \text{on (d-2)-cells}\\ \text{of dual graph}}} \prod_{\text{(d-1)-cells}} \tilde{\Omega}(\sum_{\substack{\text{(d-2)-cells}\\ \text{bordering a}\\ \text{(d-1)-cell}}} \alpha_i \underline{\ell}_i) \tag{5}$$

where the factor $M^{-\tilde{T}(d-2)}$ has been inserted to account for the many-to-one character of eq. (3). This factor is easily found by setting

$$\Omega(\underline{k}) = \delta_A[\underline{k}], \tag{6}$$

since then

$$Z(\Omega,G) = M, \ \lambda_o=1 \quad \text{and} \quad \tilde{\Omega}(\underline{\ell}) = 1 \quad \text{for all} \quad \underline{\ell} \tag{7}$$

follow. This inserted in eq. (5) gives

$$\tilde{T}(d-2) = N_0-N_1+N_2-1. \tag{8}$$

It is remarked, that the symmetry $\tilde{\Omega}(\underline{\ell})=\tilde{\Omega}(-\underline{\ell})$ implies, that the directions of the (d-1)-cells do not have to be given, except to give values to the $\alpha_i$; the result, eq. (5), is also independent of the $\alpha_i$-values, since $\alpha_i \to -\alpha_i$ can be counteracted by $\underline{\ell}_i \to -\underline{\ell}_i$. A fixed set of values of the $\alpha_i$ must, however, be chosen in order to evaluate eq. (5).

The duality relation, eq. (5), can be generalized further: define a k-gauge model on a d-dimensional graph by a partition function

$$Z(\Omega,G,d,k) = \sum_{\substack{\underline{\ell}\text{-variables} \\ \text{on k-cells}}} \ \prod_{\substack{(k+1)- \\ \text{cells}}} \Omega\Big( \sum_{\substack{\text{k-cells } i \\ \text{bordering a} \\ (k+1)\text{-cell}}} \alpha_i \ell_i \Big). \tag{9}$$

Note that a spin model is a 0-gauge model in this definition. Doing the same duality transformation as for the spin model, one obtains

$$Z(\Omega,G,d,k) = M^{N_k-N_{k+1}} \lambda_o^{N_{k+1}} M^{-\tilde{T}(d-k-2)} Z(\tilde{\Omega},\tilde{G},d,d-k-2). \tag{10}$$

The exponent $\tilde{T}(d-k-2)$ takes care of the many-to-one character of the analogue of eq. (3) for this case; use of the special interaction of eq. (6) yields

$$T(k) + \tilde{T}(d-k-2) = N_k-N_{k+1}+N_{k+2}, \tag{11}$$

which is easily seen to imply

$$T(k) = N_{k-1}-N_{k-2}+\ldots+(-1)^{k+1}N_0+(-1)^k. \tag{12}$$

Eq. (8) follows from this by going over to the dual graph and by use of the Euler relation, eq. (1).

The k-gauge model defined by eq. (9) may also be defined through the energy function $E(\underline{\ell})$, from which $\Omega(\underline{\ell})$ is derived as

$$\Omega(\underline{\ell}) = \exp(-\beta E(\underline{\ell})). \tag{13}$$

Eq. (9) is then the partition function of a model with energy function

$$E\{\underline{\ell}\} = \sum_{\substack{(k+1)-\\ \text{cells}}} E\Big(\sum_{\substack{k\text{-cells } i\\ \text{bordering}\\ \text{a } (k+1)-\\ \text{cell}}} \alpha_i \ell_i\Big). \tag{14}$$

The question as to the symmetry group of this energy function now arises naturally; one has the following results:

Theorem 1. The symmetry group of eq. (14) contains an Abelian k-gauge group $\mathcal{G}(k)$ with $M^{T(k)}$ elements; every element of this k-gauge group may be obtained by performing a sequence of local gauge transformations. Such a local gauge transformation is defined for each (k-1)-cell of the graph by the prescription

$$\underline{\ell}'_k = \underline{\ell}_k + \beta_k \underline{a}, \tag{15}$$

for each k-cell, which has this particular (k-1)-cell on its boundary. In eq. (15), $\underline{a}$ is an arbitrary element of $\mathcal{A}$ and the $\beta_k$ indicate whether the directions of the (k-1)-cell and its adjacent k-cells coincide ($\beta_k$=+1) or nor ($\beta_k$=-1).

Proof. In summing over the k-cells bordering a certain (k+1)-cell, none or exactly two k-cells with a specific (k-1)-cell as boundary are en countered. In the first case, the energy contribution of this particular (k+1)-cell is, of course, not changed; in the second case, let these k-cells be labelled by 1 and 2. These give a contribution

$$\alpha_1\underline{\ell}_1 + \alpha_2\underline{\ell}_2 \tag{16}$$

to the sum over all k-cells. The local gauge transformation changes this to

$$\alpha_1\underline{\ell}'_1 + \alpha_2\underline{\ell}'_2 = \alpha_1\underline{\ell}_1 + \alpha_2\underline{\ell}_2 + (\alpha_1\beta_1+\alpha_2\beta_2)\underline{a}. \tag{17}$$

The extra term in eq. (17) is zero, however: (i) let k-cells 1 and 2 have the same orientation with respect to the (k+1)-cell; then one points towards the (k-1)-cell, the other away from it, so that $\alpha_1=\alpha_2$ and $\beta_1=-\beta_2$ follows; (ii) similarly, $\alpha_1=-\alpha_2$ is seen to imply $\beta_1=\beta_2$. Therefore, a set of local gauge transformations does not change any of the arguments in eq. (14). On the other hand, $M^{T(k)}$ is exactly the number of different ways to have a fixed set of arguments and it is easily seen, that any two equivalent configurations of the $\underline{\ell}$'s must be connected by a gauge transformation. ¶

Remark. It might have been expected, that there are $M^{N_{k-1}}$ gauge transformations, since there are $N_{k-1}$ (k-1)-cells at which each of the M gauge transformations of eq. (15) can be performed. That this is not true is easiest to see for the case k=1; then the spins are on the edges and $\beta_k$=+1 or -1, according to whether an edge points towards or away from the vertex at which the local gauge transformation is performed. It is then clear, that performing the same gauge transformation at every vertex does not change the edge spins at all. Since there are M such possibilities, the number of elements of the gauge group is M to the power $N_0-1$, which agrees with the general result.

Theorem 2. The total symmetry group of the energy function of eq. (14) is the direct product $\mathcal{G}(k)\otimes\mathcal{L}$ of the gauge group and of a group $\mathcal{L}$ of automorphisms of $\mathcal{A}$, which have the global action

$$\underline{\ell}_i' = \alpha(\underline{\ell}_i) \quad \text{for all k-cells } i. \tag{18}$$

$\mathcal{L}$ is the intersection of $\mathrm{AUT}(\mathcal{A})$ with the group $\mathcal{H}_1$ of the letter-1-fixing permutations of the permissible group with energy function $E(\underline{k})$.

Proof. Since G is a connected graph, a symmetry operation of eq. (14) must be global if it has no local part. It can then only be given by a one-to-one mapping of $\mathcal{A}$ onto itself. The form of the energy implies that such a global symmetry must be an automorphism of $\mathcal{A}$. Those automorphisms, which keep the energy values fixed are exactly those described in the theorem. Since every such automorphism commutes with every element of the gauge group, the total symmetry group is a direct product. ¶

The above theorem shows, that the symmetry group of a k-gauge model which for k=0 reduces to a spin model with permissible group $\mathcal{G}$, is not independent of the choice of the Abelian, regular subgroup $\mathcal{A}$ of $\mathcal{G}$ with which $\Omega(i,j)$ is parametrized to give $\Omega(\underline{k})$. Indeed, the group $\mathcal{L}$ is certainly different for different choices of $\mathcal{A}$. In particular, let $\mathcal{G}=\mathcal{S}(M)$, the Potts model. Then any Abelian group with M elements is a regular subgroup of $\mathcal{G}$. The corresponding groups $\mathcal{L}$ are always the full automorphism groups of $\mathcal{A}$, but these do depend on $\mathcal{A}$; for M=4, for example, one has the two choices $\mathcal{A}_1=\mathcal{S}(2)\otimes\mathcal{S}(2)$ and $\mathcal{L}_1\simeq\mathcal{S}(3)$ as well as $\mathcal{A}_2=\mathcal{C}(4)$ and $\mathcal{L}_2\simeq\mathcal{S}(2)$. This rather unsettling feature does not show up in the duality relation, eq. (10), however: in fact, this equation can be used to show that the partition function only depends on the function $\Omega(i,j)$ and not on the particular Abelian subgroup of $\mathcal{G}$ used to parametrize this as $\Omega(\underline{k})$.

## 7.5. Duality transformations for k-gauge models in a magnetic field.

As shown in Section 1.5, an external field F(i) acting on an M-component spin model can, for a completely permissible symmetry group, be decomposed in its projections on either the eigenspaces of $\underline{\underline{\Omega}}$ or on the invariant subspaces of the carrier space. For the models with a regular, Abelian subgroup considered in this chapter, this implies, that $\exp-\beta F(i)$ can be replaced by an interaction $\Omega_f(\underline{i})=\Omega_f(\underline{i},\underline{0})$, since eqs. (1.15) and (1.5) imply

$$\Omega_f(\underline{i}) = (2M)^{-1} \sum_{\underline{t}\varepsilon A} \lambda_{\underline{t}} \{(\mu_{\underline{t}})_{\underline{i}}+(\mu_{\underline{t}})^*_{\underline{i}}\}. \tag{1}$$

In view of eq. (1), a k-gauge model in an external field may be described by the coupling of all spins to a "ghost spin", which is in a fixed state (chosen as $\underline{0}$ here):

$$Z(\Omega,G,d,k,\Omega_f) = \sum_{\substack{\underline{\ell} \text{ on k-}\\ \text{cells}}} \prod_{\substack{(k+1)-\\ \text{cells}}} (\sum_i \alpha_i \underline{\ell}_i) \prod_{\text{k-cells}} \Omega_f(\alpha_i \underline{\ell}_i), \tag{2}$$

where the factor $\alpha_i$ has been inserted in the last factor for convenience. For this model, a duality transformation is not immediately possible. Eq. (2) is, therefore, now cast into a different form, for which a duality transformation does exist. To this end, the system is embedded in (d+1)-dimensional space and the graph G is supplemented by a vertex (the "ghost vertex") and by edges extending from this ghost vertex to all other vertices of the original graph. This extended graph G' is easily seen to have $N_0'=N_0+1$ vertices, $N_1'=N_1+N_0$ edges, $N_2'=N_2+N_1$ faces, etc. Eq. (2) can now be written as a k-gauge model on G':

$$Z(\Omega,G,d,k,\Omega_f) = \sum_{\substack{\underline{\ell}_i \text{ on old k-cells}\\ \underline{0} \text{ on new k-cells}}} \prod_{\substack{\text{old (k+1)-}\\ \text{cells}}} \Omega(\sum_i \alpha_i \underline{\ell}_i) \times$$
$$\times \prod_{\substack{\text{new (k+1)-}\\ \text{cells}}} \Omega_f(\sum_i \alpha_i \underline{\ell}_i). \tag{3}$$

The condition "$\underline{0}$ on new k-cells" can be dropped, since this can always be obtained by a gauge transformation:

$$Z = M^{-N_{k-1}} \sum_{\substack{\underline{\ell}_i \text{ on all k-}\\ \text{cells of G'}}} \prod_{\substack{\text{old (k+1)-}\\ \text{cells}}} \Omega(\sum_i \alpha_i \underline{\ell}_i) \prod_{\substack{\text{new (k+1)-}\\ \text{cells}}} \Omega_f(\sum_i \alpha_i \underline{\ell}_i). \tag{4}$$

Since this is proportional to the partition function of a k-gauge model on G', the duality transformation of the previous section can be applied with the difference, that there are two types of interactions now. The result is

$$Z = M^{-N_{k+1}} \lambda_o^{N_{k+1}} (\lambda_{of})^{N_k} M^{-T'(d-k-1)} \sum_{\substack{\underline{w}_i \text{ on } (d-k-1)-\\ \text{cells of } G'}} \prod_{\substack{(d-k)\text{-cells}\\ \text{dual to new}\\ (k+1)\text{-cells}}} \times$$

$$\times \tilde{\Omega}_f(\sum_i \beta_i \underline{w}_i) \prod_{\substack{(d-k)\text{-cells}\\ \text{dual to old}\\ (k+1)\text{-cells}}} \tilde{\Omega}(\sum_i \beta_i \underline{w}_i) \ . \tag{5}$$

Now it is easy to see, that this looks like a (d-k-1)-gauge model defined on the original dual graph $\tilde{G}$ in a magnetic field; in fact, one finds

$$Z(\Omega,G,d,k,\Omega_f) = M^{-N_{k+1}} \lambda_o^{N_{k+1}} (\lambda_{of})^{N_k} Z(\tilde{\Omega}_f,\tilde{G},d,d-k-1,\tilde{\Omega}) . \tag{6}$$

Note that $\underline{\underline{\Omega}}$ and $\underline{\underline{\Omega}}_f$ exchange their roles in the dualization process.

Eq. (6) can be used to show the "triviality" of (d-1)-gauge models without field in d dimensions. Indeed, setting

$$\Omega_f(\underline{k}) = 1 \quad \text{for all} \quad \underline{k} \in A, \tag{7}$$

reduces eq. (6) to the case of no field; this gives

$$Z(\Omega,G,d,d-1) = M^{N_{d-1}-N_d} \lambda_o^{N_d} Z(\tilde{\Omega}_f,\tilde{G},d,0,\tilde{\Omega}), \tag{8}$$

where use has been made of $\lambda_{of}=M$ and $\tilde{\Omega}_f(\underline{k})=\delta_A[\underline{k}]$. Due to this form of $\tilde{\Omega}_f$, the partition function on the right-hand-side of eq. (8) represents just one spin in an external field:

$$Z(\tilde{\Omega}_f,\tilde{G},d,0,\tilde{\Omega}) = \tilde{\lambda}_o = M/\lambda_o, \tag{9}$$

so that eq. (8) gives

$$Z(\Omega,G,d,d-1) = M^{N_{d-1}-N_d+1} \lambda_o^{N_d-1} . \tag{10}$$

Such a model obviously cannot have a phase transition at a finite temperature. This result holds for 1-dimensional spin models, 2-dimensional

1-gauge models, 3-dimensional 2-gauge models, etc.

## 7.6. Self-duality for k-gauge models on hypercubic lattices.

The results of the previous two sections yield the following cases, in which a k-gauge model is dual to a k-gauge model again:
a). No external field: $k=d-k-2$ or $d=2k+2$: spin models in two dimensions, 1-gauge models in four dimensions, etc.
b) With external field: $k=d-k-1$ or $d=2k+1$: spin models in one dimension, 1-gauge models in three dimensions, etc.
In order to obtain from this information on the phase diagrams of specific models, these have to have self-dual symmetry groups and they must be considered on self-dual graphs. The simplest of these latter are the finite d-dimensional hypercubic lattices with periodic boundary conditions. Although these cannot quite be embedded in d-dimensional Euclidean space, this distinction disappears in the thermodynamic limit $N_O \to \infty$, as shown explicitly by the duality inequalities (1.20) and (1.21) for the case $d=2$. Therefore, these finite d-dimensional hypercubic lattices can, for large $N_O$, be treated as self-dual graphs $G(N_O)$. The values of the $N_k$ for these graphs are given by

$$N_k = N_{d-k} = N_k = N_{d-k} = \binom{d}{k} N_O. \tag{1}$$

The free energy (up to a factor $-\beta$) per spin in the thermodynamic limit is defined by

$$\gamma(\Omega,d,k) = \lim_{N_O \to \infty} N_O^{-1} \ln Z(\Omega,G(N_O),d,k) \tag{2}$$

for k-gauge models without a field and by

$$\gamma(\Omega,d,k,\Omega_f) = \lim_{N_O \to \infty} N_O^{-1} \ln Z(\Omega,G(N_O),d,k,\Omega_f) \tag{3}$$

for k-gauge models in an external field. Choosing $d=2k+2$ for the field-free case, eq. (4.10) yields, in the thermodynamic limit,

$$\gamma(\Omega,2k+2,k) = \frac{1}{2} \binom{2k+2}{k+1} \ln(\lambda_O{}^2/M) + \gamma(\tilde{\Omega},2k+2,k). \tag{4}$$

Similarly, eq. (5.6) gives, for the choice $d=2k+1$,

$$\gamma(\Omega,2k+1,k,\Omega_f) = \binom{2k+1}{k} \ln(\lambda_o\lambda_{of}/M) + \gamma(\tilde{\Omega}_f,2k+1,k,\tilde{\Omega}). \qquad (5)$$

Self-duality for eq. (5) implies $\Omega=\tilde{\Omega}_f$, which in turn implies $\tilde{\Omega}=\Omega_f$ and $\lambda_{of}=M/\lambda_o$; here, it is remarked, that for $\Omega=\tilde{\Omega}_f$, eq. (5) has the same form as eq. (4); upon absorption of the combinatorial factors in the $\gamma$'s, both read

$$\gamma'(\Omega) = \ln(\lambda_o^2/M) + \gamma'(\tilde{\Omega}), \qquad (6)$$

$$\gamma'(\Omega) = 2\frac{\gamma(\Omega,2k+2,k)}{\binom{2k+2}{k+1}} = \frac{\gamma(\Omega,2k+1,k,\Omega_f)}{\binom{2k+1}{k}}. \qquad (7)$$

The single eq. (6), therefore, describes k-gauge models without a field on (2k+2)-dimensional hypercubic lattices as well as k-gauge models with coupling to a ghost spin via the spin-spin interaction on (2k+1)-dimensional hypercubic lattices.

The matrices $\underline{\underline{\Omega}}$ and $\tilde{\underline{\underline{\Omega}}}$ of eq. (6) depend on the symmetry group of the model; assuming there to be $s$ graphs in the MI of the corresponding spin model, $\underline{\underline{\Omega}}$ depends on $s$ parameters $\omega_1,\omega_2,..,\omega_s$. Similarly, $\tilde{\underline{\underline{\Omega}}}$ depends on the $s$ parameters $\tilde{\omega}_1,\tilde{\omega}_2,..,\tilde{\omega}_s$:

$$\tilde{\omega}_i = \lambda_i/\lambda_o. \qquad (8)$$

Therefore, $\gamma'$ can be written as a function of an s-dimensional vector and eq. (6) is equivalent to

$$\gamma'(\omega_1,\omega_2,..,\omega_s) = \ln(\lambda_o^2/M) + \gamma'(\tilde{\omega}_1,\tilde{\omega}_2,..,\tilde{\omega}_s). \qquad (9)$$

The space of s-dimensional vectors $\underline{\omega}$ which can occur in eq. (9) is restricted by the condition that it must belong to the ferromagnetic region (see Section 1), i.e., all $\tilde{\omega}_i$ must be positive for eq. (9) to make sense. It is this ferromagnetic region, that is mapped onto itself by a duality transformation. The simple example of a duality transformation represented by eqs. (4.6) and (4.7) shows, that the duality transformation always maps points in the neighbourhood of (0,...,0) onto points in the neighbourhood of (1,...,1) and vice versa. Therefore, it is useful to define a metric in the space of s-dimensional vectors and to study those vectors, whose length is left invariant by the duality transformation. Two metrics will be used in the following:

(i) an $L_1$-type metric defined by

$$e(\underline{\omega}) = \sum_{k=1}^{s} z_k \omega_k \tag{10}$$

and (ii) a standard Euclidean-type metric:

$$d(\underline{\omega})^2 = \sum_{k=1}^{s} z_k \omega_k^2. \tag{11}$$

In eqs. (10) and (11), the $z_k$ have their usual meaning as the degrees of the regular graphs occurring in the MI of the permissible group of the corresponding spin model. By Lemma 2.3, these are also the multiplicities of the eigenvalues $\lambda_k$ of $\underline{\underline{\Omega}}$, so that eqs. (10) and (11) are also well-defined for the dual vector $\underline{\tilde{\omega}}$. The following theorem shows the meaning of the logarithmic term occurring in eq. (9):

Theorem 1. There is a hyperplane of dimension $s-1$ in the space of $\underline{\omega}$-vectors such that

$$e(\underline{\omega}) = e(\underline{\tilde{\omega}}) \quad \text{and} \quad d(\underline{\omega}) = d(\underline{\tilde{\omega}}) \tag{12}$$

both hold for all points of this plane. This hyperplane $H$ is the intersection of the plane $\lambda_o = \sqrt{M}$ with the ferromagnetic region.

Proof. One has

$$e(\underline{\omega}) = \lambda_o - 1, \tag{13}$$

since $\lambda_o$ is the eigenvalue of $\underline{\underline{\Omega}}$ corresponding to the (M-dimensional) vector (1,...,1). On the other hand,

$$M = \mathrm{Tr}\ \underline{\underline{\Omega}} = \sum_{k=0}^{s} \lambda_k = \lambda_o\{1+e(\underline{\tilde{\omega}})\} \tag{14}$$

holds, so that

$$e(\underline{\tilde{\omega}}) = (M/\lambda_o) - 1 \tag{15}$$

follows. Equality of eqs. (13) and (15) implies $\lambda_o^2 = M$.

Secondly, the structure of $\underline{\underline{\Omega}}$ implies

$$\sum_{j=1}^{M} \Omega(i,j)\Omega(j,i) = 1 + d(\underline{\omega})^2, \tag{16}$$

independent of i, or

$$d(\underline{\omega})^2 = M^{-1}\ \mathrm{Tr}\ \underline{\underline{\Omega}}^2 - 1. \tag{17}$$

On the other hand, one has

$$\mathrm{Tr}\ \underline{\underline{\Omega}}^2 = \lambda_o{}^2 + \sum_{k=1}^{s} \lambda_k{}^2 = \lambda_o{}^2\{1+d(\underline{\tilde{\omega}})^2\}, \tag{18}$$

so that $d(\underline{\omega})$ and $d(\underline{\tilde{\omega}})$ are related by

$$d(\underline{\omega})^2+1 = (\lambda_o{}^2/M)\ \{d(\underline{\tilde{\omega}})^2+1\}. \tag{19}$$

Clearly, $d(\underline{\omega})=d(\underline{\tilde{\omega}})$ only iff $M=\lambda_o{}^2$. ¶

Before considering specific models, two more points which deserve mention will be discussed. The first of these is rather trivial: the partition function for a system with interaction matrix $\underline{\underline{\Omega}}$ factorizes if this matrix is a direct product of two interaction matrices for smaller systems:

$$\underline{\underline{\Omega}} = \underline{\underline{\Omega}}_1 \otimes \underline{\underline{\Omega}}_2 \rightarrow Z(\Omega) = Z(\Omega_1)\ Z(\Omega_2). \tag{20}$$

The second point is somewhat more interesting: the duality transformation is linear, i.e., an $s\times s$ matrix, on the hyperplane $H$ defined in Theorem 1; now the partition function may have other symmetries, which leave this hyperplane invariant. In particular, the permutations of the $\omega$'s, which correspond to the elements of $\mathcal{S}(M)$ which leave the MI of a spin model invariant (described as $A_1(G)$ in Section 2.1, Lemma 2.1.2 ff.), have this property and are, therefore, also representable by $s\times s$ permutation matrices. The total group $D_s$ of the symmetries of the partition function, which leave $H$ invariant, is, therefore, generated by the duality transformation and by these permutations. There is no a priori reason for this group to be Abelian; in fact, cases do occur, for which the duality transformation is not in the center of $D_s$. In this latter case, the group will contain also conjugates of the duality transformation, which are of equal interest as the original. The subsets of $H$, which are left pointwise invariant by such a conjugate of the duality transformation, are defined to be <u>self-dual</u> <u>hyperplanes</u>. Keeping the above in mind, some specific models are considered below. These will be classified according to their names as permissible spin model groups; this does not imply, that the corresponding k-gauge models have the same symmetry group: these latter are given by Theorem 4.1 instead.

<u>Potts models</u>. These are the models corresponding to the permissible group $\mathcal{S}(M)$. The $\underline{\omega}$-space is one-dimensional, since there is only one graph in the MI of $\mathcal{S}(M)$. The hyperplane $H$ is a single point given by

$\lambda_o=\sqrt{M}$, or, explicitly,

$$\omega_c = (1+\sqrt{M})^{-1}. \qquad (21)$$

This must then also be the only self-dual point of the phase diagram; indeed, eq. (21) satisfies the equation

$$\tilde{\omega} = \frac{1-\omega}{1+(M-1)\omega} = \omega. \qquad (22)$$

The ferromagnetic region is simply $0\leq\omega\leq 1$ by eq. (22). The point given by eq. (21) is the unique phase transition point for two-dimensional spin models ([9]) and there are strong indications, that this is also the case for four-dimensional 1-gauge models (from Monte Carlo simulations). On the other hand, eq. (21) certainly does not correspond to a phase transition for a one-dimensional spin model in an external field.

If eq. (21) describes the unique phase transition point of a Potts model, some more information can be obtained from eq. (9), which now takes the form

$$\gamma'(\omega) = \ln\{[1+(M-1)\omega]^2/M\} + \gamma'(\frac{1-\omega}{1+(M-1)\omega}). \qquad (23)$$

Writing $\omega=\omega_c+\delta\omega$ with $\delta\omega$ small, gives

$$\gamma'(\omega_c+\delta\omega) \cong \{2(M-1)/\sqrt{M}\}\delta\omega + \gamma'(\omega_c-\delta\omega). \qquad (24)$$

Therefore, the singularity at $\omega_c$ is <u>symmetric</u>.

<u>$S(M)\sim S(M)$ models</u>. These are the self-dual imprimitive permissible groups with two graphs in their MI's. The eigenvalues are, for a Boltzmann factor $\omega_1$ attached to the $z=M-1$ graph and $\omega_2$ attached to the graph with $z=M(M-1)$:

$$\lambda_o = 1 + (M-1)\omega_1 + M(M-1)\omega_2, \qquad \text{(nondegenerate)}, \qquad (25)$$

$$\lambda_1 = 1 + (M-1)\omega_1 - M\omega_2, \qquad \text{((M-1)-fold degenerate)}, \qquad (26)$$

$$\lambda_2 = 1 - \omega_1, \qquad \text{( M(M-1)-fold degenerate)}. \qquad (27)$$

The ferromagnetic region is enclosed by the straight lines

$$\omega_1=1 \quad \text{and} \quad \omega_2=\{1+(M-1)\omega_1\}/M. \qquad (28)$$

The invariant hyperplane H is also a straight line:

$$\omega_2 = (1-\omega_1)/M. \tag{29}$$

This is also the self-dual line, as is easily checked; this is due to the fact, that of the original s duality conditions (here s=2), two are automatically fulfillied by $M=\lambda_o^2$ as shown by the nontrivial expressions in Theorem 1.

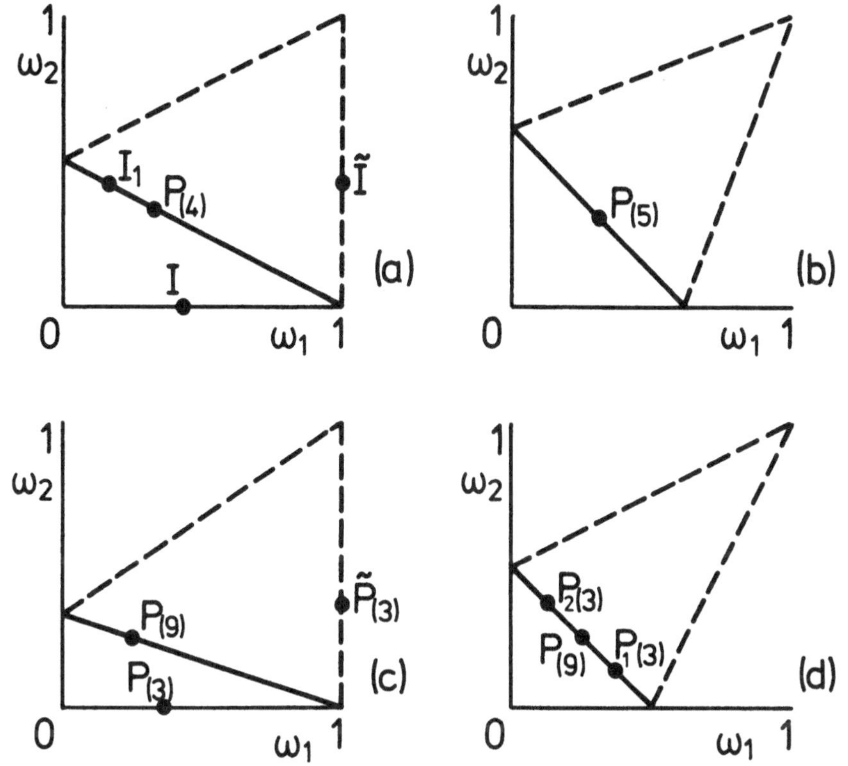


Fig. 1. Phase diagrams for permissible models with two graphs in their MI's. (a) S(2)∿S(2), (b) D(5), (c) S(3)∿S(3), (d) G($G_9$). Full straight lines are the self-dual lines; the broken straight lines are the boundaries of the ferromagnetic regions. For a discussion of the special points of these phase diagrams, see the text.

Two examples are shown in Figs. 1(a) and 1(c) for M=2 and M=3, respectively. Also shown here are the pairs of dual points I and $\tilde{I}$ (Fig. 1(a)) and P(3) and $\tilde{P}(3)$ (Fig. 1(c)); these are the self-dual points of an Ising and of a 3-state Potts model, respectively, to which the present models reduce when they factorize, which is the case for

$\omega_2=0$ and for $\omega_1=1$:

$$\underline{\underline{\Omega}}[\mathrm{S}(M)\sim\mathrm{S}(M);\omega_1,\omega_2=0] = \underline{\underline{\Omega}}[\mathrm{S}(M);\omega_1] \otimes \begin{pmatrix} 1 & 0 \\ 0 & 1 \end{pmatrix} \tag{30}$$

and

$$\underline{\underline{\Omega}}[\mathrm{S}(M)\sim\mathrm{S}(M);\omega_1=1,\omega_2] = \begin{pmatrix} 1 & 1 \\ 1 & 1 \end{pmatrix} \otimes \underline{\underline{\Omega}}[\mathrm{S}(M);\omega_2]. \tag{31}$$

The points $P(4)$ and $P(9)$, respectively, are the Potts model self-dual points obtained for $\omega_1=\omega_2$:

$$\underline{\underline{\Omega}}[\mathrm{S}(M)\sim\mathrm{S}(M);\omega_1=\omega_2] = \underline{\underline{\Omega}}[\mathrm{S}(M^2);\omega_1]. \tag{32}$$

The self-dual Ising point $I_1$ in Fig. 1(a) is exceptional: the $\mathrm{S}(2)\sim\mathrm{S}(2)$ model only also factorizes for $\omega_1=\omega_2{}^2$. If all these special points are phase transitions, it may be expected that the phase diagram contains the following features: (i) a phase transition line along the self-dual line extending from the point $\omega_1=0$, $\omega_2=1/M$ up to $P(M^2)$ and (ii) a pair of dual phase transition lines, both starting at $P(M^2)$ and ending at the points $P(M)$ and $\tilde{P}(M)$, respectively. For the general k-gauge model, this is pure conjecture; it will, however, be shown in Chapters 9 and 11, that there are strong indications for the correctness of this interpretation for spin models.

The $\mathrm{D}(5)$ model. The simplest primitive, permissible group with a 2-graph MI is $\mathrm{D}(5)$, see Section 6.1. The eigenvalues of $\underline{\underline{\Omega}}$ are:

$$\lambda_0 = 1+2\omega_1+2\omega_2, \qquad \text{(nondegenerate)}, \tag{33}$$

$$\begin{aligned} \lambda_1 &= 1+\tfrac{1}{4}(\sqrt{5}-1)\,\omega_1-\tfrac{1}{4}(\sqrt{5}+1)\,\omega_2, \qquad \text{( 2-fold degenerate)}, \\ \lambda_2 &= 1-\tfrac{1}{4}(\sqrt{5}+1)\,\omega_1+\tfrac{1}{4}(\sqrt{5}-1)\,\omega_2, \qquad \text{( 2-fold degenerate)}. \end{aligned} \tag{34}$$

The duality transformation $D$ is, therefore, given as

$$D(\omega_1,\omega_2) = (\tilde{\omega}_1,\tilde{\omega}_2) \quad \text{with} \quad \tilde{\omega}_i=\lambda_i/\lambda_0 \quad \text{for} \quad i=1,2. \tag{35}$$

The partition function is symmetric with respect to the interchange $\sigma$ of $\omega_1$ and $\omega_2$; this commutes with the duality transformation, eq. (35). $D$ reduces to the identity on the hyperplane $H$, which is, therefore, the same as the self-dual line for $D$:

$$\omega_1+\omega_2 = \tfrac{1}{2}(\sqrt{5}-1). \tag{36}$$

The mapping $D\sigma$ only keeps the five-state Potts model self-dual point $P(5)$ fixed; since the model can, of course, not factorize, this is the only special point of the phase diagram, shown as Fig. 1(b). From this, not much information on the phase transitions of this model can be extracted, except for the fact, that if there is a unique phase transition for some values of $E_1$ and $E_2$, this then must lie on the self-dual line.

The $\underset{\sim}{G}(G_9)$ model. The next primitive, permissible group with a 2-graph MI is $\underset{\sim}{G}(G_9)$. The eigenvalues are here:

$$\lambda_o = 1+4\omega_1+4\omega_2, \text{ (nondegenerate)}, \tag{37}$$

$$\begin{aligned} \lambda_1 &= 1+\omega_1-2\omega_2, \quad \text{(four-fold degenerate)}, \\ \lambda_2 &= 1-2\omega_1+\omega_2, \quad \text{(four-fold degenerate)}. \end{aligned} \tag{38}$$

Again, the duality transformation is given by eq. (35), where now eq. (38) for the eigenvalues must be used; the duality transformation also commutes with the interchange of $\omega_1$ and $\omega_2$. The model factorizes for $\omega_1=\omega_2^2$ and for $\omega_2=\omega_1^2$ into two 3-state Potts models with equal coupling constants, so that there are two special points $P_1(3)$ and $P_2(3)$ given by

$$\omega_1 = (1+\sqrt{3})^{-1}, \ \omega_2=\omega_1^2, \quad \text{and} \quad \omega_2 = (1+\sqrt{3})^{-1}, \ \omega_1=\omega_2^2. \tag{39}$$

These are shown in Fig. 1(d) to lie on the self-dual line (for $D$), which is given by $\lambda_o=3$ or

$$\omega_1+\omega_2 = \tfrac{1}{2}, \tag{40}$$

on which also the nine-state Potts model self-dual point $P(9)$ is located. It is tempting to conclude, that, if the Potts models have phase transitions, then the whole self-dual line is a line of phase transitions. This is indeed found for the spin model by means of an approximate renormalization group calculation in Chapter 11.

$\underset{\sim}{S}(M_1)\otimes\underset{\sim}{S}(M_2)$ models. These are models with an imprimitive, self-dual, permissible group having three graphs with z-values $M_1-1$, $M_2-1$ and $(M_1-1)(M_2-1)$ in their MI's. The eigenvalues of $\underline{\underline{\Omega}}$ are easily found as

$$\lambda_o = 1+(M_1-1)\omega_1+(M_2-1)\omega_2+(M_1-1)(M_2-1)\omega_3, \quad \text{(nondegenerate)}, \tag{41}$$

$$\lambda_1 = 1-\omega_1+(M_2-1)(\omega_2-\omega_3), \quad \text{( } (M_1-1)\text{-fold degenerate)},$$

$$\lambda_2 = 1-\omega_2+(M_1-1)(\omega_1-\omega_3), \quad \text{( } (M_2-1)\text{-fold degenerate)}, \tag{42}$$

$$\lambda_3 = 1-\omega_1-\omega_2+\omega_3, \quad \text{( } (M_1-1)(M_2-1)\text{-fold degenerate)}.$$

Since the phase diagrams have rather strongly different features according as to whether $M_1$ equals $M_2$ or not and also, in the first case, depend on whether $M_1=M_2=2$ or not, three different cases are considered.

(i) $\mathcal{S}(2)\otimes\mathcal{S}(2)$, the Ashkin-Teller model. The MI of this group consists of three isomorphic $z=1$ graphs, see Fig. 1.2.1(a); therefore, all permutations of $\omega_1$, $\omega_2$ and $\omega_3$ leave the partition function invariant. These do, however, not all commute with the duality transformation given for the present case by

$$\tilde{\omega}_1 = \frac{1-\omega_1+\omega_2-\omega_3}{1+\omega_1+\omega_2+\omega_3}; \quad \tilde{\omega}_2 = \frac{1+\omega_1-\omega_2-\omega_3}{1+\omega_1+\omega_2+\omega_3}; \quad \tilde{\omega}_3 = \frac{1-\omega_1-\omega_2+\omega_3}{1+\omega_1+\omega_2+\omega_3}. \tag{43}$$

Denoting this transformation by $D$, the cyclic permutation of the $\omega$'s by $(123)=g$ and the permutation of $\omega_1$ and $\omega_2$ by $(12)=\sigma$, one has

$$D\sigma=\sigma D, \quad (Dg)(g^2\sigma)=(g^2\sigma)(Dg), \quad (Dg^2)(g\sigma)=(g\sigma)(Dg^2), \quad Dg=g^2D, \quad Dg^2=gD. \tag{44}$$

There is, therefore, a group $\mathcal{D}_s$ of order $12$, abstractly isomorphic to $\mathcal{D}(6)$ or $\mathcal{F}(12)$, the elements of which all map the ferromagnetic region onto itself and which all keep the hyperplane $H$, given by

$$\omega_1+\omega_2+\omega_3 = 1, \tag{45}$$

invariant. The element $D\sigma$ has square $e$ and is in the center of $\mathcal{D}_s$; it keeps the whole of $H$ pointwise invariant. The transformations $D$, $Dg$ and $Dg^2$ are conjugates (with square $e$), which each keep a line in $H$ invariant pointwise:

$$D: \ \omega_1=\omega_2=\tfrac{1}{2}(1-\omega_3); \quad Dg: \ \omega_1=\omega_3=\tfrac{1}{2}(1-\omega_2); \quad Dg^2: \ \omega_2=\omega_3=\tfrac{1}{2}(1-\omega_1). \tag{46}$$

The other two transformations of $\mathcal{D}_s$, which map the vicinity of $(0,0,0)$ onto the vicinity of $(1,1,1)$, are $Dg\sigma$ and $Dg^2\sigma$; these have squares

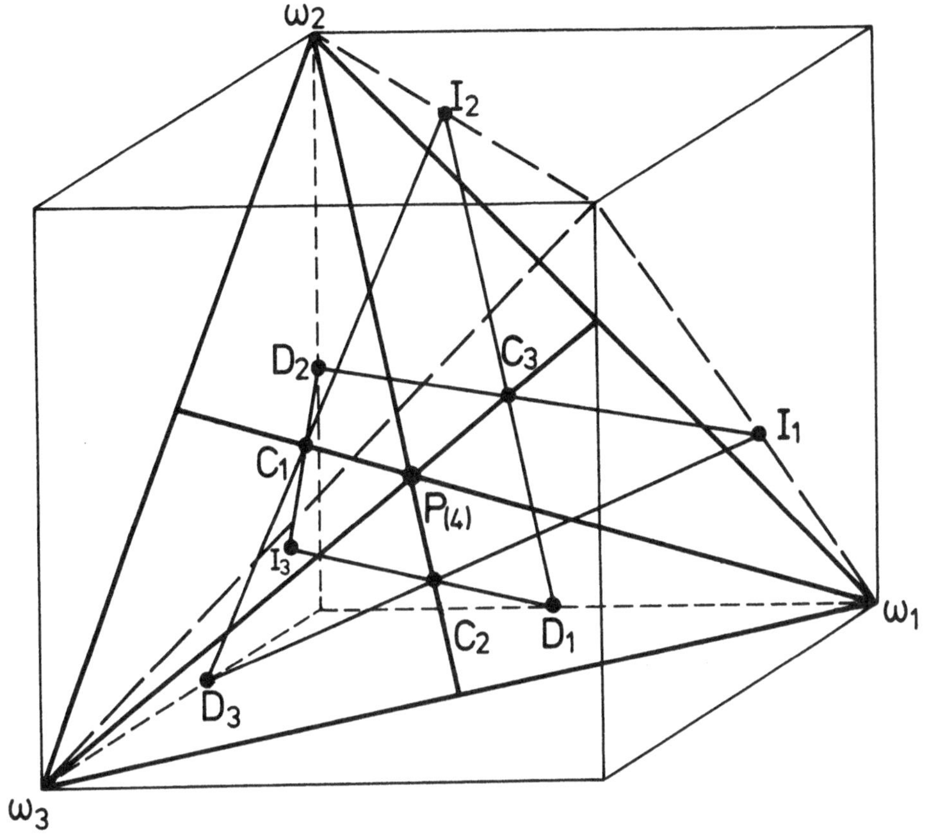


Fig. 2. The phase diagram of the $S(2)\otimes S(2)$ model from duality. The broken lines are the boundaries of the ferromagnetic region, the solid ones are boundaries of the self-dual plane or self-dual lines. The special (thin) lines and points are explained in the text.

$g$ and $g^2$, respectively, and only keep the point $\omega_1=\omega_2=\omega_3=\frac{1}{3}$ invariant, which is the self-dual point for the 4-state Potts model.

The model factorizes into two Ising models as soon as one of the $\omega$'s is the product of the other two; there are, therefore, six self-dual Ising lines in the phase diagram:

$$\begin{aligned}
&\omega_1=\sqrt{2}-1,\ \omega_3=(\sqrt{2}-1)\,\omega_2; \quad \omega_1=\sqrt{2}-1,\ \omega_2=(\sqrt{2}-1)\,\omega_3;\\
&\omega_2=\sqrt{2}-1,\ \omega_3=(\sqrt{2}-1)\,\omega_1; \quad \omega_2=\sqrt{2}-1,\ \omega_1=(\sqrt{2}-1)\,\omega_3; \qquad (47)\\
&\omega_3=\sqrt{2}-1,\ \omega_2=(\sqrt{2}-1)\,\omega_1; \quad \omega_3=\sqrt{2}-1,\ \omega_1=(\sqrt{2}-1)\,\omega_2;
\end{aligned}$$

These are the lines $D_1I_2$, $D_1I_3$, $D_2I_1$, $D_2I_3$, $D_3I_1$ and $D_3I_2$ shown in the phase diagram of Fig. 2, which also contains the ferromagnetic region, the self-dual plane of eq. (45) and the self-dual lines of eqs. (46). Pairs of these Ising lines cut each other in the points $C_1$, $C_2$ and $C_3$, which all lie on the self-dual plane in such a way that each of these lies on exactly one of the self-dual lines.

Assuming all special lines to be phase transitions, e.g., for a spin model in two dimensions, Fig. 2 suggests the existence of two (curved) phase transition planes, which touch each other and the self-dual plane along the line segments $C_iP(4)$ for $i=1,2$ and 3. In Chapters 9 and 11, more evidence for such a picture is found.

(ii) $\mathcal{S}(3)\otimes\mathcal{S}(3)$ is treated here as an example of the groups of the type $\mathcal{S}(M)\otimes\mathcal{S}(M)$ with $M\neq 2$. For these models, the partition function only has the $\omega_1\leftrightarrow\omega_2$ interchange $\sigma$ as extra symmetry. This is still enough to make the self-dual plane for $D\sigma$ the same as the hyperplane $H$:

$$\omega_1+\omega_2+2\omega_3 = 1, \tag{48}$$

if $D$ again denotes the duality transformation derived from eqs. (42) for this case. $D$ itself only leaves the line

$$\omega_1=\omega_2=\frac{1}{2}\omega_3 \tag{48'}$$

invariant pointwise.

The model factorizes into two 3-state Potts models for $\omega_3=\omega_1\omega_2$; the resulting two self-dual Potts lines are shown as $D_1D_2$ and $D_3D_4$ in Fig. 3, together with the ferromagnetic region, the self-dual plane $H$ of eq. (48) and the self-dual line $AB$ of eq. (48'). The two Potts lines still cross the self-dual plane at the same point $C$, which is on the self-dual line. If these lines are again interpreted as phase transition lines, Fig. 3 again suggests the presence of two curved phase transition planes, which touch along the self-dual line.

(iii) $\mathcal{S}(2)\otimes\mathcal{S}(3)=\mathcal{D}(6)$ is chosen as an example with $M_1\neq M_2$, so that the partition function has now no symmetries left. The hyperplane $H$ is given by

$$\omega_1+2\omega_2+2\omega_3 = \sqrt{6}-1, \tag{49}$$

but an extra condition is imposed by the (unique) self-duality now. The self-dual line is given by the relations (obtained from eqs. (42) for the present case):

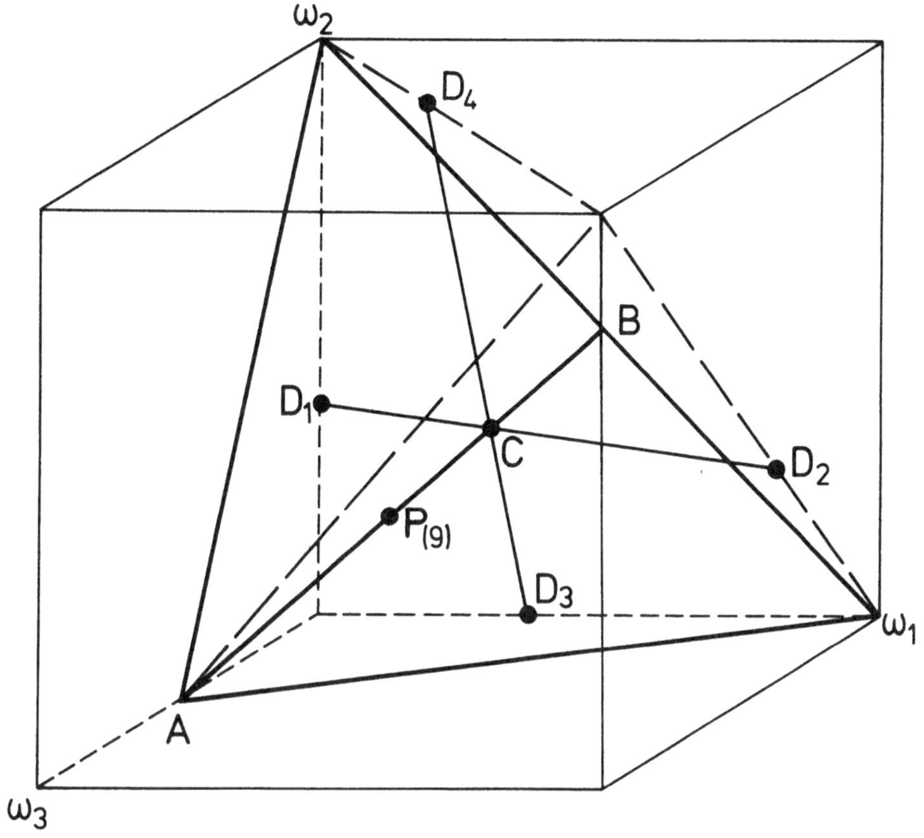


Fig. 3. The phase diagram of the model with $S(3)\otimes S(3)$ symmetry from duality. The boundaries of the ferromagnetic region are the broken lines, whereas the solid lines are the intersections of the self-dual plane with this region, except for the solid line AP(9)CB, which is the self-dual line. The thin lines indicate, where the model factorizes into self-dual Potts models.

$$\omega_2 = \tfrac{1}{4}\sqrt{6}\ \omega_1 + \tfrac{1}{4}(\sqrt{6}-2); \quad \omega_3 = \tfrac{1}{4}\sqrt{6} - \tfrac{1}{4}(\sqrt{6}+2)\,\omega_1 . \tag{50}$$

This is the line AB in Fig. 4. The present model can again factorize: for $\omega_3=\omega_1\omega_2$, it reduces to the product of an Ising model and of a 3-state Potts model. The corresponding self-dual Ising and Potts lines are $I_1I_2$ and $D_1D_2$ in Fig. 4, respectively. These lines cross the self-dual line in the common point C. The 6-state Potts model self-dual point P(6), given by $\omega_1=\omega_2=\omega_3=(\sqrt{6}-1)/5$, also must lie on the self-dual

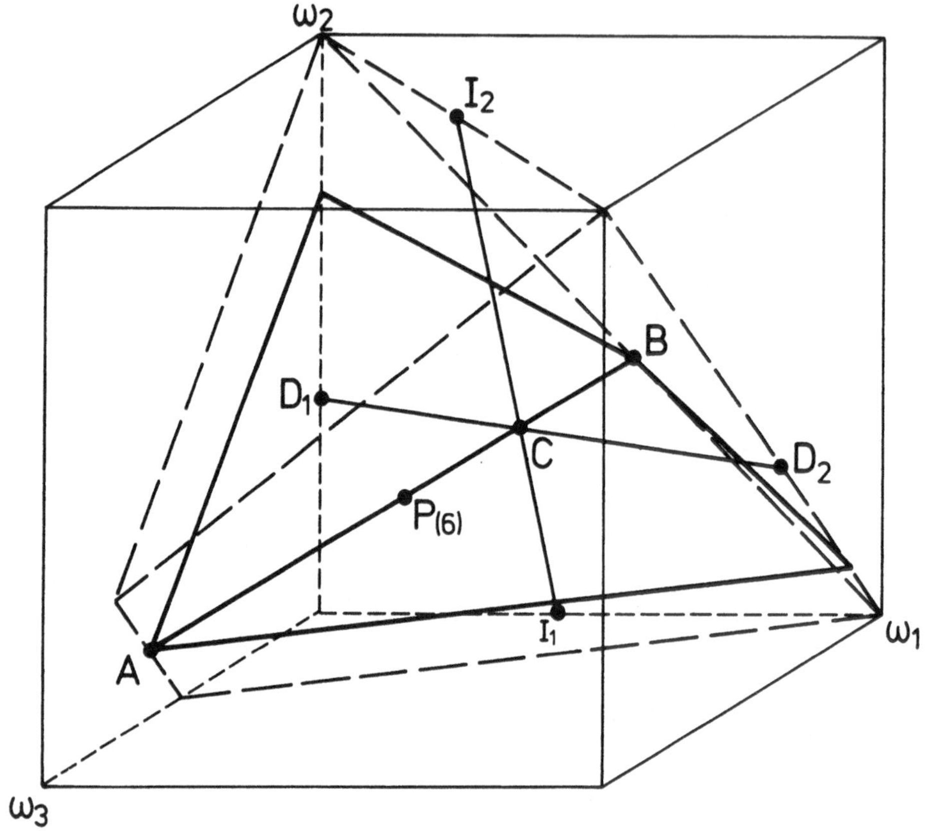


Fig. 4. The phase diagram of the $\mathcal{S}(2)\otimes\mathcal{S}(3)=\mathcal{D}(6)$ model from duality. The ferromagnetic region, the hyperplane H, and the self-dual line are designated in the same way as in the previous two figures. The special thin lines and points are explained in the text.

line, of course. Conclusions as to phase transitions are similar to the previous case, although the evidence presented by Fig. 4 is rather more scanty.

$\underline{\mathcal{S}(2)\sim\mathcal{S}(2)\sim\mathcal{S}(2)}$. This model is studied here as an example of an imprimitive, permissible group with a 3-graph MI, which is not of the direct product type. From Section 6.1, the z-values are 1, 2 and 4. The eigenvalues of the matrix $\underline{\underline{\Omega}}$ for this model are easily found to be:

$$\lambda_o = 1+\omega_1+2\omega_2+4\omega_3, \qquad \text{(nondegenerate)}, \tag{51}$$

$$\lambda_1 = 1+\omega_1+2\omega_2-4\omega_3, \qquad \text{(nondegenerate)}, \tag{52}$$

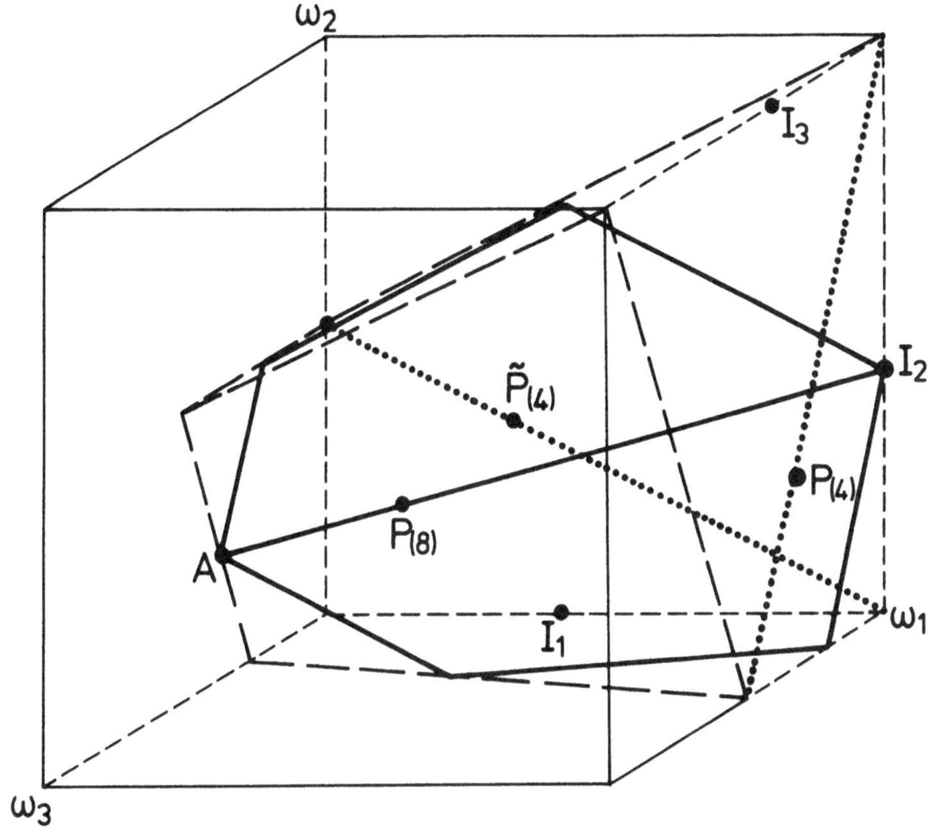


Fig. 5. The phase diagram of the $S(2)\sim S(2)\sim S(2)$ model from duality. The ferromagnetic region and the self-dual line are designated as in the previous figures. Special (dotted) lines and points are explained in the text.

$$\lambda_2 = 1+\omega_1-2\omega_2, \qquad \text{(2-fold degenerate)}, \tag{53}$$

$$\lambda_3 = 1-\omega_1, \qquad \text{(4-fold degenerate)}. \tag{54}$$

The hyperplane $H$ is, therefore, given by

$$\omega_1+2\omega_2+4\omega_3 = 2\sqrt{2}-1, \tag{55}$$

whereas the absence of extra symmetries in the partition function again yields a self-dual line only:

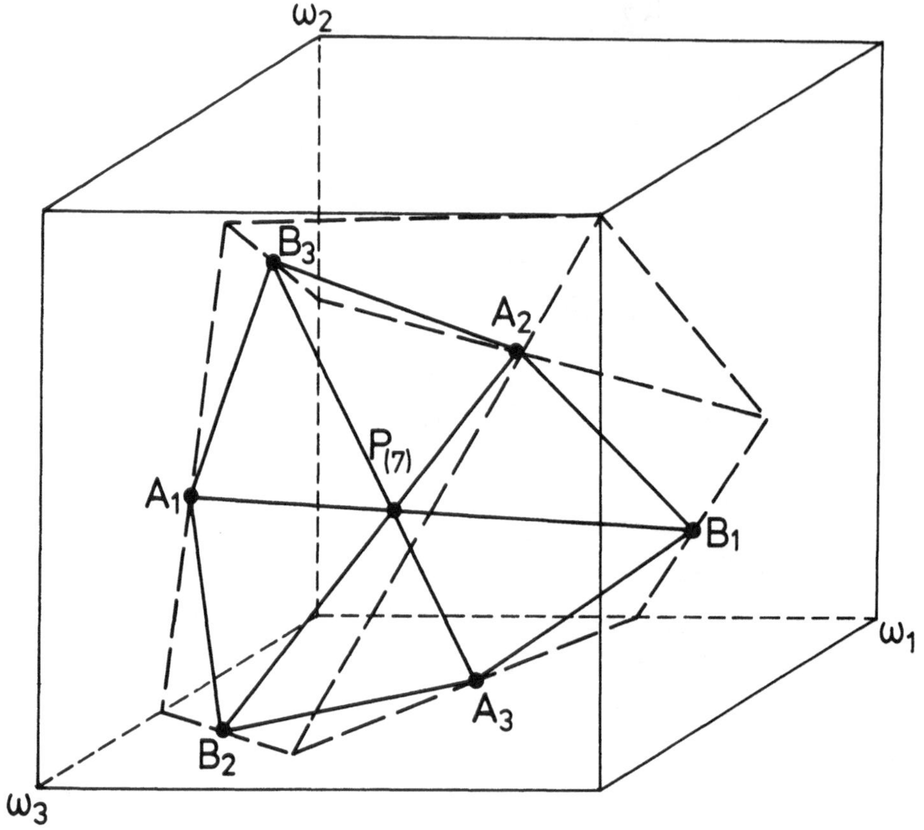


Fig. 6. The phase diagram of the model with $\mathbb{D}(7)$ symmetry from duality. There are three self-dual lines, as explained in the text.

$$\omega_2 = \frac{1}{2}(\sqrt{2}-1)(1+\omega_1); \quad \omega_3 = \frac{1}{4}\sqrt{2}(1-\omega_1). \tag{56}$$

The phase diagram is shown in Fig. 5; use has been made of the fact, that the model reduces to $\mathbb{S}(2) \sim \mathbb{S}(2)$ on the planes $\omega_2=0$ and $\omega_1=1$. Therefore, Fig. 1(a) is recovered here, yielding the dual pair of 4-state Potts points $P(4)$ and $\tilde{P}(4)$ as well as the dual pair of Ising points $I_1$ and $I_3$ and the self-dual Ising point $I_2$, which is actually the end point of the self-dual line $AI_2$. If all these special points are phase transitions, the true phase diagram must be qualitatively different from the ones for the direct product groups; in Chapter 9, this is confirmed for a special (pseudo) lattice.

The $\mathcal{D}(7)$ model. This is an example of a permissible, primitive group with a 3-graph MI. The eigenvalues are:

$$\lambda_o = 1+2\omega_1+2\omega_2+2\omega_3\ , \qquad \text{(nondegenerate)}, \tag{57}$$

$$\begin{pmatrix}\lambda_1\\ \lambda_2\\ \lambda_3\end{pmatrix} = \begin{pmatrix}1\\ 1\\ 1\end{pmatrix} + \begin{pmatrix}a & b & c\\ b & c & a\\ c & a & b\end{pmatrix}\begin{pmatrix}\omega_1\\ \omega_2\\ \omega_3\end{pmatrix}, \qquad \text{(all twofold degenerate)}, \tag{58}$$

with

$$a = \cos(2\pi/7),\quad b = \cos(4\pi/7),\quad c = \cos(6\pi/7). \tag{59}$$

The hyperplane $H$ is simply given by

$$\omega_1+\omega_2+\omega_3 = \sqrt{7}-1. \tag{60}$$

The partition function is invariant with respect to cyclic permutations of the $\omega$'s only; it is easy to see from eq. (58), that such cyclic permutations do not commute with the duality transformation $D$, so that the symmetry group $\mathcal{D}_s$ of the partition function is isomorphic to $\mathcal{S}(3)$. This group contains three conjugate elements of square $e$, which yield the three self-dual lines $A_iB_i$ (i=1,2,3) shown in Fig. 6. Any of the three duality transformations keeps one of these lines pointwise invariant and exchanges the other two; the third-order element of $\mathcal{S}(3)$ permutes these three lines cyclically. All three self-dual lines pass through the 7-state Potts model self-dual point $P(7)$.

## 7.7. General gauge models and Higgs fields.

The 1-gauge models arrived at in Section 4 may be generalized to arbitrary graphs $G$ and regular groups $\mathcal{R}$; these will be called gauge models from now on. Consider a directed graph, on every edge of which is a "spin" variable taking values in the group $\mathcal{R}$ with $M$ elements. To every closed cycle $C$ of edges of the graph, one can associate an energy $E_C$, which is a function of the product of the group elements on its edges, denoted by $r$:

$$E_C = E_C\Big(\prod_{\substack{\text{edges } i \text{ of}\\ \text{the closed}\\ \text{cycle } C}} r_i^{\alpha_i}\Big), \tag{1}$$

where $\alpha_i$ is +1 if the edge points in the direction in which the cycle is traversed and -1 otherwise. To have this independent of the direction in which the cycle is oriented and independent of the starting point, $E_C(r)$ must satisfy

$$E_C(r) = E_C(r^{-1}) \quad \text{for all} \quad r \in \mathcal{R}, \tag{2}$$

and

$$E_C(r^{-1} r_1 r) = E_C(r_1) \quad \text{for all} \quad r,\ r_1 \in \mathcal{R}. \tag{3}$$

$E_C(r)$ is, therefore, a symmetric class function on $\mathcal{R}$, so that the corresponding spin model,

$$E(i,j) = E_C(r_i^{-1} r_j), \tag{4}$$

contains $\mathcal{H}(\mathcal{R})$ as a subgroup of its permissible symmetry group $\mathcal{G}$, see Section 4.3. Defining

$$\Omega_C(r) = \exp - E_C(r), \tag{5}$$

the partition function of a genaral gauge model is

$$Z(\mathcal{R},G) = \sum_{r_e \in \mathcal{R}} \prod_{\substack{\text{cycles} \\ C \text{ of } G}} \Omega_C\Big( \prod_{\substack{\text{edges} \\ e \text{ of } C}} r_e^{\alpha_e} \Big). \tag{6}$$

A general gauge model is again invariant with respect to a local gauge transformation defined at every vertex $v$ of $G$ by

$$\begin{aligned} r_e &\to r_e a \quad \text{if } e \text{ is directed towards } v, \\ r_e &\to a^{-1} r_e \quad \text{if } e \text{ is pointing away from } v, \\ r_e &\to r_e \quad \text{if the edge } e \text{ does not have } v \text{ as end point.} \end{aligned} \tag{7}$$

The total symmetry group is a product of two subgroups, the first one being generated by all local gauge transformations of eq. (7) [ this group is easily seen to contain $M^{|V|}/|Z(\mathcal{R})|$ elements ], and the second one consisting of global symmetries induced by those (outer) automorphisms of $\mathcal{R}$, which leave $E_C(r)$ invariant. It is to be noted, that there are also global symmetries in the gauge group, for instance the transformation

$$r_e \to a^{-1} r_e a, \quad \text{for all } r_e, \tag{8}$$

obtained by performing the same local gauge transformation (7) at all vertices. Also, if the graph is <u>bipartite</u>, i.e., if its vertex set $V$ can be split into two disjoint sets $V_1$ and $V_2$, such that each edge extends from a vertex of $V_1$ towards a vertex of $V_2$, then all edges may be directed from $V_1$ towards $V_2$. (A simple example of a bipartite graph is the Kuratowski graph of Fig. 4.1.) Now performing the same local gauge transformation at all vertices of $V_1$ yields the global symmetry

$$r_e \to a^{-1} r_e. \tag{9}$$

Similarly, choosing $V_2$ to perform the local gauge transformation at, gives a global symmetry

$$r_e \to r_e a. \tag{10}$$

A general gauge model may be coupled in a gauge-invariant way to a second set of "spins" $s \varepsilon \mathcal{R}$ located at the vertices of $G$ by defining a so-called <u>Higgs</u> <u>interaction</u>

$$E_H = \sum_{\substack{\text{directed}\\ \text{edges } e}} E_H(s_{v_1}^{-1} r_e s_{v_2}), \tag{11}$$

where $e$ is directed from its end point $v_1$ towards $v_2$. The model is still gauge invariant, if one agrees to supplement eqs. (7) by

$$s_v \to a^{-1} s_v. \tag{12}$$

In order to conserve the gauge-invariance of the partition function

$$Z_H = \sum_{r_e \varepsilon \mathcal{R}} \sum_{s_v \varepsilon \mathcal{R}} \prod_{\substack{\text{cycles}\\ C \text{ of } G}} \Omega_C\Big(\prod_{\substack{\text{edges}\\ \text{of } C}} r_e^{\alpha_e}\Big) \prod_{\substack{\text{directed}\\ \text{edges } e}} \Omega_H(s_{v_1(e)}^{-1} r_e s_{v_2(e)}), \tag{13}$$

the symmetry $\Omega_H(r)=\Omega_H(r^{-1})$ must be valid for the Higgs interaction; it does, however, <u>not</u> have to be given by a class function. The Higgs model of eq. (13) is equivalent to a gauge model in a (gauge-invariance-breaking) external field, called a <u>Higgs</u> <u>field</u>. Indeed, taking new variables $r_e' = s_{v_1(e)}^{-1} r_e s_{v_2(e)}$, eq. (13) becomes

$$Z = M^{|V|} \sum_{r'_e \in R} \prod_{\substack{\text{cycles} \\ C \text{ of } G}} \Omega_C\Big( \prod_{\substack{\text{edges} \\ e \text{ of } C}} r_e'^{\alpha_e} \Big) \prod_{\substack{\text{edges} \\ e \text{ of } G}} \Omega_H(r'_e) \; . \tag{14}$$

In the same way, it can be shown, that all averages depending only on the s-variables always have their trivial, i.e., noninteracting, values. On the other hand, eq. (13) reduces to a spin model in the limit, where $\Omega_C(r)$ is given by eq. (3.3): the cycle restrictions are then satisfied for

$$r_e = t_{v_1(e)}^{-1} \, t_{v_2(e)}, \quad t_v \text{ arbitrary,} \tag{15}$$

so that, with $s'_v = t_v s_v$, eq. (13) reduces to

$$Z_{spin} = M^{|V|-1} \sum_{s'_v \in R} \prod_{e \in E} \Omega_H(s'^{-1}_{v_1(e)} s'_{v_2(e)}) . \tag{16}$$

The phase transitions of gauge models with Higgs fields are of a different nature than those for spin models in magnetic fields, since the gauge invariance cannot be broken spontaneously; this is Elitzur's theorem [10], see also Section 8.3.3 and Chapter 12.

REFERENCES.


(1). G.A. Dirac, J. London Math. Soc. 31 (1956) 460.
(2). B. Huppert, Endliche Gruppen I (Springer-Verlag, Berlin, Heidelberg, New York, 1963) p. 302.
(3). Ref. (2), p. 33 ff.
(4). Ref. (2), p. 420.
(5). J.S. Lomont, Applications of Finite Groups (Academic Press, New York, London, 1959) p. 34, 35.
(6). C. Kuratowski, Fund. Math. 15 (1930) 271.
(7). S. MacLane, Fund. Math. 28 (1937) 22.
(8). K. Wagner, Graphentheorie (Bibliographisches Institut, Mannheim, 1970) p. 15.
(9). F.Y. Wu, Rev. Mod. Phys. 54 (1982) 235.
(10). S. Elitzur, Phys. Rev. D 12 (1975) 3978.


GENERAL REFERENCES.

The derivation of the two-dimensional duality relations in Section 7.1 is patterned after
F.Y. Wu and Y.K. Wang, J. Math. Phys. 17 (1976) 439,
F.J. Wegner, Physica 68 (1973) 570.
The discussion in Section 7.2 has been inspired by some results from
M. Marcu and V. Rittenberg, J. Math. Phys. 22 (1981) 2753,
M. Marcu, Ph. D. dissertation, Bonn University (1981, unpublished).
The question of duality for nonabelian groups, Section 7.3, has been considered by a number of authors, e.g.,

J.-M. Drouffe, C. Itzykson and J.-B. Zuber, Nucl. Phys. B 147 (1979) 132.
R. Casalbuoni, V. Rittenberg and S. Yankielowicz, Nucl. Phys. B 170 (1980) 139.
M. Marcu, A. Regev and V. Rittenberg, J. Math. Phys. 22 (1981) 2740.
These are all partial results for the existence of a regular, Abelian subgroup of the PCF group.
The duality transformations in higher dimensions studied in Sections 7.4 and 7.5 have been patterned after the formalism described by
J.-M. Drouffe and J.-B. Zuber, Physics Reports 102 (1983) 1.
Gauge-invariant models were first considered by
F. Wegner, J. Math. Phys. 12 (1971) 2259,
and proposed as models for the nonabelian gauge theories of elementary particle physics by
K. Wilson, Phys. Rev. D 10 (1974) 2445.
Duality transformations have also recently been reviewed by
R. Savit, Rev. Mod. Phys. 52 (1980) 453;
see also
J.B. Kogut, Rev. Mod. Phys. 51 (1979) 1659,
J.-M. Drouffe and C. Itzykson, Physics Reports 38 (1978) 133,
as well as the review of the Potts model, Ref. ([9]) above.
The discussion of Higgs fields in Section 7.7 is again based on the above-mentioned review of Drouffe and Zuber.
Some of the models of Section 7.6 have also been studied as to their duality properties on a square lattice by a number of authors. These works cannot all be listed here; instead, the reader is referred to the review articles listed above. The $\mathbb{D}(M)$ models for $M \leq 10$ have been investigated in some detail by Zittartz and coworkers; this work remains largely unpublished, although some results can be found in
W. Wolff, Diplomarbeit, University of Cologne (1979, unpublished).
Some of the group-theoretical aspects of this chapter are detailed in
J. Klein, Diplomarbeit, University of Cologne (1984, unpublished).
In Section 7.7, the gauge symmetry is broken by an external field; the same effect can, of course, be produced by the introduction of extra pair interactions. The resulting model has been studied by
A. Weinkauf and J. Zittartz, Z. Phys. B 45 (1982) 223,
S. Wansleben and A. Weinkauf, Z. Phys. B 50 (1983) 255.

PART C : PSEUDO-LATTICES AND RENORMALIZATION.

## 8. PSEUDO-LATTICES.

### 8.1. Recursive site graph sequences.

Let $G$ be a finite, connected graph with vertex set $V$ and edge set $E$. The vertex set is considered to consist of three mutually disjoint sets:

(i) one special vertex is singled out to play the role of a "top vertex" $v_1$, sometimes called the blue vertex of $G$.

(ii) a nonempty set $V_y \subseteq V - \{v_1\}$ of yellow vertices.

(iii) the other vertices, i.e., those from $V - V_y - \{v_1\}$ are called the inner or green vertices. This set may be empty.

Some examples are shown in Fig. 1. The set of all such graphs is denoted

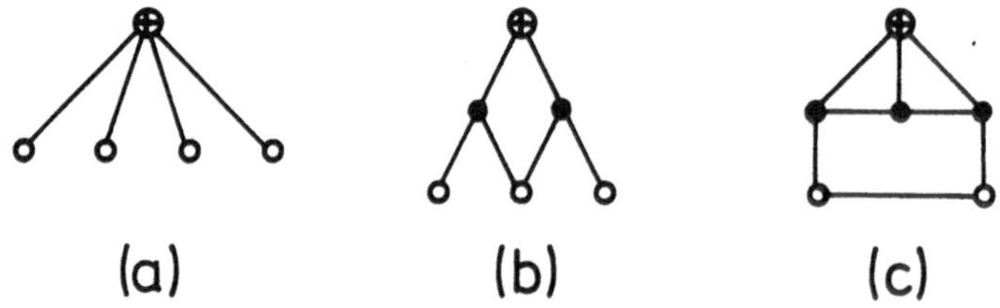

Fig. 1. Examples of graphs $G \varepsilon \Sigma$: the blue vertex is denoted by ⊕, the yellow vertices by O, the green ones by ●.

by $\Sigma$. This set is endowed with a semigroup structure by the following definition:

If $A, B \varepsilon \Sigma$, then the product $A(.)B$ is obtained by merging the yellow vertices of $A$ with the $|V_y(A)|$ blue vertices of as many copies of $B$; in this process, the merged spins become inner vertices, i.e., blue + yellow = green.

Fig. 2 shows some examples of this procedure.

The product (.) is associative:

$$A(.)[B(.)C] = [A(.)B](.)C, \tag{1}$$

but, in general, not commutative, see Fig. 2(b). $\Sigma$ is, therefore, a semigroup without a unit element. This latter fact is clear from the following formulae for the numbers of vertices and edges of $C = A(.)B$:

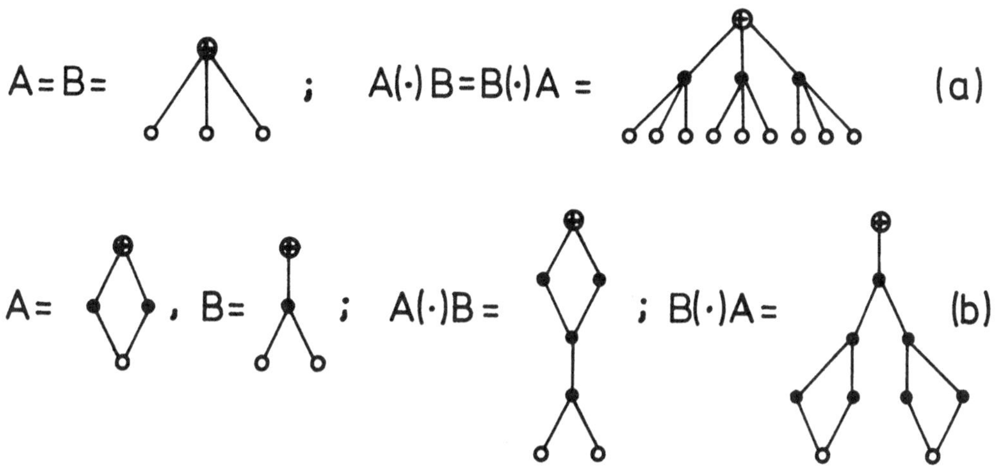


Fig. 2. Some examples of the product (.) in $\Sigma$.

$$V(C) = V(A) + \{V(B)-1\}V_y(A), \tag{2}$$

$$V_y(C) = V_y(A)V_y(B), \tag{3}$$

$$E(C) = E(A) + V_y(A)E(B). \tag{4}$$

It is expedient, to extend $\Sigma$ to a semigroup $\Sigma_o$ with a unit by adjoining the graph v consisting of a single vertex and defining

$$v(.)v = v,$$

$$v(.)A = A \quad \text{for all} \quad A\varepsilon\Sigma, \tag{5}$$

$$A(.)v = A \quad \text{for all} \quad A\varepsilon\Sigma.$$

Eqs. (2-4) automatically hold with $V(v)=V_y(v)=1$, $E(v)=0$, which are the logically expected values. The difference with the general definition is, that the merging of v with any vertex does not change the colour of this vertex: v is colourless, which is what one expects from a unit, or neutral, element.

A recursive site graph sequence $\{S_n\}$ is defined by a store of graphs from $\Sigma$, $\{G_n|n=1,2,..\}$, and by a recursive prescription:

$$S_n = G_n(.)S_{n-1}, \quad S_o=v. \tag{6}$$

$S_n$ is called the n-th generation graph of the recursive site graph sequence. The numbers of vertices and edges of $S_n$ follow immediately from eqs. (2-4) as

$$V_y(S_n) = \prod_{k=1}^{n} V_y(G_n), \tag{7}$$

$$V(S_n)/V_y(S_n) = 1+\sum_{k=1}^{n} [\{V(G_k)-V_y(G_k)\}/V_y(S_k)], \tag{8}$$

$$E(S_n)/V_y(S_n) = \sum_{k=1}^{n} E(G_k)/V_y(S_k) \ . \tag{9}$$

A recursive site graph sequence is called <u>self-similar</u> if $G_n=G$ for all $n$. Setting

$$V_y(G) = m, \quad V(G)-V_y(G) = \Delta m, \quad E(G) = f, \tag{10}$$

eqs. (7-9) reduce, for this case, to

$$V_y(S_n) = m^n; \quad V(S_n) = m^n+(\Delta m)\{(m^n-1)/(m-1)\}; \quad E(S_n) = f(m^n-1)/(m-1); \tag{11}$$

These equations are for the case $m>1$; for $m=1$, one obtains:

$$V_y(S_n) = 1; \quad V(S_n) = 1+n\Delta m; \quad E(S_n) = fn. \tag{12}$$

In this latter case, $S_n$ is just a chain of graphs $G$; for such sequences, the limits as $n\to\infty$ of eqs. (8) and (9) do not exist. In general, a recursive site graph sequence will be called essentially one-dimensional if the limits of eqs. (8) and (9) for $n\to\infty$ do not exist. If these limits <u>do</u> exist, $\{S_n\}$ is nontrivial, and the limits are $\delta^{-1}$ and $\varepsilon^{-1}$:

$$\delta= \lim_{n\to\infty} V_y(S_n)/V(S_n) = [1+\sum_{k=1}^{\infty} \{V(G_k)-V_y(G_k)\}/V_y(S_k)]^{-1}, \tag{13}$$

$$\varepsilon= \lim_{n\to\infty} V_y(S_n)/E(S_n) = [\sum_{k=1}^{\infty} E(G_k)/V_y(S_k)]^{-1}. \tag{14}$$

For a nontrivial, self-similar recursive site graph sequence, this gives

$$\delta=(m-1)/(m+\Delta m-1), \quad \varepsilon=(m-1)/f. \tag{15}$$

A recursive site graph sequence is called a Cayley-like branch if the $G_k$ of its store do not contain any green or inner vertices and if $E(G_k)=V_y(G_k)$ holds, i.e., $G_k$ looks like Fig. 1(a) or like the graphs A and B of Fig. 2(a). Denoting this kind of store-graph by $L_{m_k}$, if the common number of edges and yellow vertices is $m_k$, one has

$$V_y(S_n) = \prod_{k=1}^{n} m_k, \tag{16}$$

so that $\delta$ and $\varepsilon$ are given by

$$\delta^{-1}=\varepsilon^{-1}= 1+ \sum_{k=1}^{\infty} \left( \prod_{t=1}^{k} m_t^{-1} \right). \tag{17}$$

A Cayley-like branch is, graph-theoretically, a tree, i.e., a graph without cycles.

A self-similar Cayley-like branch is simply called a Cayley branch $\{C_n(m)\}$ with branching ratio $m$; in this case, one has

$$\delta=\varepsilon=(m-1)/m. \tag{18}$$

From the sequence of Cayley branches $\{C_n(m)\}$, the sequence $\{T_n(m+1)\}$ of Cayley trees is obtained by

$$T_n(m+1) = L_{m+1}(.)C_n(m). \tag{19}$$

Every vertex of a Cayley tree, except for those on the boundary or surface (i.e., the yellow vertices), have coordination number $m+1$. Note that $\{T_n(m)\}$ is, strictly speaking, not a recursive site graph sequence, since it is obtained from such a sequence by an extra step and cannot itself be generated by a store of graphs.

A number of Cayley-like branches, including a Cayley branch and a Cayley tree, are shown in Fig. 3. The Cayley-like branch of Fig. 3(d) is a member of the exponential branch $\{X_n\}$ defined by

$$X_n = L_n(.)X_{n-1}. \tag{20}$$

For this one has

$$V_y(X_n) = n!, \quad \delta=\varepsilon= 1/e, \tag{21}$$

where $e$ is the basis of the natural logarithm.

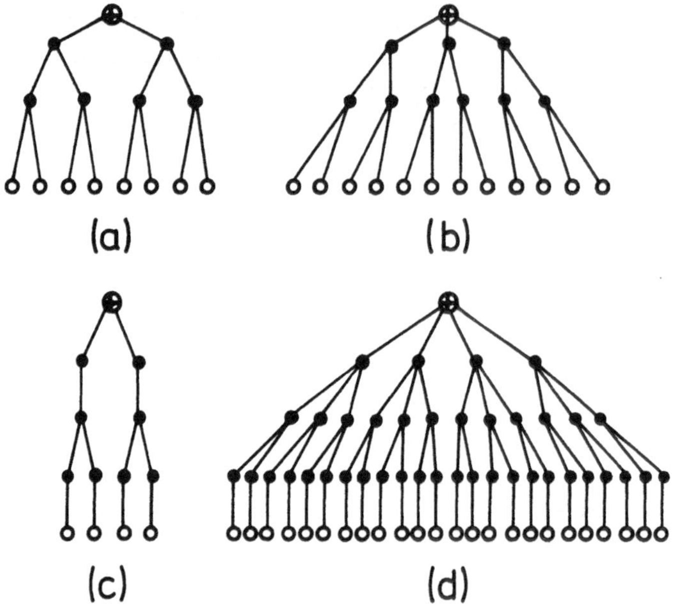


Fig. 3. (a) The Cayley branch $B_3(2)$; (b) The Cayley tree $T_2(3)$; (c) The Cayley-like branch $L_2(.)L_1(.)L_2(.)L_1$; (d) The exponential branch $X_4$.

Another class of recursive site graphs are the <u>cactus</u> <u>branches</u>, which have regular polygons as building blocks, see Fig. 4. A self-similar cactus branch with, e.g., a triangle as basic building block, is call- a triangular cactus branch, etc.

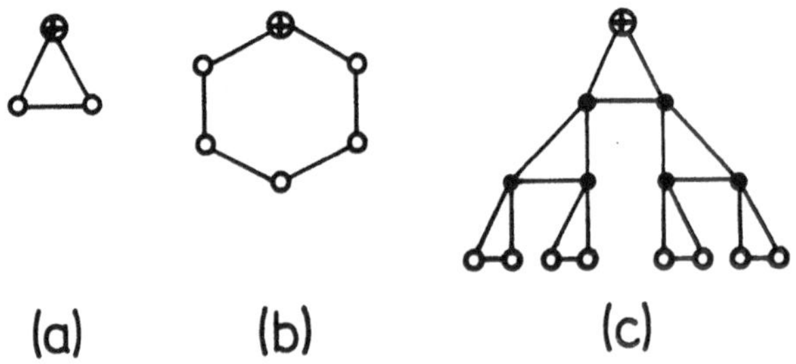


Fig. 4. Triangular (a) and hexagonal (b) building blocks for cactus branches; (c) A triangular cactus branch with three generations.

Cayley-like branches (including Cayley branches and trees) as well as cactus branches are planar graphs, so that planar dual graphs can be constructed as in Section 7.1. For the Cayley-like branches, these consist of a single vertex, which is connected to itself by as many edges as there are edges in the original graph. It is interesting to note, that all Cayley-like and cactus branches remain planar upon addition of a ghost vertex, since these graphs do not contain vertices completely surrounded by edges; therefore, the duals of these graphs with ghost vertex can still be constructed in two dimensions, in contrast to the general situation described in Section 7.5.

## 8.2. Recursive bond graph sequences.

Instead of using yellow vertices, as in the previous section, yellow edges or bonds may also be used to define a sequence of graphs, which form a pseudo-lattice. To this end, a class $\Gamma_o$ of finite, connected graphs H with the following properties is defined:
(i) The vertex set V of H contains two blue vertices $v_1$ and $v_2$; the remaining vertices are colourless.
(ii) The edge set E of H contains a nonempty subset $E_y$ of yellow edges; the edges of $E-E_y$ (which may be empty) are green.
(iii) There is an automorphism of H, which exchanges the two blue vertices $v_1$ and $v_2$ and maps $E_y$ onto itself.
Some examples of graphs $H\varepsilon\Gamma_o$ are shown in Fig. 1.

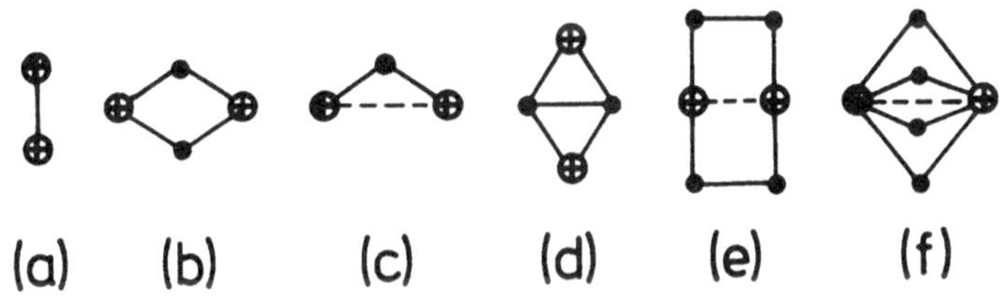

Fig. 1. Examples of graphs from $\Gamma_o$; the blue vertices are denoted by ⊕, the yellow edges by solid lines, the green edges by broken lines.

The set $\Gamma_o$ is made into a semigroup with a unit by the prescription: If $A, B\varepsilon\Gamma_o$, then the product $A(-)B$ is obtained by replacing every yellow edge of A by a copy of B, so that the blue vertices of B are merged with the end points of the removed yellow edge. In this process, the blue vertices of A remain blue, whereas a blue vertex of B merged with a colourless vertex from A becomes colourless. Fig. 2 shows some examples of this product. It is clear, that the graph e of Fig. 1(a), a single yellow edge with two blue vertices, is the unit of the semi-

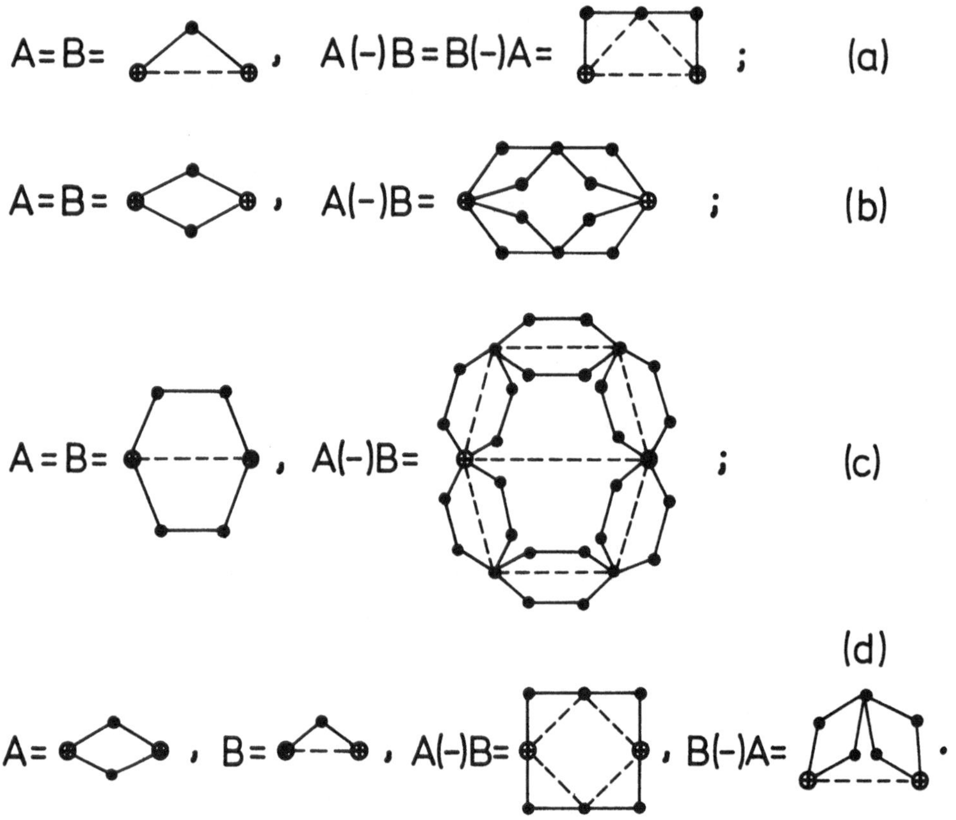


Fig. 2. Some examples of the product A(-)B.

group $\Gamma_o$; the set of graphs obtained by removing e from $\Gamma_o$ is called $\Gamma$. The numbers of vertices V, of yellow edges $E_y$ and of all edges E for the product C=A(-)B are given by

$$V(C) = V(A) + E_y(A)\{V(B)-2\}, \tag{1}$$

$$E(C) = E(A) + E_y(A)\{E(B)-1\}, \tag{2}$$

$$E_y(C) = E_y(A)\ E_y(B). \tag{3}$$

A <u>recursive bond graph sequence</u> $\{B_n\}$ can now be defined by a store of graphs $\{H_n | n=1,2,..\}$ from $\Gamma$ and by the recursion relation

$$B_n = H_n(-)B_{n-1}, \quad B_o=e. \tag{4}$$

The numbers of vertices and edges of the n-th generation graph $B_n$ are easily found from eqs. (1-3) as

$$E_y(B_n) = \prod_{k=1}^{n} E_y(H_n), \tag{5}$$

$$V(B_n)/E_y(B_n) = 2+\sum_{k=1}^{n} [\{V(H_k)-2E_y(H_k)\}/E_y(B_k)], \tag{6}$$

$$E(B_n)/E_y(B_n) = 1+\sum_{k=1}^{n} [\{E(H_k)-E_y(H_k)\}/E_y(B_k)]. \tag{7}$$

For a self-similar sequence $\{B_n\}$, all $H_n$ are equal to a fixed graph H. In this case, set

$$E_y(H) = m', \quad E(H)-E_y(H) = \Delta m', \quad V(H) = f'. \tag{8}$$

Then eqs. (5-7) become, for $m'\neq 1$,

$$E_y(B_n)=m'^n, \quad V(B_n)=\{(f'-2)m'^n+2m'-f'\}/(m-1), \quad E(B_n)=m'^n+(\Delta m')\frac{m'^n-1}{m'-1}, \tag{9}$$

whereas for $m'=1$, the result reads

$$E_y(B_n)=1, \quad V(B_n)=2+(f'-2)n, \quad E(B_n)=1+n\Delta m'. \tag{10}$$

Again, the limits of eqs. (6) and (7) as $n\to\infty$ do not exist in this latter case, since this would imply $f'=2$, $m'=0$, but then $H=e\notin\Gamma$. If one or both of the limits $n\to\infty$ of eqs. (6) and (7) do(es) not exist, the recursive bond graph sequence is called essentially one-dimensional, as for the site case. If this is not so, then the sequence is nontrivial and the limits

$$\delta' = \lim_{n\to\infty} E_y(B_n)/E(B_n) = [1+\sum_{k=1}^{\infty} \{E(H_k)-E_y(H_k)\}/E_y(B_k)]^{-1}, \tag{11}$$

$$\varepsilon' = \lim_{n\to\infty} E_y(B_n)/V(B_n) = [2+\sum_{k=1}^{\infty} \{V(H_k)-2E_y(H_k)\}/E_y(B_k)]^{-1}, \tag{12}$$

exist. For the self-similar case, these are given as

$$\delta'=(m'-1)/(m'+ m'-1), \quad \varepsilon'=(m'-1)/(f'-2). \tag{13}$$

Many of the nontrivial recursive bond graph sequences define a type of pseudo-lattice known as a hierarchical model [1,2], for which a particular kind of (real-space) renormalization group prescription is exact, see Chapter 11. For example, the self-similar recursive bond graph sequence generated by the graph of Fig. 1(b) is the diamond lattice, on which the approximate Migdal-Kadanoff renormalization transformation is exact [3]. Fig. 2(b) shows the first step of this construction.

Another interesting type of sequence is obtained if the graphs $H_n$ are restricted to those, which have a unique, colourless edge with the two blue vertices as end points; this class of graphs is denoted by $\Gamma_1$. Special cases of these are the q-plaquette branches, defined as those recursive bond graph sequences for which $H_n$ consists of $r_n$ regular polygons with q vertices (and edges) each; q-1 edges of each of these polygons are yellow, whereas the $r_n$ copies of this have exactly one colourless edge in common, the end points of which are the blue vertices of $H_n$. Such q-plaquette branches are useful for a study of gauge models, see Chapter 12. Examples are afforded by Fig. 1(e), r=2, q=4, and by Fig. 1(f), r=4, q=3; Fig. 2(c) shows the first step in the construction of a self-similar 4-plaquette branch.

## 8.3. Free energies, thermodynamic limits and renormalization.

### 8.3.1. Spin systems on recursive site graph sequences.

A spin model with permissible symmetry group $\mathcal{G}$ can be defined on the graphs $S_n$ of a recursive site graph sequence as in eq. (1.1.4). If one considers the partition function $Z_n(i)$ with top (blue) spin in state i, the following recursion relation is easily obtained:

$$Z_n(i) = \sum_{\substack{i_v \\ v\in V(G_n)-\{v_1\}}} \prod_{e\in E(G_n)} \Omega_e^{(n)}(i_{v_1(e)}, i_{v_2(e)}) \times$$

$$\times \prod_{v\in V(G_n)-V_y(G_n)} A_v^{(n)}(i_v) \prod_{v'\in V_y(G_n)} Z_{n-1}(i_{v'}), \quad Z_o(i)=A^{(o)}(i). \tag{1}$$

Here the most general case with generation- and edge-dependent interactions as well as generation- and vertex-dependent external fields is considered. However, all different branches are supposed to be identical, so that, e.g., the $Z_{n-1}(i_{v'})$ do not depend on the special vertex v'. These partition functions play the role of effective external fields on the yellow vertices of $G_n$, so that eq. (1) describes a <u>recursive renormalization of the fields</u>. Defining a field-distribution function by

$$\rho_n(i) = Z_n(i)/Z_n = Z_n(i)/\{\sum_{j=1}^{M} Z_n(j)\}, \tag{2}$$

eq. (1) may be cast in the form of a recursion relation for this quantity:

$$\rho_n(i) = c_n^{-1} \sum_{\substack{i_v \\ v\varepsilon V(G_n)-\{v_1\}}} \prod_{e\varepsilon E(G_n)} \Omega_e^{(n)}(i_{v_1(e)}, i_{v_2(e)}) \times$$

$$\times \prod_{v\varepsilon V(G_n)-V_y(G_n)} A_v^{(n)}(i_v) \prod_{v'\varepsilon V_y(G_n)} \rho_{n-1}(i_{v'}),$$

$$\rho_o(i) = A^{(o)}(i)/\{\sum_{j=1}^{M} A^{(o)}(j)\}. \tag{3}$$

The normalization constant $c_n$ is the partition function of the spin system on the graph $G_n$ with its yellow vertices in a normalized field $\rho_{n-1}(i)$:

$$c_n = \sum_{\substack{i_v \\ v\varepsilon V(G_n)}} \prod_{e\varepsilon E(G_n)} \Omega_e^{(n)}(i_{v_1(e)}, i_{v_2(e)}) \prod_{v\varepsilon V(G_n)-V_y(G_n)} A_v^{(n)}(i_v) \times$$

$$\times \prod_{v'\varepsilon V_y(G_n)} \rho_{n-1}(i_{v'}), \quad c_o = \sum_{j=1}^{M} A^{(o)}(j) = Z_o. \tag{4}$$

The free energy per spin on $S_n$ follows from the recursion relation

$$Z_n = c_n Z_{n-1}^{V_y(G_n)}, \quad Z_o = c_o, \tag{5}$$

as

$$\gamma_n = V(S_n)^{-1} \ln Z_n = \{V_y(S_n)/V(S_n)\} \sum_{k=0}^{n} (\ln c_k)/V_y(S_k) . \tag{6}$$

The thermodynamic limit of the free energy exists if $\{\gamma_n\}$ converges to a quantity $\gamma$ in the limit $n\to\infty$. The following theorem holds:

Theorem 1. Let quantities $\Omega$ and $A$ be defined by

$$\Omega(\max) = \sup_n \max_{e \in E(G_n)} \max_{i,j} \Omega_e^{(n)}(i,j),$$

$$\Omega(\min) = \inf_n \min_{e \in E(G_n)} \min_{i,j} \Omega_e^{(n)}(i,j),$$

$$A(\max) = \sup_n \max_{v \in V(G_n)-V_y(G_n)} \sum_{j=1}^{M} A_v^{(n)}(j),$$

$$A(\min) = \inf_n \min_{v \in V(G_n)-V_y(G_n)} \sum_{j=1}^{M} A_v^{(n)}(j), \tag{7}$$

$$\Omega = \max\{\Omega(\max), \Omega(\min)^{-1}\},$$

$$A = \max\{A(\max), A(\min)^{-1}\}.$$

A nontrivial recursive site graph sequence has a thermodynamic limit of the free energy, given by the absolutely convergent series

$$\gamma = \varepsilon \sum_{k=0}^{\infty} (\ln c_k)/V_y(S_k), \tag{8}$$

if both $\Omega$ and $A$ are finite.

Proof. The quantity $c_n$ of eq. (4) satisfies

$$\Omega(\min)^{E(G_n)} A(\min)^{V(G_n)-V_y(G_n)} \leqq c_n \leqq \Omega(\max)^{E(G_n)} A(\max)^{V(G_n)-V_y(G_n)}. \tag{9}$$

This implies

$$|\ln c_n| \leqq E(G_n) \ln \Omega + \{V(G_n)-V_y(G_n)\} \ln A. \tag{10}$$

Eq. (8) is then an absolutely convergent series, since the limits $\delta$ and $\varepsilon$ exist for a nontrivial sequence, eqs. (1.13) and (1.14). ¶

Remark. The conditions of Theorem 1 are certainly fulfilled, if the spin-spin and spin-field interactions are bounded for $n \to \infty$. These conditions can be weakened ([4]), but the above yields the easier proof.

The equipartition distribution function,

$$\rho^{(o)}(i) = M^{-1} \quad \text{for all } i, \tag{11}$$

is the unique solution of the recursion relation (3) for all $n$ in the absence of external fields, since then $\rho_n(i)$ is invariant with respect to the transformations $i \to g(i)$ of the transitive symmetry group $\mathcal{G}$. For small external fields, $\rho_n(i)$ may still converge to equipartition in the limit $n \to \infty$:

$$\lim_{\text{fields}\to 0} \lim_{n\to\infty} \rho_n(i) = \rho^{(o)}(i), \text{ for all } i; \tag{12}$$

if this is the case, the zero-field fixed point $\rho^{(o)}$ is stable and the symmetry $\mathcal{G}$ is unbroken. Small-field phase transitions occur for such values of the interaction parameters for which eq. (12) breaks down; these will be studied in detail in Chapters 9 and 10. There, a restriction to Cayley-like branches will be made; it can, however, be shown, that the general eq. (3) reduces, in the neighbourhood of $\underline{\rho}^{(o)}$, to the Cayley-like case with suitably redefined couplings, see Section 9.8 for an explicit example. Other phase transitions induced by nonzero critical values of the external fields are called high-field phase transitions.

The small-field phase transitions are not accompanied by singularities in the field-free free energy. This is clear from eq. (8), which in zero field reads

$$\gamma = \varepsilon \sum_{k=0}^{\infty} (\ln c_k^{(o)})/V_y(S_k), \tag{13}$$

where $\ln c_k^{(o)}$ is simply the free energy of the spin model on the graph $G_k$ in the absence of a field. This is, obviously, an analytic function of the temperature for $T>0$; the convergence of the series (13) is uniform for $\Omega$ finite (see eq. (7)), so that $\gamma$ is analytic too. This is particularly clear for a self-similar recursive site graph sequence, since eq. (13) then reduces to

$$\gamma = \ln c^{(o)}, \tag{14}$$

which is the free energy of the spin system on the finite graph $G$ from which the sequence is constructed.

### 8.3.2. Spin systems on recursive bond graph sequences.

A spin model defined on a sequence of recursive bond graphs gives some peculiar problems if there are fields acting on the spins: several spins, which already have fields in (n-1)-generation graphs have to be merged for the n-generation graph. The simplest prescription, which is simply to add these fields, leads to unphysical behaviour, see, e.g.,

Ref. ([5]). Fixing the fields anew in each generation, on the other hand, gives "extra step" recursion relations, which are rather cumbersome. For these reasons, only the field-free case is considered here. Then a recursion relation for the n-generation partition function with both blue spins $v_1$ and $v_2$ fixed in states $i$ and $j$, respectively, is easily found:

$$Z_n(i,j) = \sum_{\substack{i_v \\ v\in V(H_n)-\{v_1,v_2\}}} \prod_{e\in E(H_n)-E_y(H_n)} \Omega_e^{(n)}(i_{v_1(e)},i_{v_2(e)}) \times$$

$$\times \prod_{e'\in E_y(H_n)} Z_{n-1}(i_{v_1(e')},i_{v_2(e')}),\quad Z_o(i,j)=\Omega^{(o)}(i,j). \tag{15}$$

By condition (iii) on the $H_n$, see Section 2, and by the invariance of the spin-spin interactions, one has

$$Z_n(i,j) = Z_n(j,i);\quad Z_n(g(i),g(j)) = Z_n(i,j)\quad \text{for all}\quad g\in G, \tag{16}$$

so that the <u>interaction</u> is <u>recursively</u> <u>renormalized</u> to give, in the n-th generation:

$$\Omega^{(n)}(i,j) = Z_n(i,j)/Z_n(1,1). \tag{17}$$

Now the procedure is the same as in the recursive site case: a distribution function $\rho_n(i,j)$ and a normalization constant $d_n$ are given as:

$$\rho_n(i,j) = Z_n(i,j)/\{\sum_{k,m=1}^{M} Z_n(k,m)\} = Z_n(i,j)/Z_n, \tag{18}$$

$$\rho_n(i,j) = d_n^{-1} \sum_{\substack{i_v \\ v\in V(H_n)-\{v_1,v_2\}}} \prod_{e\in E(H_n)-E_y(H_n)} \Omega_e^{(n)}(i_{v_1(e)},i_{v_2(e)}) \times$$

$$\times \prod_{e'\in E_y(H_n)} \rho_{n-1}(i_{v_1(e')},i_{v_2(e')}),\quad \rho_o(i,j)=\Omega^{(o)}(i,j)/M\lambda_o^{(o)}, \tag{19}$$

$$d_n = \sum_{\substack{i_v \\ v\in V(H_n)}} \prod_{e\in E(H_n)-E_y(H_n)} \Omega_e^{(n)}(i_{v_1(e)},i_{v_2(e)}) \prod_{e'\in E_y(H_n)} \times$$

$$\times\ \rho_{n-1}(i_{v_1(e')},i_{v_2(e')}),\quad d_o=M\lambda_o^{(o)}. \tag{20}$$

The free energy per spin is now given by

$$\gamma_n' = V(B_n)^{-1} \ln Z_n = \{E_y(B_n)/V(B_n)\} \sum_{k=0}^{n} (\ln d_k)/E_y(B_k), \tag{21}$$

in complete analogy with eq. (6) for the site case. The thermodynamic limit $\lim_{n\to\infty} \gamma_n' = \gamma'$ exists for weak restrictions on the interactions [1]:

Theorem 2. Let $\Omega(\max)$ be defined as in the first of eqs. (7):

$$\Omega(\max) = \sup_n \max_{e \in E(H_n) - E_y(H_n)} \max_{i,j} \Omega_e^{(n)}(i,j) < \infty. \tag{22}$$

Then the thermodynamic limit of the free energy for a nontrivial recursive bond graph sequence is given by the convergent series

$$\gamma' = \varepsilon' \sum_{k=0}^{\infty} (\ln d_k)/E_y(B_k). \tag{23}$$

Proof. The proof is slightly more intricate than for Theorem 1. First of all, it is remarked, that the invariance of $\rho_n(i,j)$ with respect to the permutations of the permissible symmetry group $\mathcal{G}$ implies

$$\sum_{j=1}^{M} \rho_n(i,j) = M^{-1}, \quad \text{independent of } i, \tag{24}$$

so that the inequality

$$\rho_n(i,j) \leqq M^{-1}, \quad \text{for all } i \text{ and } j, \tag{25}$$

holds. Now if $H_n$ is not (graph-theoretically) a tree, this can be used to obtain the inequality

$$d_n \leqq \Omega(\max)^{E(H_n)-E_y(H_n)} M^{V(H_n)-E_y(H_n)} \tag{26}$$

from eq. (20). Since $H_n$ is not a tree, $E(H_n) \geqq V(H_n)$ must hold; therefore, eq. (26) implies

$$d_n \leqq [\Omega(\max)\ M]^{E(H_n)-E_y(H_n)}. \tag{27}$$

If $H_n$ is a tree, then one uses first eq. (20) to obtain

$$d_n \leqq \Omega(\max)^{E(H_n)-E_y(H_n)} \sum_{i_v, v\in V(H_n)} \prod_{e'\in E_y(H_n)} \rho_{n-1}(i_{v_1(e')}, i_{v_2(e')}). \tag{28}$$

The sum on the right-hand-side of eq. (28) is easily seen to be equal to $M^{N_c - E_y(H_n)}$, where $N_c$ is the number of connected subtrees, in which $H_n$ splits up upon deletion of the non-yellow edges. Since there is at least one yellow edge in $H_n$, $N_c$ must be less than the number of vertices $V(H_n)$, which number equals $E(H_n)+1$ for a tree:

$$N_c \leqq V(H_n) - 1 = E(H_n). \tag{29}$$

Therefore, eq. (27) also holds for $H_n$ a tree.

Now, one considers the sequence $\{R_n\}$ defined by

$$R_n = \sum_{k=0}^{n} (\ln d_k)/E_y(B_k) + \sum_{k=n+1}^{\infty} [\{E(H_k)-E_y(H_k)\}/E_y(H_k)] \ln M\Omega(\max). \tag{30}$$

This is well-defined due to the nontriviality of the sequence of graphs, which implies that $\delta'$ exists, see eq. (2.11). $\{R_n\}$ is monotonically decreasing by eq. (27); on the other hand, it is bounded from below, since eq. (15) implies

$$Z_n = \sum_{i,j}^{M} Z_n(i,j) \geqq MZ_n(i,i) \geqq M\, Z_{n-1}(i,i)^{E_y(H_n)}; \; Z_o(i,i)=1, \tag{31}$$

which shows that $Z_n \geqq 1$ holds for all $n$, so that $R_n$ cannot be negative. $\{R_n\}$ is then a convergent sequence, which shows that $\gamma'$ exists and is given by eq. (23). ¶

The recursion relation (19) for the distribution functions $\rho_n(i,j)$ has the <u>high-temperature fixed point</u>

$$\rho^{(H)}(i,j) = M^{-2} \quad \text{for all } i,j, \tag{32}$$

if there is no path of green edges connecting the blue vertices of all $H_n$. If there is a path of yellow edges with this property, then there is also a <u>low-temperature fixed point</u>:

$$\rho^{(L)}(i,j) = M^{-1}\, \delta(i,j), \tag{33}$$

where $\delta(i,j)=0$ for $i \neq j$, $\delta(i,j)=1$ for $i=j$. It is expected, that, if both of these fixed points exist, then the sequence $\{\rho_n(i,j)\}$ converges to one of these fixed points for almost all initial values $\Omega^{(o)}(i,j)$, the few interactions, for which this is not the case, corresponding to phase transitions in the coupling constant space. This problem will be returned to in Chapter 11.

### 8.3.3. Gauge models on recursive bond graphs of class $\Gamma_1$.

The graphs of $\Gamma_1$ contain a unique "top edge", which is the green edge with the two blue vertices as end points. On such graphs, a general gauge model as described in Section 7.7 can be defined (including Higgs fields). Since cycles of edges should not "double up" on themselves, each cycle of a graph $B_n$ belongs completely to one generation, having at most one edge in common with cycles belonging to a particular graph in the preceding or following generation. A recursive bond graph sequence $B_n$ will be called <u>gauge-nontrivial</u> if, in addition to $\delta'$ as defined in eq. (2.11), the limit

$$\zeta' = \lim_{n\to\infty} C(B_n)/E_y(B_n) \tag{34}$$

exists and is nonzero. Here $C(B_n)$ is the number of cycles of the graph $B_n$; it satisfies the recursion relation

$$C(B_n) = C(H_n) + E_y(H_n)C(B_{n-1}), \quad C(B_o)=C(H_o)=0. \tag{35}$$

Therefore, eq. (34) may also be written as

$$\zeta' = \sum_{k=1}^{\infty} C(H_k)/E_y(B_k). \tag{36}$$

The partition function of a gauge model on $B_n$ is now easily seen to obey the recursion relation (spin $r$ on top edge $e_1$ fixed):

$$Z_n(r) = \sum_{\substack{r_e \in R \\ e\in E(H_n)-\{e_1\}}} \ \prod_{\substack{\text{cycles} \\ C \text{ of } H_n}} \Omega_C^{(n)}\Big( \prod_{\substack{\text{edges } e \\ \text{of cycle} \\ C}} r_e^{\alpha_e}\Big) \times$$

$$\times \prod_{e\in E(H_n)-E_y(H_n)} A_e^{(n)}(r_e) \prod_{e'\in E_y(H_n)} Z_{n-1}(r_{e'}). \tag{37}$$

Here $A_e^{(n)}(r)$ is a generation- and edge-dependent Higgs field, whereas $\Omega_C^{(n)}(r)$ is a cycle- and generation-dependent interaction of the type described in Section 7.7. All $r_e$ belong to a regular group $R$ with $M$ elements. Eq. (37) is completely analogous to eq. (1) for a spin model on a recursive site graph; therefore, eqs. (2), (3), (4) and (6) are practically unchanged:

$$\tilde{\rho}_n(r) = Z_n(r)/Z_n, \quad Z_n=\Sigma_{r'\in R} Z_n(r'); \tag{38}$$

$$\tilde{\rho}_n(r) = \tilde{c}_n^{-1} \sum_{\substack{r_e \varepsilon \mathcal{R} \\ e\varepsilon E(H_n)-\{e_1\}}} \prod_{C \leq H_n} \Omega_C^{(n)}\Big(\prod_{e\varepsilon C} r_e^{\alpha_e}\Big) \prod_{\substack{\text{green} \\ \text{edges} \\ e}} A_e^{(n)}(r_e) \times$$

$$\times \prod_{\substack{\text{yellow} \\ \text{edges } e'}} \tilde{\rho}_{n-1}(r_{e'}), \quad \tilde{\rho}_o(r)=A^{(o)}(r)/\Big\{\sum_{r'\varepsilon\mathcal{R}} A^{(o)}(r')\Big\}; \tag{39}$$

$$\tilde{c}_n = \sum_{\substack{r_e \varepsilon \mathcal{R} \\ e\varepsilon E(H_n)}} \prod_{C \leq H_n} \Omega_C^{(n)}\Big(\prod_{e\varepsilon C} r_e^{\alpha_e}\Big) \prod_{\substack{\text{green} \\ \text{edges} \\ e}} A_e^{(n)}(r_e) \prod_{\substack{\text{yellow} \\ \text{edges} \\ e'}} \tilde{\rho}_{n-1}(r_{e'}),$$

$$\tilde{c}_o = \sum_{r'\varepsilon\mathcal{R}} A^{(o)}(r') = Z_o; \tag{40}$$

$$\tilde{\gamma}_n = E(B_n)^{-1} \ln Z_n = \{E_y(B_n)/E(B_n)\} \sum_{k=0}^{n} (\ln c_k)/E_y(B_k). \tag{41}$$

These equations describe the <u>recursive renormalization of the Higgs field</u>. The thermodynamic limit of the free energy is given by

<u>Theorem 3</u>. Let $\Omega$ and $A$ be defined by analogy with eq. (7). A gauge-nontrivial recursive bond graph sequence of class $\Gamma_1$ has a thermodynamic limit of the free energy given by the absolutely convergent series

$$\tilde{\gamma} = \delta' \sum_{k=0}^{\infty} (\ln c_k)/E_y(B_k), \tag{42}$$

in case $\Omega$ and $A$ are both finite.

<u>Proof</u>. Completely analogous to the proof of Theorem 1, except that the existence of the limits $\delta'$ and $\zeta'$ instead of $\delta$ and $\varepsilon$ has to be invoked. ¶

The <u>zero-Higgs-field fixed point</u>

$$\tilde{\rho}^{(o)}(r) = M^{-1}, \text{ for all } r\varepsilon\mathcal{R}, \tag{43}$$

should <u>always</u> be stable according to Elitzur's theorem, see the end of Section 7.7. Phase transitions, therefore, only occur at nonzero values of the Higgs fields: they are the analogues of the high-field phase transitions for spin models alluded to in subsection 3.1. Phase transitions for Higgs-field models are studied in Chapter 12; there, the stability of the fixed point of eq. (43) is also proved.

## 8.4. Infinite pseudo-lattices (Bethe lattices) and the Bethe-Peierls approximation.

The recursive site and bond graph sequences discussed in the first two sections of this chapter do not, in general, "converge" in some way to a unique infinite graph. For some special cases, however, infinite pseudo-lattices can be defined in such a way, that all members of a recursive graph sequence are finite portions of the infinite graph. Examples are afforded by the Bethe lattices, which are infinite trees in which every vertex has the same coordination number $m+1$; these obviously contain all Cayley trees $T_n(m+1)$, see Fig. 1(a) for the example $m=2$ and compare this with Fig. 1.3(b). A similar graph obtained by welding triangles together (triangular Bethe lattice) is shown in Fig. 1(b); this contains all triangular cactus trees, compare Fig. 1.4(c).

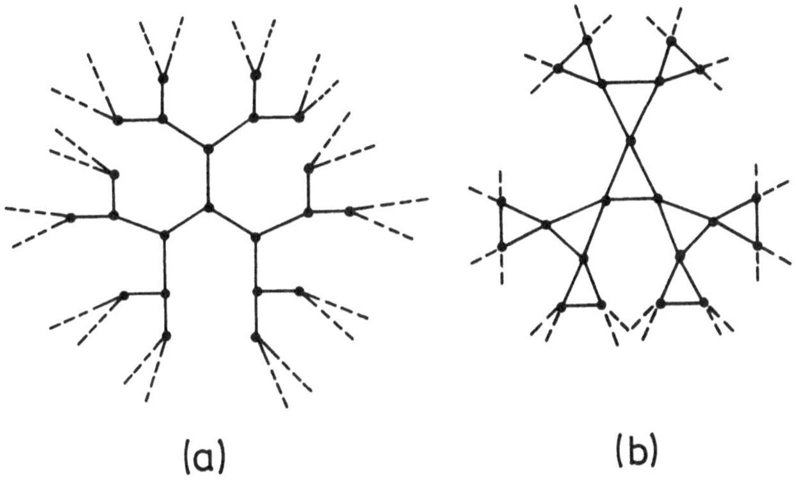


Fig. 1. (a) Portion of the Bethe lattice with $m+1=3$; (b) Portion of the triangular Bethe lattice.

These infinite pseudo-lattices are translation-invariant, i.e., every vertex is equivalent to any other vertex, and the same holds true for the edges. (The automorphism groups of these infinite graphs are then transitive on the sets of vertices and edges.) This is in marked contrast to the "corresponding" recursive graphs, since these have a surface of yellow edges or vertices, and the number of edges or vertices in this surface is a finite fraction of the total number if the graph is nontrivial in the thermodynamic limit. It is, therefore, not to be expected, that the thermodynamic limit of the free energy for a spin or gauge model on a sequence of recursively defined graphs has much to do with the corresponding quantity for the "corresponding" infinite graph, if this latter entity exists at all. On these grounds, the recursive approach has been criticized ([6]). There is, of course, no reason for critique, if one is interested in the properties of very large, but finite graphs.

Nonetheless, the problem of defining a free energy for a spin or gauge model on an infinite pseudo-lattice and its relation (if any) with the corresponding finite graphs is interesting in itself. To study it, a different starting point has to be taken, namely the thermodynamic limit of probability measures on infinite graphs ([7]). Below, a simplified version of this proper method will be given, which does not claim to be exact, but is in close analogy with the Bethe-Peierls approximation for real lattices. In this latter approximation, one considers a spin together with its nearest neighbours; the influence of the rest of the lattice is supposed to be given by a mean field. Self-consistency arguments then fix the value of this mean field.

In order to start the investigation, consider first a Cayley branch $C_n(m)$ with branching ratio $m$, on which a spin model with Boltzmann factors $\Omega(i,j)$ for every edge is defined. Further, all spins are supposed to feel the same external field $A(i)$. Then eqs. (3.3), (3.4) and (3.6) reduce to:

$$\rho_n(i) = c_n^{-1} A(i) \left[\sum_{j=1}^{M} \Omega(i,j)\, \rho_{n-1}(j)\right]^m, \quad \rho_o(i)=A(i)/\{\sum_{j=1}^{M} A(j)\}, \tag{1}$$

$$c_n = \sum_{i=1}^{M} A(i) \left[\sum_{j=1}^{M} \Omega(i,j)\, \rho_{n-1}(j)\right]^m, \quad c_o = \sum_{j=1}^{M} A(j), \tag{2}$$

$$\gamma_n = \frac{m^n(m-1)}{m^{n+1}-1} \sum_{k=0}^{n} (\ln c_k)/m^k. \tag{3}$$

It will be assumed, that $\{\rho_n(i)\}$ converges (in $L_1$-sense) to a distribution $\rho(i)$, so that one has

$$\lim_{n\to\infty} c_n = c = \sum_{i=1}^{M} A(i) \left[\sum_{j=1}^{M} \Omega(i,j)\, \rho(j)\right]^m. \tag{4}$$

The corresponding sequence of Cayley trees $\{T_n(m+1)\}$ has, as free energies of its members, $\gamma_n^{(t)}$,

$$N^{(t)}(n)\, \gamma_n^{(t)} = m^n(m+1) \left[\sum_{k=0}^{n} (\ln c_k)/m^k + (\ln c'_{n+1})/\{m^n(m+1)\}\right], \tag{5}$$

with

$$\lim_{n\to\infty} c'_{n+1} = c' = \sum_{i=1}^{M} A(i) \sum_{j=1}^{M} \Omega(i,j)\ \rho(j)^{m+1}, \tag{6}$$

and $N^{(t)}(n)$ the number of vertices of $T_n(m+1)$:

$$N^{(t)}(n) = (m+1)\frac{m^{n+1}-1}{m-1} + 1. \tag{7}$$

In order to obtain a free energy, which does not contain contributions from the surface, the following "tree subtraction" procedure ([9]) can be used:

A tree of $n+\ell$ generations contains $m^{n+\ell}(m+1)$ surface vertices; $m^{\ell}$ n-generation trees contain exactly as many. The difference in free energies of the two systems is, from eq. (5),

$$\Delta\gamma(n,\ell) = N^{(t)}(n+\ell)\gamma_{n+\ell}^{(t)} - m^{\ell}N^{(t)}(n)\gamma_n^{(t)} =$$

$$= m^{n+\ell}(m+1)\ \Big[\sum_{k=n+1}^{n+\ell} (\ln c_k)/m^k + (\ln c'_{n+\ell+1})/\{m^{n+\ell}(m+1)\} - (\ln c'_{n+1})/ /\{m^n(m+1)\}\Big]. \tag{8}$$

The number of vertices "left over" is

$$\Delta v(n,\ell) = N^{(t)}(n+\ell) - m^{\ell}N^{(t)}(n) = 2(m^{\ell}-1)/(m-1). \tag{9}$$

In the limit $n\to\infty$ (for fixed $\ell$), the difference free energy per left-over vertex is

$$\Gamma = \lim_{n\to\infty} \Delta\gamma(n,\ell)/\Delta v(n,\ell) = \tfrac{1}{2}(m+1)\ \ln c - \tfrac{1}{2}(m-1)\ \ln c'. \tag{10}$$

Note, that although the surface does not contribute to this free energy, the type of surface field still determines the fixed point, for which $c$ and $c'$ are calculated. Therefore, if no reference to the surface field is to be made, eq. (10) must be evaluated for all fixed points $\rho(i)$ and the thermodynamical criterion of lowest free energy is needed to select the proper fixed points in each phase (or in each point of the space of coupling constants). Such an approach has recently been applied to the gauge model case with excellent results ([10]), see also Section 12.5.

The same spin model is now considered on the Bethe lattice with m+1 next nearest neighbours per site. For a ferromagnetic type of ordering, it is expected, that the distribution of spin states for a disconnected branch can be described by a single distribution function $\sigma_d(i)$ for every site. Now consider the combined probability $\sigma(i;j_1,..,j_{m+1})$ that a spin is in state $i$ and that its nearest neighbours are in states $j_1,..,j_{m+1}$. Since these neighbours all belong to independent branches, one expects this combined probability to be of the form

$$\sigma(i;j_1,..j_{m+1}) = A(i)[\prod_{k=1}^{m+1} \Omega(i,j_k)\sigma_d(j_k)]/C', \tag{11}$$

where $C'$ is the normalization constant

$$C' = \sum_{i=1}^{M} A(i)[\sum_{j=1}^{M} \Omega(i,j)\sigma_d(j)]^{m+1} . \tag{12}$$

The pair distribution function $\sigma(i,j)$ must be symmetric, due to the isotropy of the Bethe lattice; this implies

$$\sum_{j=1}^{M} \sigma(i,j) = \sum_{j_1,..,j_{m+1}=1}^{M} \sigma(i;j_1,..j_{m+1}) =$$

$$= \sum_{j_1,..,j_{m+1}=1}^{M} \sigma(j_1;i,j_2,..,j_{m+1}). \tag{13}$$

Using eq. (11), this yields

$$\sigma_d(i) = C^{-1} A(i) [\sum_{j=1}^{M} \Omega(i,j)\sigma_d(j)]^{m}, \tag{14}$$

with $C$ the appropiate normalization constant. Comparison with eqs. (1), (4) and (6) now yields the identifications

$$\sigma_d(i) = \rho(i), \text{ the limiting distribution for a Cayley branch,}$$
$$C' = c', \quad C = c. \tag{15}$$

The pair distribution function is then given as

$$\sigma(i,j) = (c/c')\, \rho(i)\Omega(i,j)\rho(j), \tag{16}$$

so that the normalization for this quantity gives the useful identity

$$c'/c = \sum_{k,m=1}^{M} \rho(k)\Omega(k,m)\rho(m). \tag{17}$$

The pair distribution of eq. (17) can now be used to calculate the average energy per spin from the equation

$$\langle E\rangle = \frac{1}{2}(m+1)\sum_{i,j=1}^{M}\sigma(i,j)E(i,j) + \sum_{i,j=1}^{M}\sigma(i,j)F(i); \tag{18}$$

here the forefactor of the first term is due to the fact, that there are $(m+1)/2$ edges pro vertex. Eq. (18) yields, with $\Omega(i,j)=\exp-\beta E(i,j)$ and $A(i)=\exp-\beta F(i)$,

$$\langle E\rangle = -\frac{1}{2}(m+1)(c/c')\sum_{i,j=1}^{M}\frac{\partial\Omega(i,j)}{\partial\beta}\rho(i)\rho(j) - (c/c')\sum_{i,j=1}^{M}\frac{\partial\ln A(i)}{\partial\beta}\times$$

$$\times\ \Omega(i,j)\rho(i)\rho(j). \tag{19}$$

In order to obtain the free energy $\Gamma_B$ per spin, this has to be integrated:

$$\Gamma_B(\beta)-\Gamma_B(\beta_o) = -\int_{\beta_o}^{\beta}\langle E\rangle(\beta')\,d\beta'. \tag{20}$$

Eq. (19) can be cast into the form of a derivative; to this end, quantities a, b and d are defined by

$$a = (c/c')\sum_{i,j=1}^{M}\frac{\partial\Omega(i,j)}{\partial\beta}\rho(i)\rho(j),$$

$$b = (c/c')\sum_{i,j=1}^{M}\Omega(i,j)\rho(i)\frac{\partial\rho(j)}{\partial\beta}, \tag{21}$$

$$d = (c/c')\sum_{i,j=1}^{M}\frac{\partial\ln A(i)}{\partial\beta}\Omega(i,j)\rho(i)\rho(j).$$

With these abbreviations, eq.(19) reads

$$\langle E\rangle = -\frac{1}{2}(m+1)a - d. \tag{22}$$

The derivative of $(c/c')$ with respect to $\beta$ yields, using eq. (17):

$$\frac{\partial}{\partial\beta}\ln(c/c') = a+2b, \tag{23}$$

whereas differentiation of eq. (14), for $\rho(i)=\sigma_d(i)$, yields, after some manipulation,

$$b = -\frac{\partial}{\partial\beta}(\ln c) + d + m(a+b). \tag{24}$$

Combination of eqs. (22), (23) and (24) gives

$$\langle E\rangle = -\frac{\partial}{\partial\beta}\Gamma = -\frac{\partial}{\partial\beta}\Gamma_B, \tag{25}$$

with $\Gamma$ as derived for the "subtracted Cayley trees" in eq. (10). The free energy of a spin model on a Bethe lattice is, therefore,

$$\Gamma_B = \Gamma + w, \tag{26}$$

with $w$ a temperature-independent constant in each phase. In the high-temperature phase, e.g., $\Gamma_B$ must equal the noninteracting spin value $\ln M$; since $\Gamma$ has this same high-temperature limit (here $c=c'=M$ holds), $w$ equals zero in this phase. In the other phases, one has to use the procedure outlined following eq. (10).

The above result, eq. (26), can be extended to cases in which the external field is not hohmogeneous and also to cases, for which the sequence of distribution functions tends to a limit cycle. It is clear, that in all these cases all interesting information on symmetry-breaking phase transitions, etc., can be obtained by studying recursive graph sequences. The study of the infinite pseudo-lattices does not give any new points of view and will, therefore, not be pursued further here.

## 8.5. Other recursively defined graph sequences: Fractals.

Mandelbrot ([11]) has discussed many self-similar structures built up by recursion relations, which are fractals, i.e., which must be assigned non-integer dimensions. On such structures, spin (or gauge) models can also be defined ([12,13,14]). Here only a few simple examples can be given:

(a) Koch curves.([12]). These are actually self-similar recursive bond graph sequences, for which the graph $H$ is quasi-linear, i.e., the blue vertices are at the end points of (different) yellow edges. Some examples are shown in Fig. 1.

An intrinsic (graph fractal) dimension can be defined generally by ([15]):

$$D = (\ln N)/(\ln B), \tag{1}$$

where $N$ is the number of (n-1)-generation graphs needed to built an n-generation graph and $B$ is the ratio of the distances between the blue vertices in (n-1)- and n-generation graphs (in the limit $n\to\infty$, if

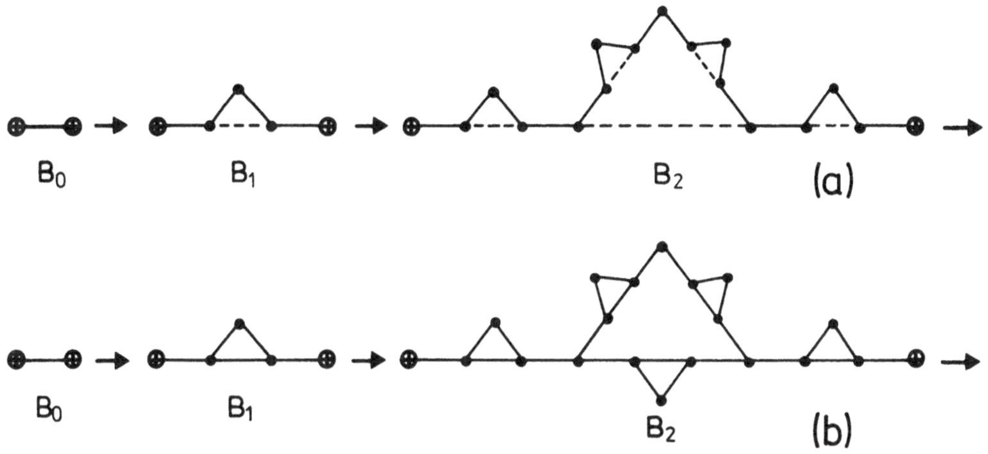


Fig. 1. Two examples of Koch curves as recursive bond graph sequences.

necessary). Therefore, the Koch curve of Fig. 1(a) has $N=4$, $B=2$ and $D=2$, whereas the curve of Fig. 1(b) has $N=5$, $B=3$ and $D=(\ln 5)/(\ln 3)$.

The definition of eq. (1) is applicable to all recursive bond graph sequences, which are not of class $\Gamma_1$; the diamond lattice of Fig. 2.2(b), for example, also having $D=(\ln 4)/(\ln 2)=2$.

(b) Sierpinski gaskets. ([13]). These are constructed recursively from triangles as shown in Fig. 2. Since $N=3$ and the analogue of $B$, the

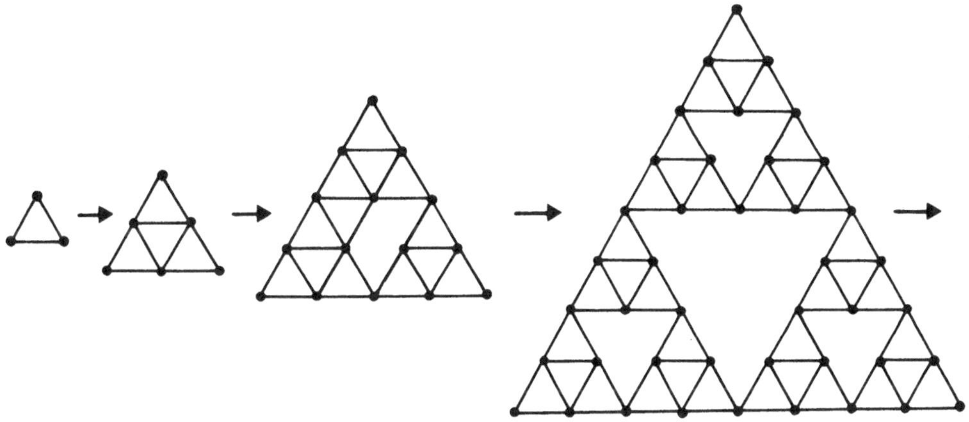

Fig. 2. Recursive construction of Sierpinski gaskets.

ratio of the distances between the extreme vertices, equals 2, the graph fractal dimension is here $(\ln 3)/(\ln 2)$. This might be called a triangular site recursive bond graph sequence, since at every step three $B_{n-1}$-graphs are used to construct one $B_n$-graph by the merging of vertices. It is not difficult, to extend this type of construction to higher dimensions, using tetrahedra etc., but this will not be shown here, see Ref. ([13]).

(c) Sierpinski carpets. ([14]). The recursive prescription uses here the merging of edges to obtain "carpets with many holes". Fig. 3 shows an example with $N=8$, $B=3$, $D=(\ln 8)/(\ln 3)$. Such a procedure can also be generalized, for which the reader is referred to ([14]).

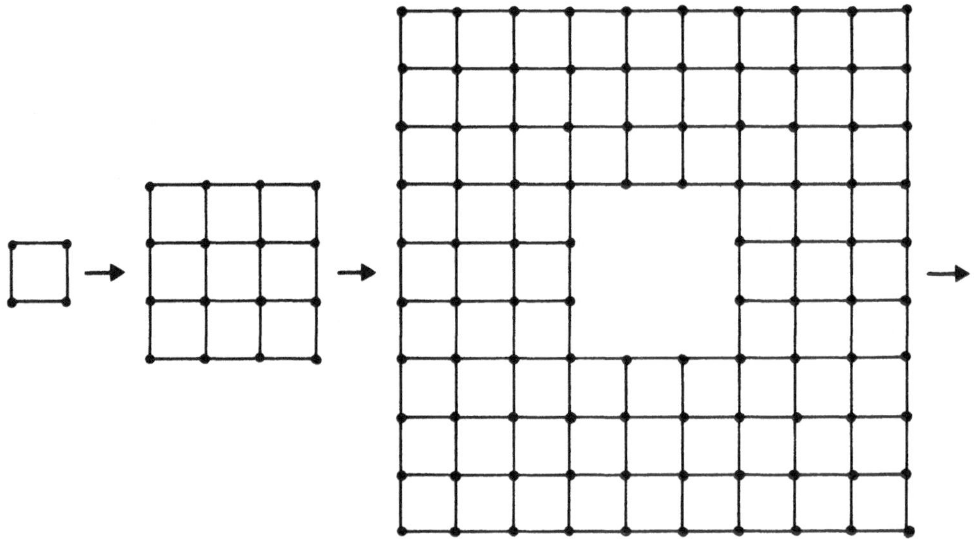

Fig. 3. Recursive construction of a Sierpinski carpet.

It might be asked, whether it is possible to assign a graph fractal dimension to self-similar recursive site graph sequences and to recursive bond graph sequences of class $\Gamma_1$. For the latter, one has $B=1$, so that eq. (1) may be interpreted as

$$D(\Gamma_1) = \begin{cases} 1 & \text{if the sequence is essentially one-dimensional,} \\ \infty & \text{if the sequence is nontrivial.} \end{cases} \qquad (2)$$

If one defines $B$ for a recursive site graph sequence by the limit as $n \to \infty$ of the ratio of the distances from the top spin to the surface for two successive graphs, $B=1$ here as well, so that the same result holds:

$$D(\text{site}) = \begin{cases} 1 & \text{if the sequence is essentially one-dimensional,} \\ \infty & \text{if the sequence is nontrivial.} \end{cases} \quad (3)$$

With these definitions, the graph fractal dimensions of nontrivial Cayley- and q-plaquette branches is always $\infty$; this reflects the fact, that in the limit $n\to\infty$, these graphs demand spaces with $d\to\infty$ in order to embed them in d-dimensional hypercubic lattices.

The existence of the thermodynamic limit for spin or gauge models on fractal lattices has not been proved generally. Some results can be found in Ref. ([1]); there is no reason to assume, that the type of proof used in Section 3 must be modified for this more general case.

REFERENCES.


([1]). R.B. Griffiths and M. Kaufman, Phys. Rev. B 26 (1982) 5022.
([2]). M. Kaufman and R.B. Griffiths, preprint (1984).
([3]). A.N. Berker and S. Ostlund, J. Phys. C 12 (1979) 4961.
([4]). H. Moraal, Physica 85 A (1976) 457.
([5]). M. Kaufman and R.B. Griffiths, J. Phys. A 15 (1982) L239.
([6]). F. Peruggi, J. Phys. A 16 (1983) L713.
([7]). D. Ruelle, Thermodynamic Formalism (Addison-Wesley, Reading, 1978).
([8]). H.A. Bethe, Proc. Roy. Soc. A 150 (1935) 122.
R. Peierls, Proc. Camb. Phil. Soc. A 32 (1936) 471.
([9]). H.G. Baumgärtel and E. Müller-Hartmann, Z. Physik B 46 (1982) 227.
([10]). J.-B. Zuber, Nucl. Phys. B 235 [FS 11] (1984) 435.
([11]). B.B. Mandelbrot, Fractals: Form, Chance and Dimension (Freeman, San Francisco, 1977).
B.B. Mandelbrot, The fractal Geometry of Nature (Freeman, San Francisco, 1982).
([12]). Y. Gefen, A. Aharony and B.B. Mandelbrot, J. Phys. A 16 (1983) 1267.
([13]). Y. Gefen, A. Aharony, Y. Shapir and B.B. Mandelbrot, J. Phys. A 17 (1984) 435.
([14]). Y. Gefen, A. Aharony and B.B. Mandelbrot, J. Phys. A 17 (1984) 1277.
([15]). B.B. Mandelbrot, J. Stat. Phys. 34 (1984) 895.

## 9. HOMOGENEOUS SPIN MODELS ON CAYLEY BRANCHES WITH SURFACE FIELD.

### 9.1. Small-field phase transitions.

In this section, a general multicomponent spin model with permissible symmetry group $\mathcal{G}$ on a Cayley branch sequence $\{C_n(m)\}$ with branching ratio $m$ is considered. All interactions between pairs of spins are taken equal and a field is applied to the surface or boundary, consisting of the yellow spins, only. Then the recursion relations (8.3.3) and (8.3.4) for the n-th generation field-distribution function read

$$\rho_n(i) = c_n^{-1}\{\sum_{j=1}^{M}\Omega(i,j)\rho_{n-1}(j)\}^m, \quad \rho_o(i)=A(i)/\{\sum_{j=1}^{M}A(j)\}, \tag{1}$$

$$c_n = \sum_{i=1}^{M}\{\sum_{j=1}^{M}\Omega(i,j)\rho_{n-1}(j)\}^m, \quad c_o=\sum_{j=1}^{M}A(j). \tag{2}$$

The thermodynamic limit of the free energy exists by Theorem 8.3.1 and is given by eq. (8.3.8), which for this case reads

$$\gamma = \frac{m-1}{m}\sum_{k=0}^{\infty}(\ln c_k)/m^k. \tag{3}$$

The stability of the field-free fixed point, $\rho^{(o)}(i)=M^{-1}$ for all $i$, is studied by setting

$$\rho_n(i) = \rho^{(o)}(i) + \delta_n(i), \tag{4}$$

with $\underline{\delta}_n$ small. Noting, that

$$\sum_{i=1}^{M}\delta_n(i) = 0 \tag{5}$$

must hold, since $\rho_n(i)$ is normalized, eqs. (1) and (2) give, up to linear terms in $\underline{\delta}$,

$$\delta_n(i) = (m/\lambda_o)\sum_{j=1}^{M}\Omega(i,j)\delta_{n-1}(j). \tag{6}$$

Since $\underline{\underline{\Omega}}$ is a real symmetric matrix, $\underline{\delta}_n$ can be expanded in its eigenvectors $\underline{\mu}_k$; eq. (5) implies, that only those orthogonal to $\mu_o(i)=1$ (for all $i$), which has the eigenvalue $\lambda_o$, have to be taken into account:

$$\underline{\delta}_n = \sum_{k=1}^{M-1} \alpha_n(k)\ \underline{\mu}_k. \tag{7}$$

This yields

$$\alpha_n(k) = (m\lambda_k/\lambda_o)\ \alpha_{n-1}(k), \quad \text{for} \quad k=1,..,M-1. \tag{8}$$

Therefore, the zero-field fixed point becomes unstable with respect to the eigenvector $\underline{\mu}_k$ as soon as the phase transition

$$|m\lambda_k/\lambda_o|=1 \tag{9}$$

is reached. For $|m\lambda_k/\lambda_o|>1$, the eigenvectors belonging to $\lambda_k$ are said to <u>propagate</u> on the tree. The phase transition for $\lambda_k>0$, i.e., in the ferromagnetic region, is called a ferromagnetic phase transition; it is expected, that $\{\underline{\rho}_n\}$ converges always to a fixed point in this region. The phase transition for $\lambda_k<0$ leads to alternating signs for the sequence $\{\alpha_n(k)\}$ and is called antiferromagnetic; the sequence of distribution functions converges possibly to a two-point limit cycle here.

Eqs. (9) lead, for a model with $q$ different eigenvalues ($q=s$, the number of graphs in the MI of $\mathcal{G}$, if this group has a P-algebra), to $3^q$ different phases. The transition hyperplanes separating these phases are of two different kinds: (i) symmetry-breaking phase transitions and (ii) dimensionality-changing phase transitions. In symmetry-breaking phase transitions, the subgroups $\mathcal{G}_1$ and $\mathcal{G}_2$ of $\mathcal{G}$, which leave the phases on different sides of the transition hyperplane invariant, are not equal; here these subgroups are defined in terms of the subgroups $\mathcal{G}(k)$, which leave each vector of the eigenspace of the eigenvalue $\lambda_k$ invariant, by

$$\mathcal{G}(\text{phase } \alpha) = \bigcap_{\substack{\text{values of } k,\\ \text{such that } \underline{\mu}_k\\ \text{propagates}\\ \text{in phase } \alpha}} \mathcal{G}(k). \tag{10}$$

(Technically speaking, the propagating eigenvectors of phase $\alpha$ carry a representation of $\mathcal{G}$; the kernel of this representation is the unbroken symmetry group $\mathcal{G}(\text{phase } \alpha)$). If both phases separated by a transition hyperplane have the <u>same</u> symmetry group, $\mathcal{G}_1=\mathcal{G}_2$, the dimensionality of the space of propagating eigenvectors changes, in general. Since in the symmetry-breaking phase transitions, the fixed point or limit cycle to which $\{\underline{\rho}_n\}$ converges, changes qualitatively, whereas this is not the case in dimensionality-changing phase transitions, the singularities associated with the former are, supposedly, stronger than those associated with the latter, if these possess such singularities at all. Therefore, only the symmetry-breaking phase transitions will, in what follows, be considered of importance as far as conclusions for real lattices are concerned.

## 9.2. The Ising and Potts models.

### 9.2.1. The small-field phase transitions.

For the Potts models with symmetry groups $S(M)$, of which the Ising model is the special case $M=2$, eq. (9) yields, in general, two small-field phase transition points:

$$\omega_f = \frac{m-1}{m+M-1}, \quad \omega_{af} = \frac{m+1}{m-M+1} . \tag{1}$$

The antiferromagnetic one, $\omega_{af}$, is absent for $m \leq M-1$, indicating that this may also be the case on real lattices with low coordination number. Note that this is not the case for the Ising model: $M=2$ implies $\omega_f = \omega_{af}^{-1}$.

Since every vector orthogonal to $\underline{\mu}_o$ is an eigenvector with eigenvalue $1-\omega=\lambda_1$, no symmetries are left in the low-temperature phases: these are (complete) symmetry-breaking phase transitions.

### 9.2.2. Absence of high-field phase transitions for the Ising model.

For the Ising model, the recursion relations (1.1) and (1.2) can be rewritten as a recursion relation for the ratio

$$x_n = \rho_n(1)/\rho_n(2) . \tag{2}$$

This has the form

$$x_n = \left(\frac{x_{n-1}+\omega}{\omega x_{n-1}+1}\right)^m, \quad x_o=A(1)/A(2) . \tag{3}$$

Since the partition function $Z_n$ can be written as

$$Z_n = Z_n(2)\ (1+x_n), \tag{4}$$

this can, for general complex values of $x_o$, be zero only if $x_n=-1$ holds. Now the nonlinear mapping (3) keeps the unit circle in the complex x-plane invariant: writing $x_n=y_n+iz_n$, one finds

$$|x_n|^2=y_n^2+z_n^2=\left(\frac{|x_{n-1}|^2+2\omega y_{n-1}+\omega^2}{\omega^2|x_{n-1}|^2+2\omega y_{n-1}+1}\right)^m, \tag{5}$$

which has the fixed point $|x|=1$. Therfore, $Z_n$ can have zeroes in the complex $x_o$-plane only for $|x_o|=1$. By the Yang-Lee theory of phase transitions ([1]), this implies, that a phase transition can only occur for $x_o=1$, i.e., for zero field. This shows, that high-field phase transitions are excluded for the Ising model. This result also holds in case $\omega$ and/or $m$ are functions of the generation number $n$; further, the values of $\omega$ are not restricted to the ferromagnetic region $\omega<1$, so that the result holds for antiferromagnetic interactions as well. In this way, the standard Yang-Lee theorem on the zeroes of the partition function ([2]) for ferromagnetic Ising systems is extended here to the case of arbitrarily interacting Ising spins on (possibly inhomogeneous) Cayley-like branches.

### 9.2.3. Explicit solution of the recursion relation for Ising systems.

The recursion relation (3) is easily seen to converge to the fixed point $x_o=1$ for values of $\omega$ satisfying $\omega_f<\omega<\omega_{af}$. For $\omega\leqq\omega_f$, $\{x_n\}$ converges to the fixed point $x_1$ given by

$$x_1 = \left(\frac{x_1+\omega}{\omega x_1+1}\right)^m, \tag{6}$$

which may be calculated explicitly for low values of $m$. For $\omega\geqq\omega_{af}$, $\{x_n\}$ converges to a two-point limit cycle; one has

$$\lim_{n\to\infty} x_{2n} = x_1, \quad \lim_{n\to\infty} x_{2n+1} = x_1^{-1}. \tag{7}$$

In Fig. 1, the average top spin magnetization $\langle\sigma\rangle$, which is given by

$$\langle\sigma\rangle = \frac{x_1-1}{x_1+1}, \text{ for } \omega\leqq\omega_f, \quad \langle\sigma\rangle=0 \text{ for } \omega_f\leqq\omega\leqq 1, \tag{8}$$

(in the thermodynamic limit), is plotted as a function of $\omega$. The behaviour of $\langle\sigma\rangle$ for $\omega\uparrow\omega_f$ is, from eq. (6), always of the form

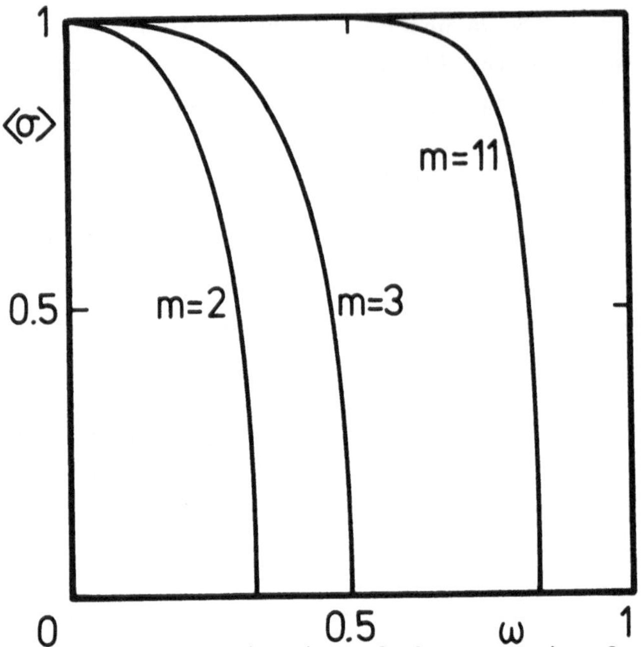


Fig. 1. The average magnetization of the top spin of a Cayley branch as a function of the interaction strength for several values of the branching ratio; the curves for $m=2$ and for $m=3$ have been calculated from eq. (6); the curve for $m=11$ has been obtained numerically by iterating eq. (3).

$$\langle\sigma\rangle \sim \sqrt{(\omega_f-\omega)}, \tag{9}$$

which is the classical or mean field exponent. This is a general feature of Cayley branches with boundary field.

Finally, the free energy of the Ising model on a Bethe lattice with coordination number $m+1$ follows from eq. (8.4.10) as

$$\begin{aligned} \Gamma &= m\ \ln(1+\omega x_1) - \tfrac{1}{2}(m-1)\ \ln(1+2\omega x_1+x_1{}^2), \quad \text{for} \quad \omega \leqq \omega_f, \\ \Gamma &= \tfrac{1}{2}(m+1)\ \ln(1+\omega) - \tfrac{1}{2}(m-1)\ \ln 2, \quad \text{for} \quad 1 \geqq \omega \geqq \omega_f. \end{aligned} \tag{10}$$

For $m=3$, for example, one finds the explicit expressions

$$\begin{aligned} \Gamma &= \ln\ (1+\omega)^2/2 \quad \text{for} \quad \tfrac{1}{2} \leqq \omega \leqq 1, \\ \Gamma &= \ln\ (1-\omega^2)^2/(1-2\omega^2) \quad \text{for} \quad 0 \leqq \omega \leqq \tfrac{1}{2}. \end{aligned} \tag{11}$$

These functions have the same values and first derivatives at $\omega=\omega_f=\frac{1}{2}$; the second derivatives, however, are not equal here, so that the specific heat has a discontinuity at the phase transition. This is shown in Fig. 2 below.

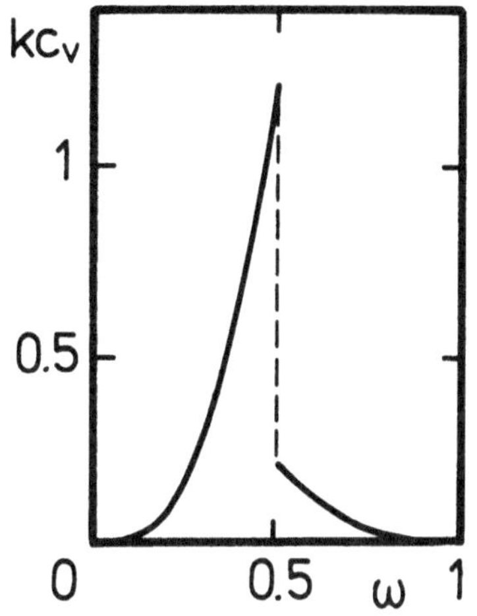


Fig. 2. The discontinuity in the specific heat per spin, $kc_v$, for the Ising model on a Bethe lattice with coordination number 4.

## 9.2.4. High-field phase transitions for the Potts model.

The recursion relations (1.1) and (1.2) for the Potts model with $M>2$ can be expressed in terms of the ratios

$$x_n(i) = \rho_n(i)/\rho_n(1), \quad i=2,3,..,M, \tag{12}$$

as

$$x_n(i) = \left(\frac{x_{n-1}(i)+\omega+\omega\Sigma_{k\neq i}x_{n-1}(k)}{1+\omega x_{n-1}(i)+\omega\Sigma_{k\neq i}x_{n-1}(k)}\right)^m, \quad x_o(i)=A(i)/A(1). \tag{13}$$

This is similar to eq. (3). It is useful to define new parameters $z(i)$ by the expression taken to the m-th power in eq. (13). This gives:

$$z_{n-1}(i)^m = \frac{\{1+(M-1)\omega\}z_n(i)-\omega(1+\Sigma_z^{(n)})}{1+(M-2)\omega-\omega\Sigma_z^{(n)}}, \quad \Sigma_z^{(n)} = \sum_{j=2}^{M} z_n(j). \tag{14}$$

For $n\to\infty$, the fixed point in the ferromagnetic region is given by parameters $z(i)$, which are solutions of the equations

$$z(i)^m-1 = \{z(i)-1\}\{1+(M-1)\omega\}\{1+\omega(M-2)-\omega\Sigma_z\}^{-1} \tag{15}$$

Since $\Sigma_z$ does not depend on $i$, this has either the solution 1 or else another solution $\neq 1$, but independent of $i$; the fixed point then has the structure: $z(i)=1$ holds for $M-r-1$ values of $i$; the other $r$ values are all equal to a solution $\neq 1$ of

$$\frac{z^m-1}{z-1} = \{1+(M-1)\omega\}\{1+(r-1)\omega-r\omega z\}^{-1}, \tag{16}$$

which can also be written as a polynomial in $z$:

$$\omega r z^m-(1-\omega)(z^{m-1}+z^{m-2}+\ldots+z)+(M-r)\omega = 0. \tag{17}$$

For $m=2$, for instance, eq. (17) has two real positive solutions for all values of $M$ and $r$ as soon as $\omega<\omega_c$ holds; at $\omega_c$, these solutions coincide:

$$\omega_c = \{1+2\sqrt{r(M-r)}\}^{-1}. \tag{18}$$

One has $\omega_c \geqq \omega_f$, where equality holds iff $r=M/2$.

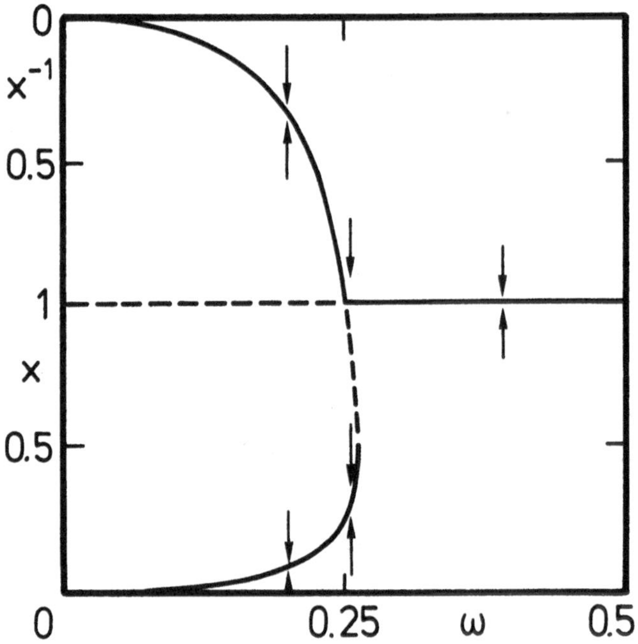


Fig. 3. The high-field and small-field phase transitions for the 3-state Potts model on a Cayley branch with $m=2$. Solid curves are stable fixed points, broken curves are unstable ones; these latter are also the phase boundaries. The arrows indicate, in which direction the recursion goes.

The different solutions of eq. (17) are plotted in Figs. 3 and 4 for the case $m=2$ for the 3- and 4-state Potts models, respectively. The unstable fixed points (broken lines) are phase transition curves, since they separate parts of the phase diagram, in which the recursion approaches different stable fixed points (solid curves). These stable fixed points are selected by the type of boundary field also and not only by the strength: although the cases $r=1$ and $r=2$ are equivalent for the 3-state Potts model, the cases $r=1$ (or 3) and $r=2$ are <u>not</u> equivalent for the 4-state case. In general, only the cases $r$ and $M-r$ are equivalent due to the normalization of the field distribution function. It is clear from Figs. 3 and 4, that the phase transitions are of two types: the broken lines at the value 1 are the small-field phase transitions, those not at 1 are the high-field ones.

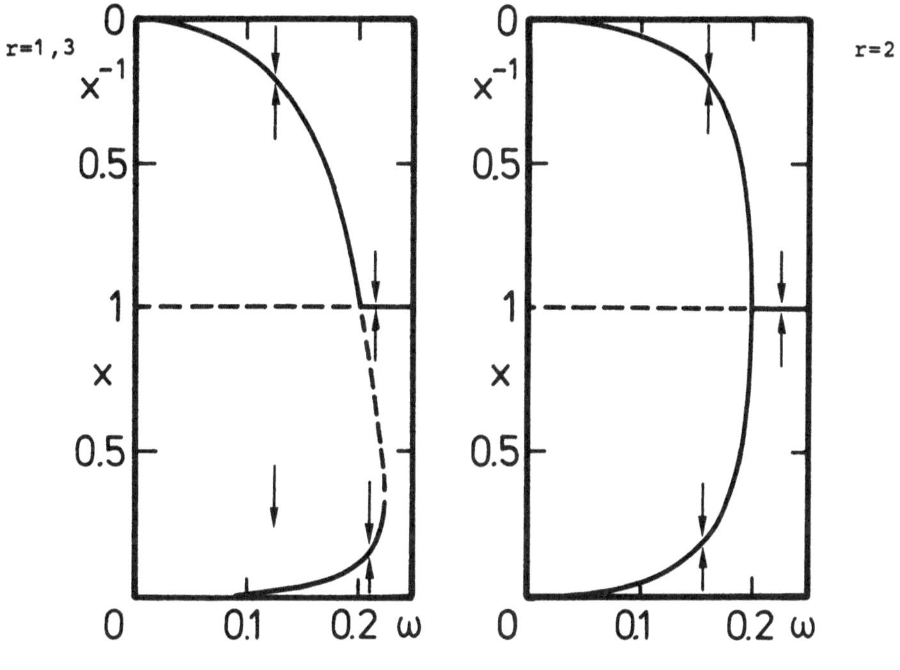


Fig. 4. The phase transitions for the 4-state Potts model on a Cayley branch with $m=2$. There are now two nonequivalent types of boundary field, one with $r=1$ or 3, one with $r=2$. The different lines have the same meaning as in Fig. 3.

## 9.3. Phase diagrams for models with permissible groups with two-graph maximal interactions.

There are, by Theorem 3.1.1, two types of models as mentioned in the title of this section: the imprimitive ones are the ones with a wreath product $S(M_1)\sim S(M_2)$ as symmetry group, whereas the primitive ones on $M\leq 10$ letters are, from Section 6.1, $D(5)$, $G(G_9)$ and $G(G_{10})$.

These will all be studied in what follows.

a) $S(M_1) \sim S(M_2)$. The MI of this group consists of two graphs, one consisting of $M_2$ complete $K(M_1)$-graphs, the other the complement of this; the first of these graphs has Boltzmann factor $\omega_1$, the second one $\omega_2$. It is not difficult to see, that the eigenvectors of the resulting $\underline{\underline{\Omega}}$-matrix are direct product of the eigenvectors for Potts models. Let $\underline{\mu}_o(M)$ be the eigenvector with eigenvalue $\lambda_o$ for the M-state Potts model and let the $M-1$ vectors $\underline{\mu}_k(M)$ be the corresponding ones for the eigenvalue $\lambda_1$. Then one has: the direct product $\underline{\mu}_o(M_1) \otimes \underline{\mu}_o(M_2)$ is the eigenvctor of $\underline{\underline{\Omega}}$ with eigenvalue

$$\lambda_o = 1+(M_1-1)\omega_1+M_1(M_2-1)\omega_2, \qquad \text{(nondegenerate)}; \tag{1}$$

the direct products $\underline{\mu}_o(M_1) \otimes \underline{\mu}_k(M_2)$ are eigenvectors with eigenvalue

$$\lambda_1 = 1+(M_1-1)\omega_1-M_1\omega_2, \qquad \text{( $(M_2-1)$-fold degenerate )}; \tag{2}$$

these eigenvectors are all invariant with respect to permutations of the form

$$\begin{pmatrix} D(g_1) & & & \\ & D(g_2) & & \\ & & \ddots & \\ & & & D(g_{M_2}) \end{pmatrix}, \qquad g_i \varepsilon S(M_1) \quad \text{for} \quad i=1,..,M_2 , \tag{3}$$

so that the group $G_1$ of preserved symmetries in the phase, in which these vectors propagate, is given as

$$G_1 \simeq S(M_1) \otimes S(M_1) \otimes \ldots \otimes S(M_1), \quad M_2 \text{ direct factors.} \tag{4}$$

All vectors orthogonal to all of the above, i.e., $\underline{\mu}_k(M_1) \otimes \underline{\mu}_o(M_2)$ and $\underline{\mu}_k(M_1) \otimes \underline{\mu}_{k'}(M_2)$, have eigenvalue

$$\lambda_2 = 1-\omega_1, \qquad \text{( $M_2(M_1-1)$-fold degenerate )}; \tag{5}$$

In this subspace, there is no symmetry left over.

In view of the general theory of Section 1, the phase diagram for the small-field case for the present models contains the following phase transition lines:

(i) The lines given by

$$m\lambda_2/\lambda_o=1 \quad \text{(fm)} \quad \text{and} \quad m\lambda_2/\lambda_o=-1 \quad \text{(afm)} \tag{6}$$

are always (completely) symmetry-breaking phase transition lines, since there are no symmetries left over in the low-temperature phases. These phases are denoted by 2 for the ferromagnetic (fm) case and by $\bar{2}$ for the antiferromagnetic (afm) case.

(ii) The lines given by

$$m\lambda_1/\lambda_o=1 \quad \text{(fm, phase } 1\text{)}, \quad \text{and} \quad m\lambda_1/\lambda_o=-1 \quad \text{(afm, phase } \bar{1}\text{)}, \tag{7}$$

are symmetry-breaking phase transition lines as long as they do not extend into the phases 2 or $\bar{2}$. If this is indeed not the case, then the phases 1 and $\bar{1}$ have the $\mathcal{G}_1$-symmetry of eq. (4).

Explicit expressions for these phase trnsition lines are as follows:

$$\begin{aligned}
&\text{phase } 1 \text{ boundary: } \omega_2=\frac{m-1}{M_1(m+M_2-1)}\{1+(M_1-1)\omega_1\},\\
&\text{phase } 2 \text{ boundary: } \frac{m+M_1-1}{m-1}\omega_1+\frac{M_1(M_2-1)}{m-1}\omega_2=1,\\
&\text{phase } \bar{1} \text{ boundary: } \omega_2=\frac{m+1}{M_1(m+1-M_2)}\{1+(M_1-1)\omega_1\}, \text{ for } m>M_2-1,\\
&\text{phase } \bar{2} \text{ boundary: } \omega_2=\frac{m+1}{M_1(M_2-1)}\left(\frac{m+1-M_1}{m+1}\omega_1-1\right), \text{ for } m\ M_1-1.
\end{aligned} \tag{8}$$

Two typical examples are shown in Fig. 1. The points $P(M_1)$, $P(M_2)$ and $P(M_1M_2)$ are the phase transition points of the Potts models, to which the present model reduces for $\omega_2=0$, $\omega_1=1$ and $\omega_1=\omega_2$, respectively, see also Section 7.6. This information strongly suggests the correctness of the conjectures concerning the phase diagrams of the $\mathcal{S}(M)\sim\mathcal{S}(M)$ models on the square lattice in Section 7.6, if the dimensionality-changing phase transitions are disregarded.

b) $\mathcal{D}(5)$. The eigenvalues for this model have already been given in eqs. (7.6.33,34); all eigenvectors of $\lambda_1$ and $\lambda_2$ break the full $\mathcal{D}(5)$-symmetry. Eqs. (1.9) yield the type of phase diagram shown in Fig. 2 for the case $m=7$.

c) $\mathcal{G}(G_9)$. The eigenvalues have been given as eqs. (7.6.37,38). Again, the general eigenvectors do not have any remaining symmetries, so that the phase diagram looks like the one in Fig. 3 for the case $m=5$. The

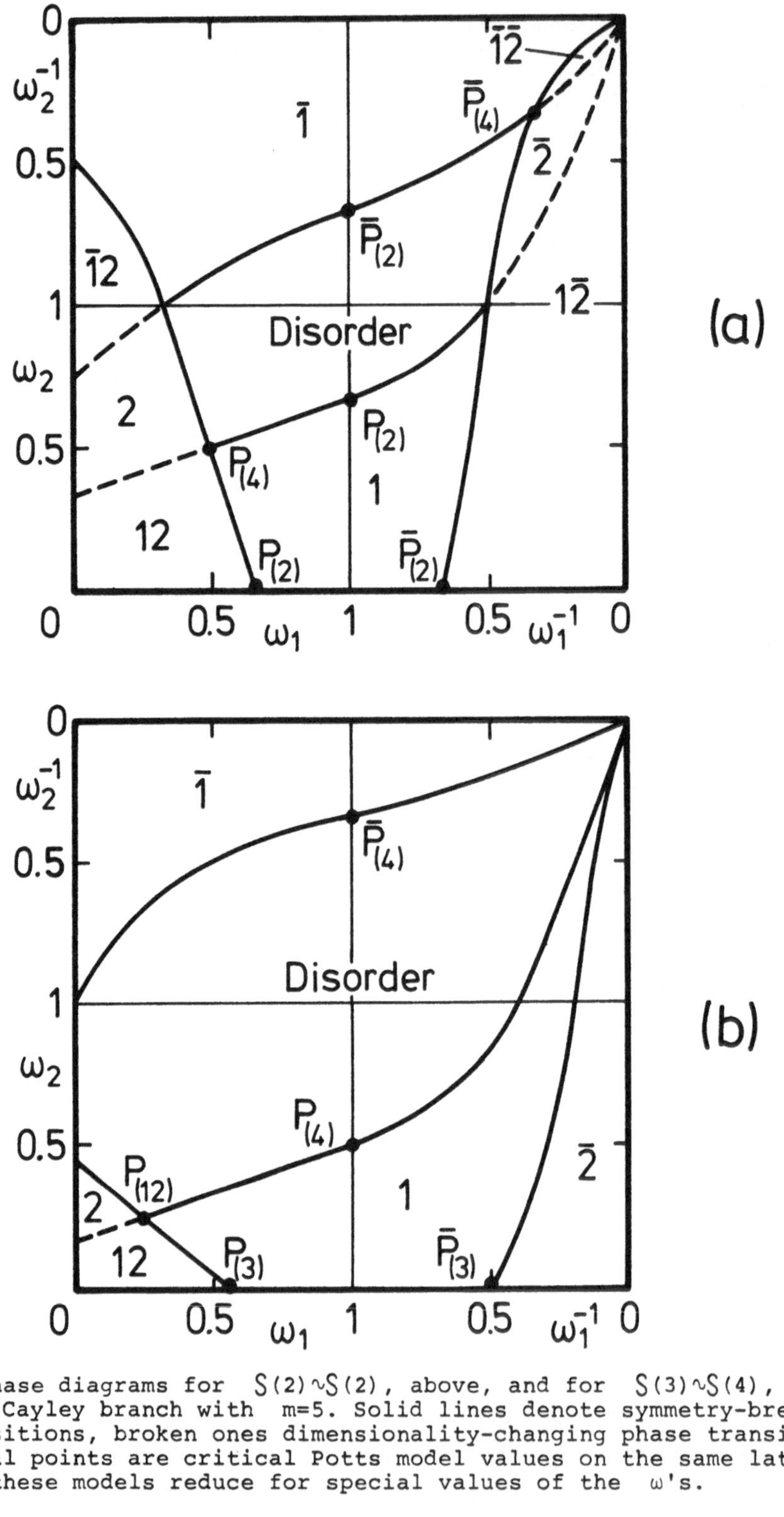


Fig. 1. Phase diagrams for $S(2)\sim S(2)$, above, and for $S(3)\sim S(4)$, below, both on a Cayley branch with $m=5$. Solid lines denote symmetry-breaking phase trnsitions, broken ones dimensionality-changing phase transitions. The special points are critical Potts model values on the same lattice, to which these models reduce for special values of the $\omega$'s.

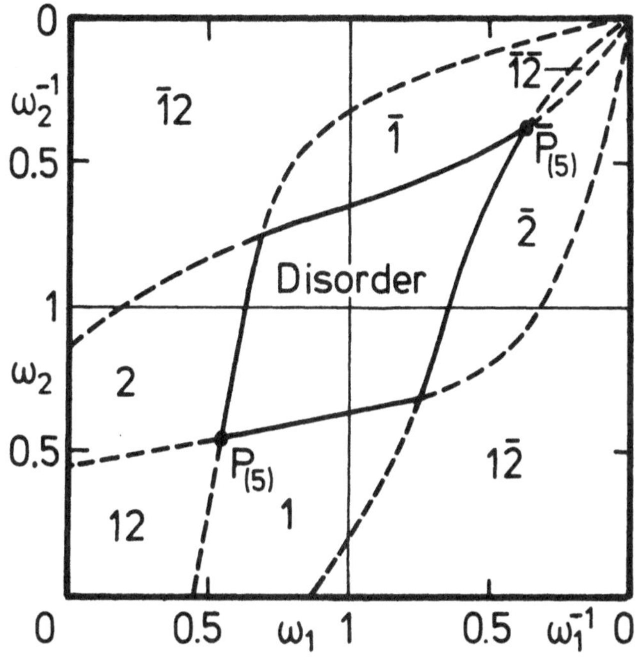


Fig. 2. Phase diagram for the D(5) model on a Cayley branch with m=7. Phases and phase transition lines as in the text and Fig. 1.

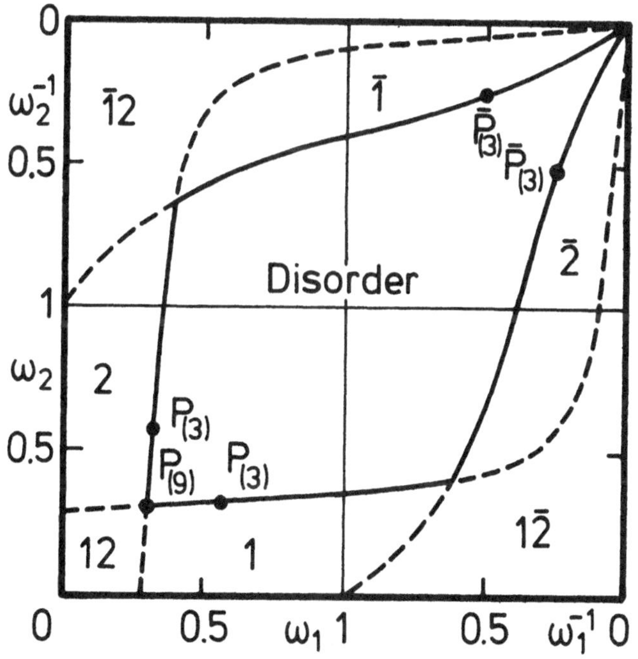


Fig. 3. Phase diagram of the G($G_9$) model on a Cayley branch with m=5. Phase transition lines as in Fig. 1, special points as in the text.

points $P_1(3)$ and $P_2(3)$ are phase transition points for the 3-state Potts model, to which the present model reduces for $\omega_1=\omega_2^2$ and for $\omega_2=\omega_1^2$.

The phase diagrams of Figs. 2 and 3 suggest, that for these models on a square lattice, the exact ferromagnetic transition line is the self-dual line of Figs. 7.6.1(b) and (d), respectively. This seems again to show, that the dimensionality-changing phase transitions are not real phase transitions at all, at least not for real lattices.

d) $\mathcal{G}(G_{10})$. Although this group is completely permissible, it does not contain a regular, Abelian subgroup (see Section 6.1), so that the eigenvalues and eigenvectors of $\underline{\underline{\Omega}}$ cannot be obtained by the method of Chapter 7. Even though $\underline{\underline{\Omega}}$ is a 10×10 matrix, it is rather easily diagonalized by observing, that it consists of four blocks of cyclic 5×5 matrices, as follows from Fig. 5.1.3 for the Petersen graph:

$$\underline{\underline{\Omega}} = \begin{pmatrix} \underline{\underline{A}} & \underline{\underline{B}} \\ \underline{\underline{B}} & \underline{\underline{C}} \end{pmatrix}, \quad \underline{\underline{A}} = \begin{pmatrix} 1 & \omega_1 & \omega_2 & \omega_2 & \omega_1 \\ & & \text{cyclic}\rightarrow & & \end{pmatrix}, \quad \underline{\underline{B}} = \begin{pmatrix} \omega_1 & \omega_2 & \omega_2 & \omega_2 & \omega_2 \\ & & \text{cyclic}\rightarrow & & \end{pmatrix},$$

$$C = \begin{pmatrix} 1 & \omega_2 & \omega_1 & \omega_1 & \omega_2 \\ & & \text{cyclic}\rightarrow & & \end{pmatrix}. \tag{9}$$

Now if one sets the first five entries of an eigenvector equal to a $\underline{\mu}_k$ of $\mathbb{D}(5)$ and the last five equal to $\alpha\underline{\mu}_k$, then the problem reduces to a 2×2 one:

$$\begin{pmatrix} \lambda_k(\underline{\underline{A}}) & \lambda_k(\underline{\underline{B}}) \\ \lambda_k(\underline{\underline{B}}) & \lambda_k(\underline{\underline{C}}) \end{pmatrix} \begin{pmatrix} 1 \\ \alpha \end{pmatrix} = \lambda \begin{pmatrix} 1 \\ \alpha \end{pmatrix}. \tag{10}$$

Using the explicit expressions for the eigenvalues of $\underline{\underline{A}}$ (given by eqs. (7.6.33,34)), of $\underline{\underline{B}}$ (these are $\omega_1+4\omega_2$ for $k=0$, $\omega_1-\omega_2$ otherwise), and of $\underline{\underline{C}}$ (these are obtained from those of $\underline{\underline{A}}$ by exchanging $\omega_1$ and $\omega_2$), one finds

$$\begin{aligned} \lambda_o &= 1+3\omega_1+6\omega_2, && \text{(nondegenerate)}, \\ \lambda_1 &= 1+\omega_1-2\omega_2, && \text{(five-fold degenerate)}, \\ \lambda_2 &= 1-2\omega_1+\omega_2, && \text{(four-fold degenerate)}. \end{aligned} \tag{11}$$

Also, there are no nontrivial symmetries in the eigenspaces of $\lambda_1$ and

$\lambda_2$. The phase diagram obtained from eq. (1.9) and eqs. (11) above, is very similar to the phase diagrams of the other two models with primitive symmetry groups treated before, except for a slight asymmetry with respect to the $\omega_1 \leftrightarrow \omega_2$ interchange, see Fig. 4.

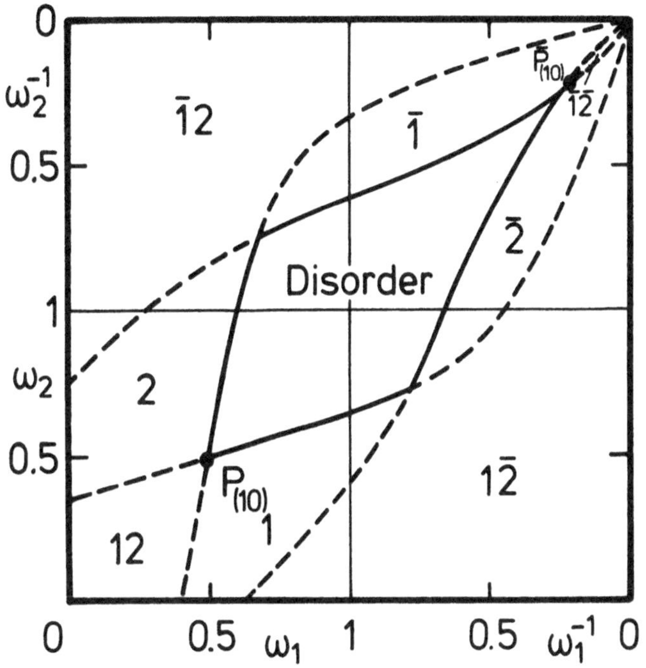


Fig.4. Phase diagram of the $G(G_{10})$ model on a Cayley branch with m=11. Phase transition lines are as in the previous figures of this section.

## 9.4. Phase diagrams for models with permissible groups with 3-graph MI's.

The models mentioned in the title of this section have 27 different phases. Therefore, only symmetry-breaking phase transitions of the ferromagnetic type will be considered here. Also, the phase diagrams will be restricted to the unit cube $0 \leq \omega_i \leq 1$, for i=1,2,3. The same models as the ones studied in Section 7.6 on duality will also be considered here: $S(2)\otimes S(2)$, $S(3)\otimes S(3)$, $S(2)\otimes S(3)$, $S(2)\sim S(2)\sim S(2)$ and $D(7)$. In addition, the icosahedral group $G(I)$ is treated as an example of a primitive group with a 3-graph MI of type (f) of Theorem 3.1.2.

a) $S(2)\otimes S(2)$. The phase transition planes are given by eq. (1.9) and by the eigenvalues of eqs. (7.6.41,42) for $M_1=M_2=2$. Each simple phase, i.e., each phase in which only one eigenvalue has a propagating eigenspace, which in the present case consists of one eigenvector only, has

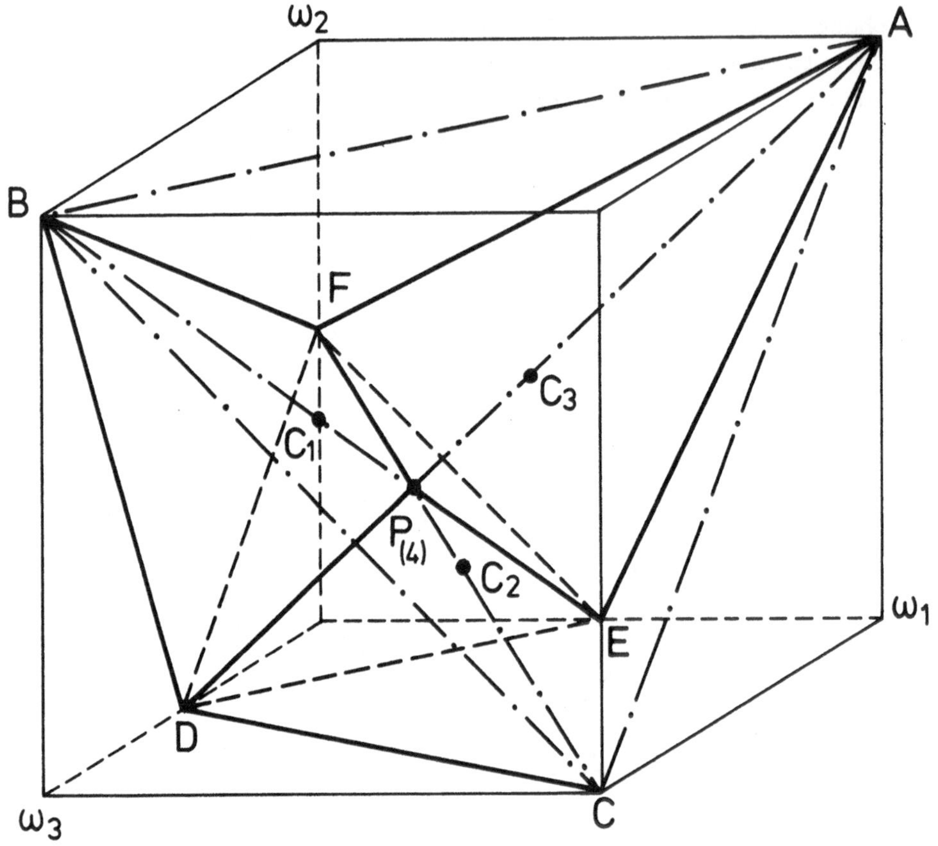


Fig. 1. The phase diagram of the Ashkin-Teller model on a Cayley branch with m=3. The different phase transition planes are described in the text.

(one of the three possible) $S(2)$ symmetry left over. All nonsimple phases have no nontrivial symmetries. This yields the phase diagram of Fig. 1 for the case m=3. There are really, as conjectured in Section 7.6, two symmetry-breaking phase transition planes: (i) the first one, which separates the disordered region around (1,1,1) from the phases with $S(2)$-symmetries, consists of the three triangles ABP(4), ACP(4) and BCP(4); these boundaries are marked —— . —— in Fig. 1. (ii) The second plane separates the $S(2)$-symmetric phases from the completely ordered phase around (0,0,0): triangles AFP(4), AEP(4), BDP(4), BFP(4), CDP(4) and CEP(4), solid and dot-dashed lines from the first plane. These planes touch along the lines AP(4), BP(4) and CP(4), on which also the Ising "crossover" points $C_3$, $C_1$ and $C_2$ are located, see also Section 7.6. Inside the region bounded by the triangles DEP(4), DFP(4)

and EFP(4) (extra broken lines), all three eigenvectors propagate. It is concluded, that, for the Ashkin-Teller model on a square lattice (and, presumably, also for the corresponding gauge model in four dimensions), there are indeed two symmetry-breaking phase transition planes, which touch exactly along the three self-dual lines in Fig. 7.6.2. These planes are dual to each other with respect to all duality transformations.

b) $S(3)\otimes S(3)$. Here the phase transition planes are obtained from eqs. (7.6.41,42) with $M_1=M_2=3$ and from eqs. (1.9). Now, only the simple phases 1 and 2 still have a $S(3)$-symmetry left over; in phase 3, all symmetry is completely broken already. The phase diagram for m=3 is shown in Fig. 2. The planes, at which the symmetry is reduced to $S(3)$ are $P(9)B_iE_iF$ for i=1,2; these have the line P(9)F in common, on which also the point C, where the model reduces to a product of two

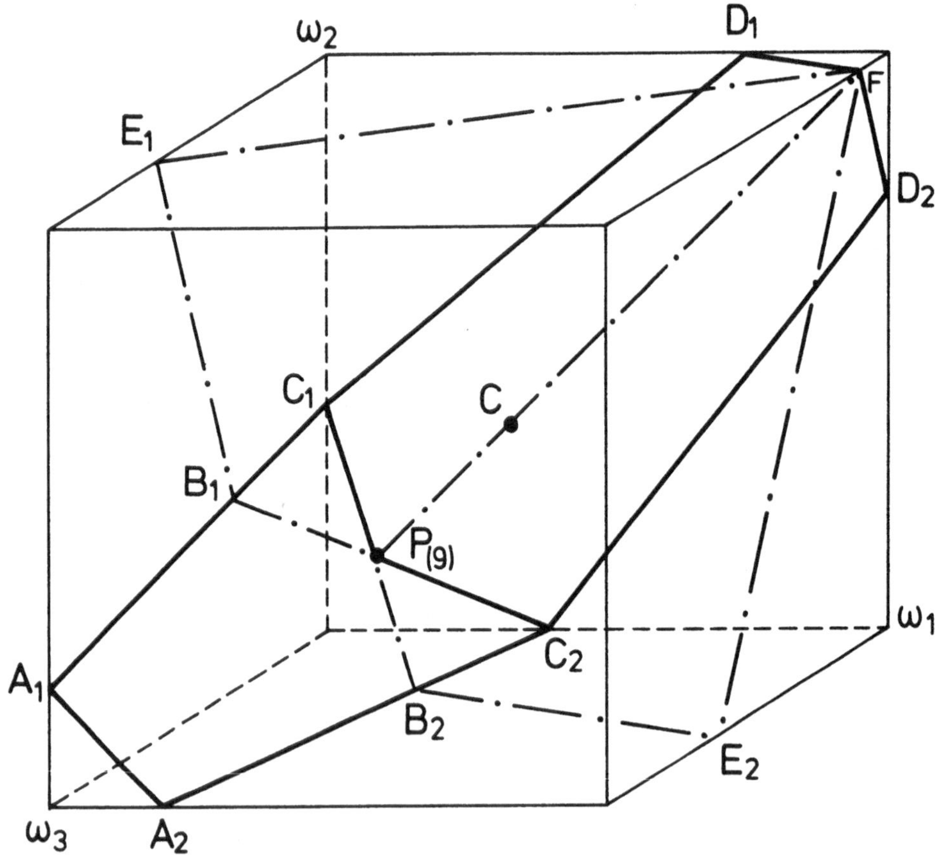


Fig. 2. The phase diagram of the $S(3)\otimes S(3)$ model on a Cayley branch with m=3. The different phase trnsition planes are described in the text.

critical Potts models, is located; at the points $C_1$ and $C_2$, the model reduces to one critical Potts model only. Complete symmetry breaking takes place at the planes $A_1C_1P(9)C_2A_2$ and $P(9)C_iD_iF$, for $i=1,2$. Such a picture is completely consistent with the duality arguments put forth in Section 7.6, see Fig. 7.6.3. It is, therefore, expected, that the phase diagram looks similar for the corresponding models on hypercubic lattices: part of the self-dual plane is a phase transition plane with complete symmetry breaking; further, there are two dual planes, at which the symmetry is broken successively; these two planes have the self-dual line $P(9)C$ in common.

c) $\mathcal{S}(2)\otimes\mathcal{S}(3)=\mathcal{D}(6)$. Here a similar phase diagram is expected, except for the absence of the symmetry $\omega_1\leftrightarrow\omega_2$ in the present case. This is indeed found, see Fig. 3, which has been derived in the same way as the previous two figures. In the point C, the model reduces to the product

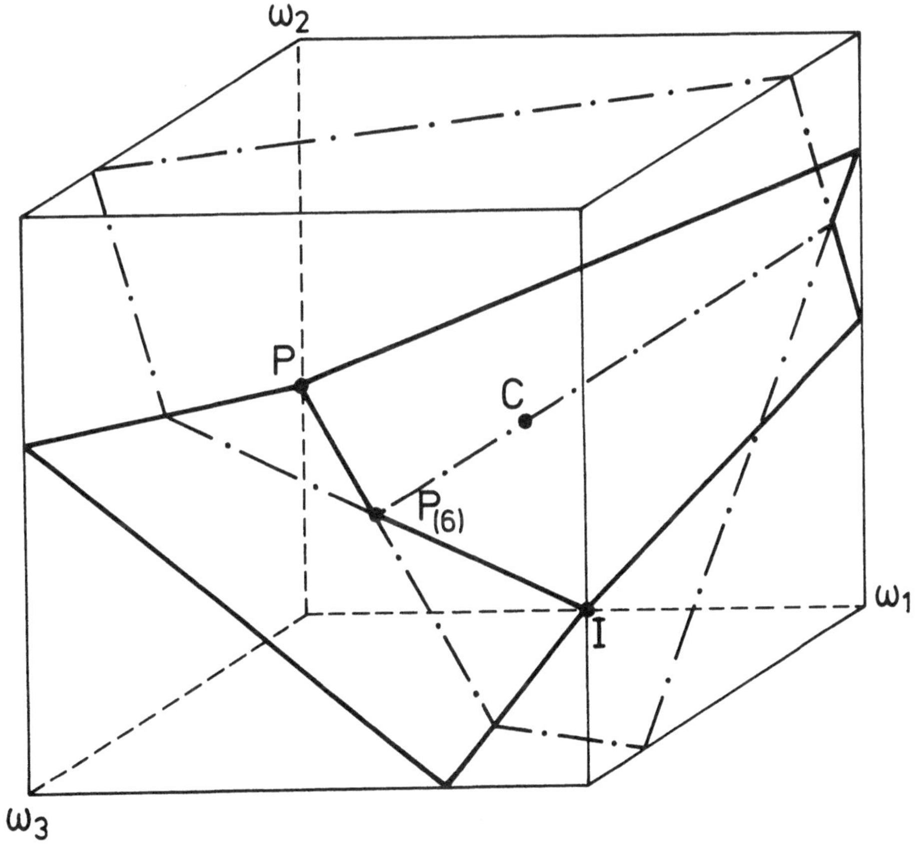


Fig. 3. The phase diagram of the $\mathcal{S}(2)\otimes\mathcal{S}(3)$ model on a Cayley branch with $m=3$. The different phase transition planes are similar to the ones in Fig. 2, see also the text.

of a critical Ising and of a critical 3-state Potts model. Further, P is a 3-state Potts critical point and I an Ising critical point, so that the Potts symmetry remains upon passing the upper plane, whereas the Ising symmetry remains upon passing the lower plane.

d) $S(2)\sim S(2)\sim S(2)$. For this model, the symmetries of the different phases are highly nontrivial. The unique eigenvector corresponding to the eigenvalue $\lambda_1$, eq. (7.6.52), still has a $S(2)\sim S(2)$ symmetry, whereas both eigenvectors corresponding to $\lambda_2$, eq. (7.6.53), have $S(2)$ symmetry; the eigenspace of $\lambda_3$, eq. (7.6.54), does not have any symmetry left over, so that one has as nontrivial symmetries:

$$G_1=S(2)\sim S(2); \quad G_2=S(2); \quad G_1\cap G_2=S(2).$$

The resulting phase diagram is shown in Fig. 4. Here the special points

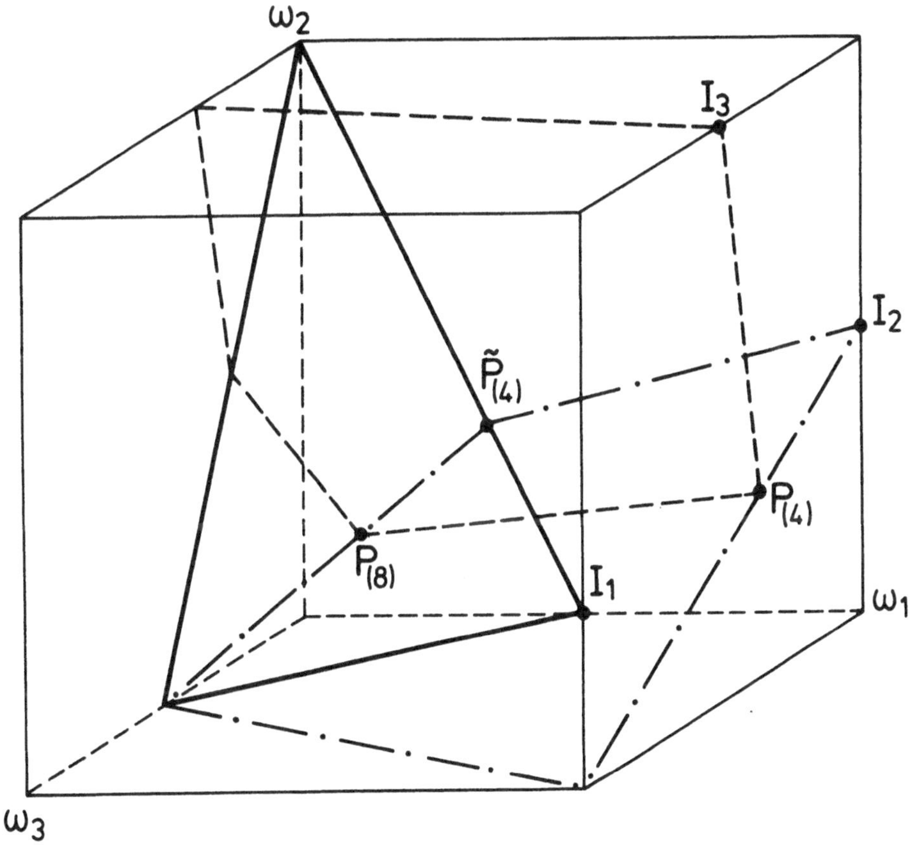


Fig. 4. Phase diagram of the $S(2)\sim S(2)\sim S(2)$ model on a Cayley branch with m=3. The phase transition planes and special points are explained in the text.

are: P(4) and $\tilde{P}(4)$: critical 4-state Potts points; $I_1$, $I_2$ and $I_3$: critical Ising model points. The 1-phase transition plane is denoted by broken lines, the 2-phase transition one by dash-dotted lines and the 3-phase transition plane by solid ones. The self-dual line found in Section 7.6 corresponds to the line $P(8)I_2$ here; the duality transformation did not give enough information to infer the presence of three phase transition planes for the present model.

e) $\mathbb{D}(7)$. Since for this model, all symmetry is completely broken for all phases, which are not completely disordered, the phase diagram obtained from eqs. (7.6.57,58) and (1.9) is relatively simple, see Fig. 5, with a unique symmetry-breaking phase transition plane. If dimensionality-changing phase transitions are unobservable, this plane ought to correspond to the hyperplane H in Fig. 7.6.6.

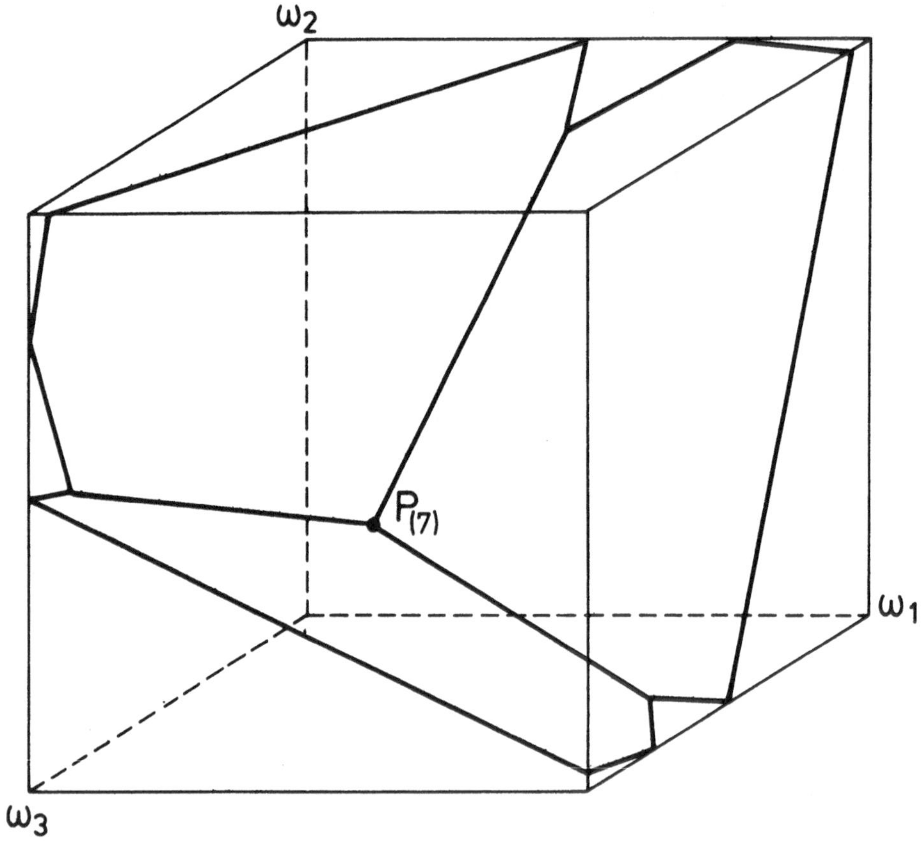


Fig. 5. Phase diagram of the $\mathbb{D}(7)$ model on a Cayley branch with m=3. The unique complete symmetry breaking phase transition plane consists of three pieces, meeting at the critical Potts model point P(7).

f) $\mathcal{G}(I)$. This group, the automorphism group of the icosahedron, has been derived in Section 5.2 as the smallest group with an MI of type (f) of Theorem 3.1.2. This MI is shown as Fig. 5.2.1; from this, the interaction matrix $\underline{\underline{\Omega}}$ is seen to be given by

$$\underline{\underline{\Omega}} = \begin{pmatrix} \underline{\underline{A}} & \underline{\underline{B}} \\ \underline{\underline{B}} & \underline{\underline{A}} \end{pmatrix}, \ A= \begin{pmatrix} 1 & \omega_1 & \omega_1 & \omega_2 & \omega_2 & \omega_2 \\ \omega_1 & 1 & \omega_1 & \omega_2 & \omega_1 & \omega_1 \\ \omega_1 & \omega_1 & 1 & \omega_1 & \omega_1 & \omega_2 \\ \omega_2 & \omega_1 & \omega_1 & 1 & \omega_1 & \omega_2 \\ \omega_2 & \omega_1 & \omega_1 & \omega_1 & 1 & \omega_1 \\ \omega_2 & \omega_1 & \omega_2 & \omega_2 & \omega_1 & 1 \end{pmatrix}, \tag{1}$$

and $\underline{\underline{B}}$ is obtained from $\underline{\underline{A}}$ by substituting $\omega_3$ for 1 and by exchanging $\omega_1$ and $\omega_2$. Since the third graph of the MI is Ising-like. the eigenvectors of $\underline{\underline{\Omega}}$ must be of the forms

$$\begin{pmatrix} \underline{a} \\ +\underline{a} \end{pmatrix}, \ \begin{pmatrix} \underline{a} \\ -\underline{a} \end{pmatrix}, \quad \text{with } \underline{a} \text{ a six-dimensional vector.} \tag{2}$$

Taking the plus sign, the eigenvalue equation becomes:

$$(\underline{\underline{A}}+\underline{\underline{B}})\,\underline{a} = \lambda \underline{a}, \tag{3}$$

the solutions of which are:

$$\lambda_o = 1+5\omega_1+5\omega_2+\omega_3, \quad \text{(nondegenerate, full symmetry)}, \tag{4}$$

$$\lambda_1 = 1-\omega_1-\omega_2+\omega_3, \quad \text{(five-fold degenerate, eigenspace symmetry } \mathcal{S}(2) \text{ left over).} \tag{5}$$

The minus sign in eq. (2) gives

$$(\underline{\underline{A}}-\underline{\underline{B}})\,\underline{a} = \underline{a}, \tag{6}$$

where $\underline{\underline{A}}-\underline{\underline{B}}$ follows from eq. (1) ff. as

$$\underline{\underline{A}}-\underline{\underline{B}} = (1-\omega_3)\underline{\underline{I}}_6+(\omega_1-\omega_2)\underline{\underline{D}}, \tag{7}$$

where $\underline{\underline{I}}_6$ is the $6\times6$ unit matrix and $\underline{\underline{D}}$ is obtained from $\underline{\underline{A}}$ by the substitutions: $1\to0$, $\omega_1\to+1$ and $\omega_2\to-1$; this matrix then is such, that

$$\underline{\underline{D}}^2 = 5\underline{\underline{I}}_6 \tag{8}$$

holds. From this, the eigenvalues and eigenspaces easily follow:

$$\begin{aligned} \lambda_3 &= 1+\sqrt{5}(\omega_1-\omega_2)-\omega_3, \\ \lambda_4 &= 1-\sqrt{5}(\omega_1-\omega_2)-\omega_3, \end{aligned} \quad \text{(both three-fold degenerate, no eigenspace symmetries).} \tag{9}$$

Eqs. (4), (5) and (9) give, for a Cayley branch with $m=3$, the phase diagram shown in Fig. 6. The points I, P(6) and P(12) have their usual meaning; these all lie on the plane $\omega_1=\omega_2$, on which the present model reduces to $\mathcal{S}(2)\sim\mathcal{S}(6)$.

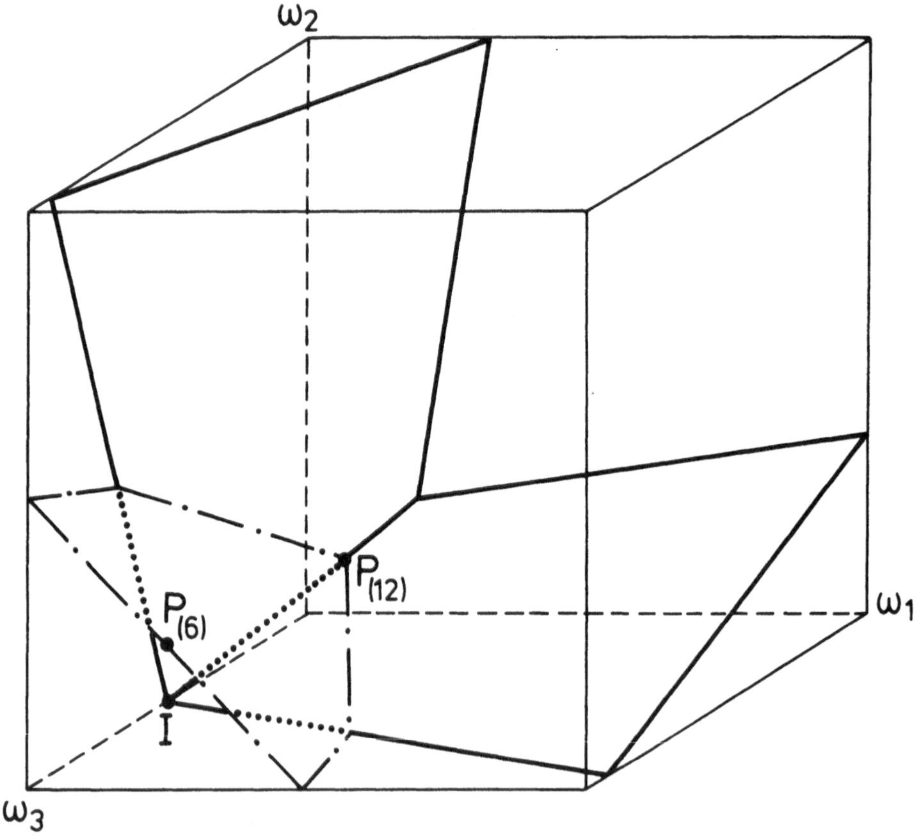


Fig. 6. Phase diagram of the $\mathcal{G}(I)$ model on a Cayley branch with $m=3$. Solid lines delineate the complete symmetry-breaking phase transition planes, dash-dotted ones the 1-phase transition plane.

## 9.5. Phase diagrams for models with permissible groups, which do not have P-algebras.

All groups treated in the preceding sections have P-algebras, even if they are not completely permissible, e.g., $\mathcal{G}(G_{10})$. For such groups, all phase transition hyperplanes are flat. In this section, two examples of groups without a P-algebra will be considered. Since all such groups have MI's with four or more graphs, it is necessary to consider non-maximal interactions in order to give three-dimensional phase diagrams.

a) $\mathcal{F}(6)$. The MI of this group is shown in Fig. 2.5.1. Here, the non-maximal interaction consisting of the graphs $G_1$ and $G_2$ of Fig. 1 below, and of the complement $G_3$ of $G_1+G_2$, will be studied. The matrix $\underline{\underline{\Omega}}$ has the form

$$\begin{pmatrix} 1 & \omega_3 & \omega_3 & \omega_1 & \omega_3 & \omega_2 \\ \omega_3 & 1 & \omega_2 & \omega_3 & \omega_1 & \omega_3 \\ \omega_3 & \omega_2 & 1 & \omega_3 & \omega_3 & \omega_1 \\ \omega_1 & \omega_3 & \omega_3 & 1 & \omega_2 & \omega_3 \\ \omega_3 & \omega_1 & \omega_3 & \omega_2 & 1 & \omega_3 \\ \omega_2 & \omega_3 & \omega_1 & \omega_3 & \omega_3 & 1 \end{pmatrix} \qquad (1)$$

as follows immediately from Fig. 1. Two of the eigenvalues and eigen-

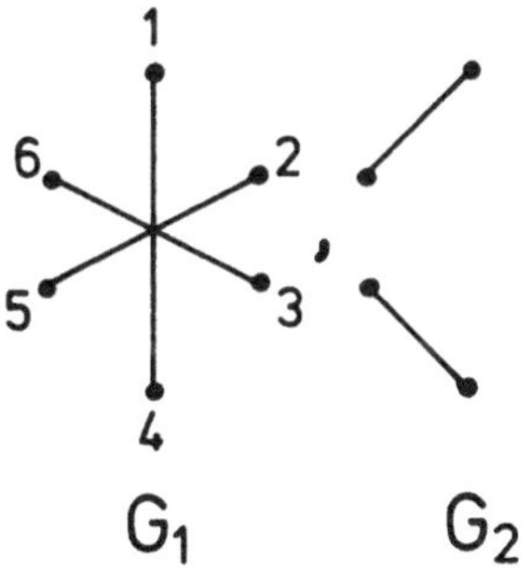


Fig. 1. Two graphs $G_1$ and $G_2$ from the MI of $\mathcal{F}(6)$. These, together with the complement $G_3$ of their sum, make up the nonmaximal interaction with $\mathcal{F}(6)$ symmetry leading to eq. (1).

vectors are obvious:

$$\lambda_o = 1+\omega_1+\omega_2+3\omega_3, \quad \text{belonging to} \quad (1,1,1,1,1,1), \text{ full symmetry,}$$
$$\lambda_1 = 1-\omega_1-\omega_2+\omega_3, \quad \text{belonging to} \quad (1,-1,1,-1,1,-1), \ \mathcal{S}(3) \text{ symmetry.} \qquad (2)$$

By going to the four-dimensional subspace orthogonal to these two eigenvectors, the others are easily found:

$$\lambda_{2,\pm} = 1-\omega_3 \pm [\omega_1{}^2+\omega_2{}^2+\omega_3{}^2-\omega_1\omega_2-\omega_1\omega_3-\omega_2\omega_3]^{\frac{1}{2}} ,$$ (both twofold degenerate and without nontrivial symmetries in their eigenspaces). (3)

Since $\lambda_{2,+}$ is always larger than $\lambda_{2,-}$, the complete symmetry breaking occurs for $m\lambda_{2,+}=\lambda_o$ in the ferromagnetic region. The phase diagram for the present model on a Cayley branch with $m=3$ is shown in Fig. 2, where the boundaries of the complete symmetry-breaking phase transition plane are solid curves, whereas phase 1 is bounded by the plane with dash-dotted boundaries.

The model reduces to a model with a completely permissible symmetry group in several cases: (i) $\omega_1=\omega_2$: nonmaximal $\mathfrak{D}(6)$ symmetry: straight lines AP(6) and DP(6); (ii) $\omega_1=\omega_3$: MI of $\mathfrak{S}(2)\sim\mathfrak{S}(3)$: straight lines $B_1P(6)$ and $C_2P(6)$; (iii) $\omega_2=\omega_3$: MI of $\mathfrak{S}(2)\sim\mathfrak{S}(3)$ as well: this gives

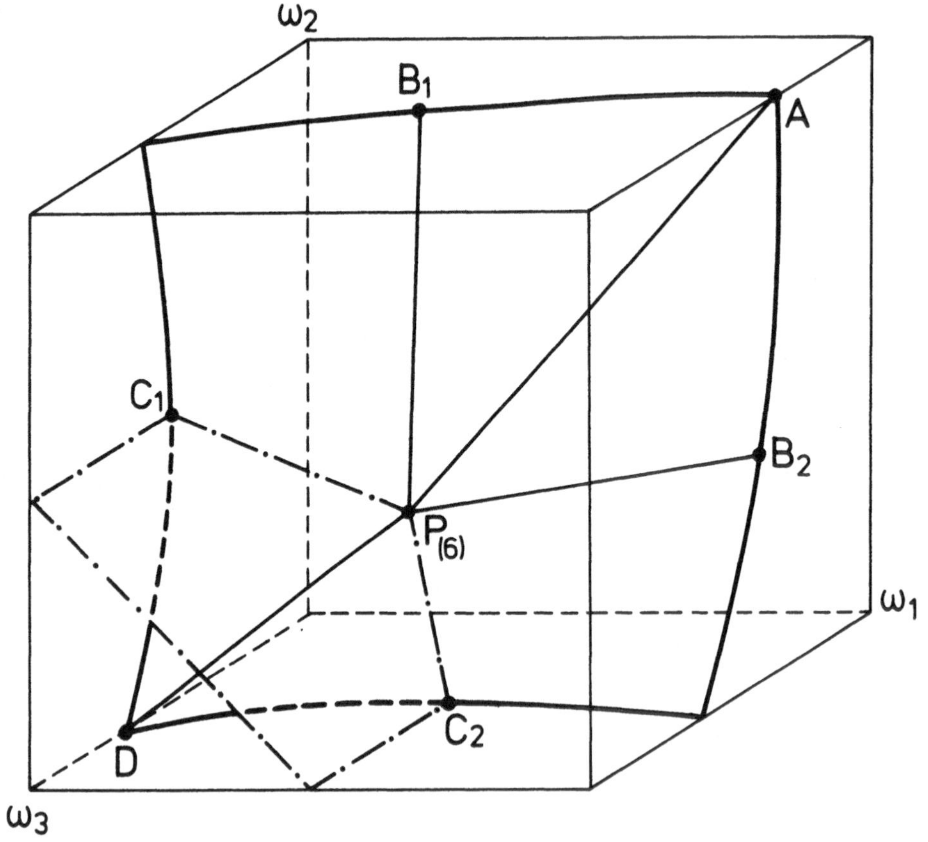


Fig. 2. Phase diagram of a nonmaximal interaction with $\mathfrak{F}(6)$ symmetry, given by Fig. 1, on a Cayley branch with $m=3$. For explanations, see the text.

$B_2P(6)$ and $C_1P(6)$. Note that the 1-phase transition plane cuts the 2-phase transition plane across the straight lines $C_1P(6)$ and $C_2P(6)$.

b) $\mathcal{R}(10)$. The graphs $G_1$ and $G_2$ of Fig. 3, together with the complement $G_3$ of their sum, define a nonmaximal interaction with permissible symmetry group $\mathcal{R}(10)$. [The MI of $\mathcal{R}(10)$ contains, in addition to $G_1$ and $G_2$, the graph $(G_2)^{(2)}$ (distance 2 graph of $G_2$), and the complement of the sum of these three graphs.] The eigenvalues and eigenvectors can be found in the same way as those for the symmetry group

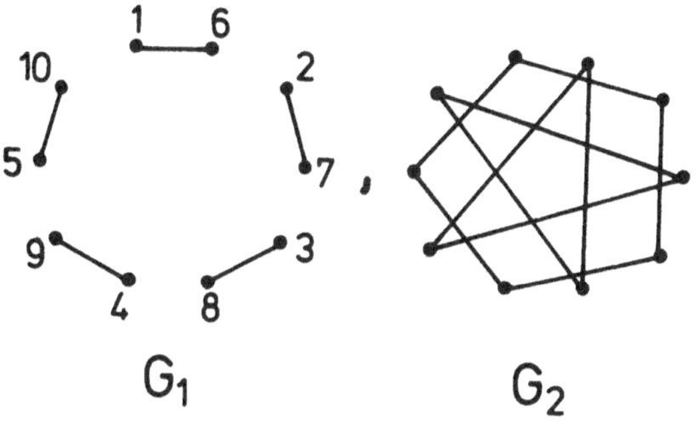


Fig. 3. Two graphs $G_1$ and $G_2$ from the MI of $\mathcal{R}(10)$. These, together with the complement of their sum, define a nonmaximal interaction with the same symmetry group.

$\mathcal{G}(G_{10})$; the results are

$$\lambda_o = 1+\omega_1+2\omega_2+6\omega_3, \quad \text{(nondegenerate, full symmetry)},$$

$$\lambda_1 = 1-\omega_1+2\omega_2-2\omega_3, \quad \text{(nondegenerate, } \mathcal{D}(5)\text{-symmetric eigenvector)},$$

$$\lambda_{2,\pm} = \tfrac{1}{2}(1-\omega_2-\omega_3)\pm[5(\omega_2-\omega_3)^2+4(\omega_1-\omega_3)^2]^{\frac{1}{2}}, \quad \text{(both four-fold degenerate with no nontrivial eigenspace symmetries)}. \tag{4}$$

The phase diagram for this model as derived from eqs. (4) and (1.9) is shown in Fig. 4 for a Cayley branch with $m=3$. The model reduces to one with a P-algebra in a number of cases again: (i) $\omega_1=\omega_2$: MI of $\mathcal{G}(G_{10})$: straight lines AP(10) and BP(10); (ii) $\omega_1=\omega_3$: nonmaximal $\mathcal{D}(5)\sim\mathcal{S}(2)$ interaction: straight lines CP(10) and DP(10); (iii) $\omega_2=\omega_3$: MI of $\mathcal{S}(2)\sim\mathcal{S}(5)$: straight lines EP(10) and FP(10). Again, the 1-phase transition plane cuts the curved 2-phase transition plane at two of these special lines, EP(10) and BP(10).

As is clear from Figs. 2 and 4, the qualitative features of these phase diagrams do not differ much from the ones for the direct product

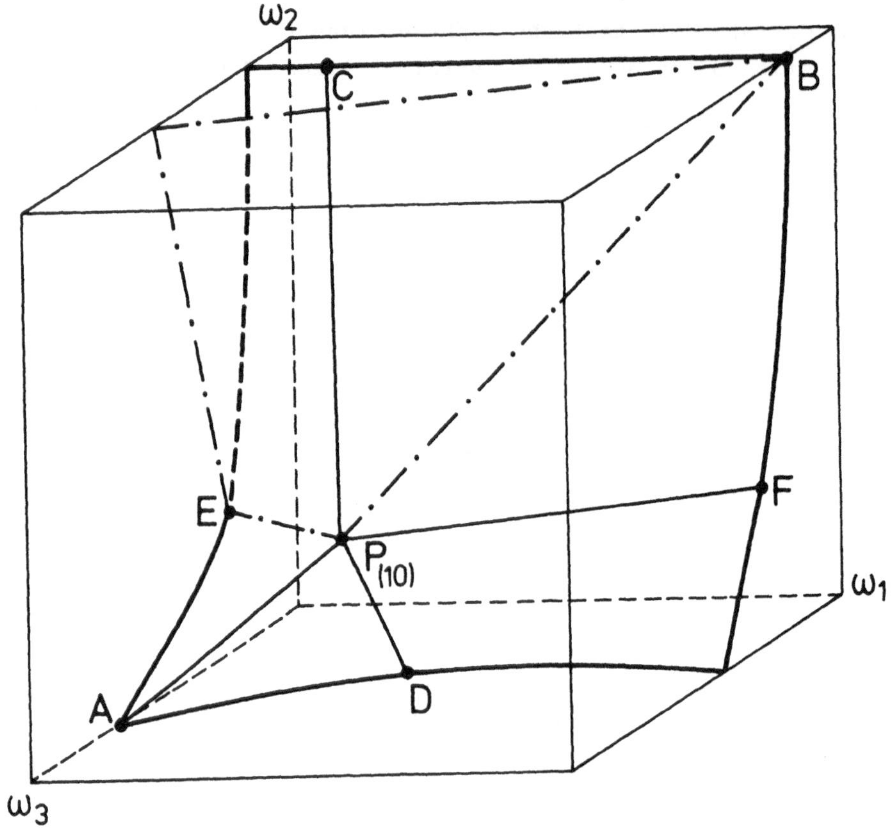


Fig. 4. Phase diagram of the nonmaximal $\mathcal{R}(10)$ model defined by the graphs of Fig. 3; the Cayley branch has m=3 again. For explanations of the special lines, see the text.

groups of the preceding section. In particular, the number of different phase transition planes is the same.

## 9.6. The closed Cayley branch and the $\mathcal{G}\otimes\mathcal{G}\rightarrow\mathcal{G}$ symmetry-breaking phase transition.

Since a Cayley branch is a tree, it does not contain any closed circuits. It has been proposed [3,4], to include such closed circuits in a simple way by joining two Cayley branches together at their surfaces by extra bonds, see Fig. 1, where these extra edges are denoted by dottes lines. Let a spin model with permissible symmetry group $\mathcal{G}$ on M letters be defined on such a graph; the interactions along the "solid bonds" are the given by a Boltzmann factor matrix $\Omega(i,j)$, those along the "dotted bonds" by $\Omega'(i,j)$. (This latter matrix may be equal to $\delta(i,j)$, as in the original work of Jelitto ([3]).) By folding the two

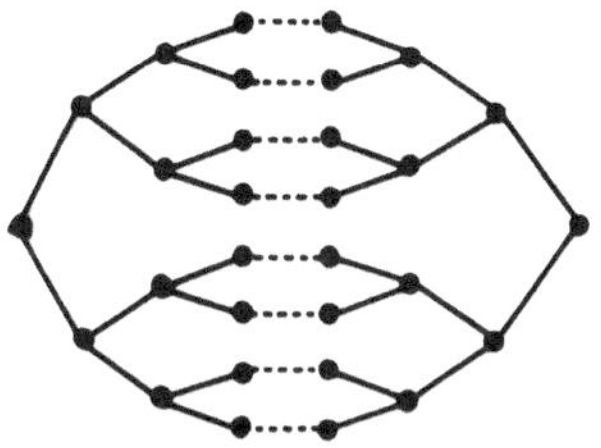

Fig. 1. A closed Cayley branch with two generations; the dotted lines are the extra bonds.

parts of the closed Cayley branch together, one sees, that one so defines a model with permissible symmetry group $\mathcal{G}^c$ on $M^2$ letters: every double solid bond carries the interaction

$$\Omega(ij,k\ell) = \Omega(i,k)\Omega(j,\ell), \tag{1}$$

whereas the boundary spins feel a field:

$$A(ij) = \Omega'(i,j). \tag{2}$$

It follows, that the symmetry group $\mathcal{G}^c$ contains the direct product $\mathcal{G}\otimes\mathcal{G}$, and that this type of boundary field cannot reduce this symmetry below a subgroup $\mathcal{G}$. A nontrivial fixed point for the $\mathcal{G}^c$ model obviously corresponds to nonvanishing correlations between the two top spins of the closed Cayley branch.

If the group $\mathcal{G}$ has a P-algebra, then $\underline{\underline{\Omega}}$ and $\underline{\underline{\Omega}}'$ have the same eigenvectors, so that eq. (2) can be written as

$$A(ij) = \sum_{k=0}^{M-1} \lambda'_k (\mu_k)_i (\mu_k)_j, \tag{3}$$

with real eigenvectors $\underline{\mu}_k$. This field is then small, if all $\lambda'_k$ are small for $k\neq 0$; this implies, that all $\omega'$ parameters are close to 1. The criterion for a small-field phase transition is, therefore,

$$m(\lambda_k/\lambda_o)^2 = 1, \tag{4}$$

since the relevant eigenvalues of (1) are the squares of those for $\underline{\underline{\Omega}}$.

Eq. (4) is equivalent to the criterion for ferromagnetic or antiferromagnetic small-field phase transitions of $\lambda_k$-type of the model with symmetry group $\mathcal{G}$ on a Cayley branch with branching ratio $\sqrt{m}$. For the Ising model, one has $\mathcal{G}^C = \mathcal{S}(2) \sim \mathcal{S}(2)$, and the phase transition points are, from eq. (2.1),

$$\omega_1 = \frac{\sqrt{m}-1}{\sqrt{m}+1}, \quad \omega_2 = \frac{\sqrt{m}+1}{\sqrt{m}-1} . \tag{5}$$

This is the same result, as obtained in ([3]) by more involved procedures. For the Potts model, one has $\mathcal{G}^C = \mathcal{S}(M) \otimes \mathcal{S}(M)$, $M > 2$, and eqs. (2.1) with $\sqrt{m}$ instead of $m$ give again the small-field phase transition points. For $\omega'$ not in the neighbourhood of 1, the high-field phase transitions studied in subsection 2.4 are also reproduced on the closed Cayley branch. This has been shown explicitly in ([4]) and will not be reproduced here, since no new information can be obtained in this way.

## 9.7. The exponential branch.

As a curiosity, the exponential branch defined by eq. (8.1.20) is considered here briefly. For such a branch, the branching ratio increases as the generation number $n$. Although the thermodynamic limit exists by Theorem 8.3.1, the behaviour of such a branch in a small boundary field is anomalous. This is most easily seen by writing down the recursion relation for the $\alpha_n(k)$ as in Section 1; the result is

$$\alpha_n(k) = n(\lambda_k/\lambda_o)\alpha_{n-1}(k), \tag{1}$$

with solution

$$\alpha_n(k) = n!\,(\lambda_k/\lambda_o)^n \alpha_o(k). \tag{2}$$

As $n \to \infty$, this always diverges for nonzero initial values, unless $\lambda_k = 0$ holds. For Ising and Potts models, this implies, that there is a small-field phase transition for all finite temperatures, even though the field-free free energy is given by

$$\gamma_o = e^{-1} \ln \lambda_o, \tag{3}$$

which is analytic.

For more complicated models, there may be finite temperatures, for which not all eigenvectors propagate on the exponential branch; for the

$S(M_1) \sim S(M_2)$ model, e.g., eq. (3.2) shows, that eigenvectors in the $\lambda_1$-eigenspace do not propagate for

$$\omega_2 = \{1+(M_1-1)\omega_1\}/M_1 . \tag{4}$$

Such "phase transitions" are always of the dimensionality-changing type.

## 9.8. Reduction of cactus branches to Cayley branches for small-field phase transitions.

In Section 8.3.1, it has been remarked, that the study of small-field phase transitions for a general recursive site graph sequence reduces to the case of a Cayley-like branch with suitably redefined couplings. Here, this will be made explicit for the cactus branches defined in Section 8.1, see Fig. 8.1.4, for example. A self-similar cactus branch with a polygon with $m+1$ vertices (and edges) as building block will be considered; if every edge carries a Boltzmann factor matrix $\underline{\underline{\Omega}}$, then the recursion relation for the partition function with fixed top spin reads, for an applied boundary field $A(i)$,

$$Z_n(i) = \sum_{i_1,\ldots,i_m=1}^{M} \Omega(i,i_1)\Omega(i_1,i_2)\ldots\Omega(i_m,i) \prod_{t=1}^{m} Z_{n-1}(i_t),$$

$$Z_o(i)=A(i). \tag{1}$$

Writing again $\rho_n(i)=Z_n(i)/Z$ and linearizing, $\rho_n(i)=M^{-1}+\delta_n(i)$, yields as recursion relation for the $\underline{\delta}_n$:

$$\delta_n(i) = \sum_{j=1}^{M} [\sum_{k=1}^{m} \Omega^k(i,j)\Omega^{m-k+1}(j,i)][M^{-1}\sum_{k=1}^{M} \Omega^{m+1}(k,k)]^{-1} \delta_{n-1}(j). \tag{2}$$

Here $\Omega^k(i,j)$ denotes the $(i,j)$-element of the k-fold matrix product of $\underline{\underline{\Omega}}$ with itself. Clearly, eq. (2) has, formally, the same form as eq. (1.6) for the Cayley branch with the renormalized interaction matrix

$$\Omega_r(i,j) = \frac{1}{m}\sum_{k=1}^{m} \Omega^k(i,j)\Omega^{m-k+1}(j,i); \tag{3}$$

$$\delta_n(i) = [m/\lambda_o(\underline{\underline{\Omega}}_r)] \sum_{j=1}^{M} \Omega_r(i,j)\delta_{n-1}(j). \tag{4}$$

This renormalized matrix is invariant with respect to the symmetry group

$G$ of $\underline{\underline{\Omega}}$ and is also symmetric; it is not yet normalized so that the diagonal elements are 1, but this is unimportant, since such a normalization factor disappears from eq. (4) by virtue of the presence of the eigenvalue $\lambda_o(\underline{\underline{\Omega}}_r)$ in the denominator. If $G$ has an associated P-algebra, the eigenvectors of $\underline{\underline{\Omega}}_r$ are also the same as those of $\underline{\underline{\Omega}}$.

As a simple example, consider the Ising model on a triangular cactus tree, m=2. The matrix $\underline{\underline{\Omega}}_r$ is then:

$$\begin{pmatrix} 1+\omega^2 & 2\omega^2 \\ 2\omega^2 & 1+\omega^2 \end{pmatrix}, \tag{5}$$

which amounts to a renormalized $\omega_r$ given as

$$\omega_r = 2\omega^2/(1+\omega^2). \tag{6}$$

The Ising model is critical on a Cayley branch with m=2 for $\omega_r = \frac{1}{3}$ (fm) and for $\omega_r=3$ (afm). From eq. (6), no antiferromagnetic transition occurs on the triangular cactus branch, whereas the ferromagnetic one is at $\omega=\frac{1}{5}\sqrt{5}$.

REFERENCES.


(1). C.N. Yang and T.D. Lee, Phys. Rev. 87 (1952) 404.
(2). T.D. Lee and C.N. Yang, Phys. Rev. 87 (1952) 410.
(3). R.J. Jelitto, Physica 99A (1979) 268.
J.E. Krizan, P.F. Barth and M.L. Glasser, Physica 119A (1983) 230.
M.L. Glasser and M.K. Goldberg, Physica 117A (1983) 670.
(4). P.L. Christiano and S. Goulart Rosa, Jr., Phys. Lett. 101A (1984) 275.
K. De'Bell, D.J.W. Geldart and M.L. Glasser, preprint (1983).


GENERAL REFERENCES.

The results of this chapter are extensions of a series of papers:
H. Moraal, Physica 85A (1976) 457; 92A (1978) 305; 105A (1981) 472; 113A (1982) 44 and Z. Phys. B 45 (1982) 237.
These were inspired by the work of Müller-Hartmann and Zittartz on Cayley trees in a homogeneous field:
E. Müller-Hartmann and J. Zittartz, Phys. Rev. Lett. 33 (1974) 893 and Z. Phys. B 22 (1975) 59,
see also Chapter 13.
The Potts model has also been considered in this vein by
L. Turban, Phys. Lett. 78A (1980) 404,
and, in the context of the random cluster model, by
H.G. Baumgärtel and E. Müller-Hartmann, Z. Phys. B 46 (1982) 227.
The special role played by the surface of a Cayley branch or tree was first discerned by

L.K. Runnels, J. Math. Phys. 8 (1967) 2081,
for a hard-core lattice gas. Some years later, the analyticity of the field-free free energy was stressed by
T.P. Eggarter, Phys. Rev. B 9 (1974) 2928.
These results then led to the discovery of the phase transitions of continuous order by Müller-Hartmann and Zittartz, see also Chapter 13. These have also been studied on a cactus branch by
N. Grewe and W. Klein, Z. Phys. B 23 (1976) 193.
The idea, that phase transitions ought to be accompanied by symmetry breaking, thus excluding dimensionality-changing phase transitions as such, has always been put forth forcefully by J. Zittartz in a number of talks on duality for spin models.
Other interesting applications of the surface-field approach are:
J. Vannimenus, Z. Phys. B 43 (1981) 141,
K. Fesser and H.J. Herrmann, J. Phys. A 17 (1984) 1493.
These authors studied systems with competing interactions, resulting in large limit cycles and chaotic behaviour. A similar study with a uniform external field has been performed by
T. Morita, Phys. Lett. 94A (1983) 232.

## 10. RANDOM SPIN MODELS ON CAYLEY BRANCHES WITH SURFACE FIELD.

### 10.1. Bond-random spin models: cumulant expansion.

In this section, a spin model with a permissible symmetry group on $M$ letters, which has a P-algebra, is considered. The Boltzmann factor matrix is taken to be a random variable for each bond of the Cayley branch in the sense, that if the model has $s$ different energy parameters $E_1, E_2, .., E_s$, then these are randomly distributed with distribution function $p(\underline{E})=p(E_1,..,E_s)$, which is normalized to 1. As in the previous chapter, a (small) boundary field is assumed, which is non-random. Then, for a particular configuration $\Psi$ of the $\underline{\underline{\Omega}}$'s, the recursion relations (8.3.3) and (8.3.4) read:

$$\rho_n(i;\Psi) = c_n(\Psi)^{-1} \prod_{t=1}^{m} \{ \sum_{j=1}^{M} \Omega^{(t)}(i,j)\rho_{n-1}(j;\Psi_t')\},\ \rho_o(i) \text{ given by the surface field,} \tag{1}$$

$$c_n(\Psi) = \sum_{i=1}^{M} \prod_{t=1}^{m} \{ \sum_{j=1}^{M} \Omega^{(t)}(i,j)\rho_{n-1}(j;\Psi_t')\}. \tag{2}$$

Here $\underline{\underline{\Omega}}^{(t)}$ is the interaction matrix for the edge connecting the top spin of $C_n(m)$ to the t-th $C_{n-1}(m)$-branch and $\Psi_t'$ is the configuration of the $\underline{\underline{\Omega}}$'s for the t-th branch, corresponding to $\Psi$, symbolically:

$$\Psi = \bigcup_{t=1}^{m} \{\underline{\underline{\Omega}}^{(t)}, \Psi_t'\}. \tag{3}$$

Since $G$ has an associated P-algebra, $\rho_n(i;\Psi)$ can be expanded in terms of the eigenvectors common to all $\underline{\underline{\Omega}}$'s:

$$\rho_n(i;\Psi) = M^{-1}\{1+\sum_{k=1}^{M-1} \alpha_k^{(n)}(\Psi)(\mu_k)_i\}, \tag{4}$$

$$\rho_o(i;\Psi) = \rho_o(i) = M^{-1}\{1+\sum_{k=1}^{M-1} \alpha_k^{(o)}(\mu_k)_i\}. \tag{5}$$

For the expansion coefficients, eqs. (1) and (2) imply the recursion

$$\alpha_k^{(n)}(\Psi) = \sum_{t=1}^{m} (\lambda_k^{(t)}/\lambda_o^{(t)})\ \alpha_k^{(n-1)}(\Psi_t'), \tag{6}$$

where $\lambda_k^{(t)}$ is the k-th eigenvalue of $\underline{\underline{\Omega}}^{(t)}$.

In contrast to the nonrandom case, $\alpha_k^{(n)}(\Psi)$ is now, of course, a random variable and not only its average (over the configurations $\Psi$), but its whole probability distribution in the limit $n\to\infty$ is of interest. Therefore, the moment generating function $f_n(x_1,..,x_s)$, defined by:

$$f_n(x_1,..,x_s) = \sum_{r_1,..,r_s=0} \prod_{\ell=1}^{s} (x_\ell^{r_\ell}/r_\ell!)\, M_n(r_1,..,r_s) , \tag{7}$$

$$M_n(r_1,..,r_s) = \int \prod_{\ell=1}^{s} \{\alpha_\ell^{(n)}(\Psi)\}^{r_\ell}\, p(\Psi)\, d\Psi, \tag{8}$$

$$p(\Psi)d\Psi = \prod_{\substack{\text{all}\\ \text{edges } e}} p(\underline{E}_e)d\underline{E}_e, \tag{9}$$

is considered in the following. Eq. (6) implies, by the independence of the random variables $(\lambda_k^{(t)}/\lambda_o^{(t)})$ and $\alpha_k^{(n-1)}(\Psi_t')$ for different branches, the following simple recursion for this function:

$$f_n(x_1,..,x_s) = [\int f_{n-1}(x_1\lambda_1/\lambda_o,x_2\lambda_2/\lambda_o,...,x_s\lambda_s/\lambda_o)\, p(\underline{E})d\underline{E}]^m , \tag{10}$$

where the eigenvalues are functions only of the $\underline{E}$-vector over which is integrated.

Since the moments defined in eq. (8) also contain "trivial" parts, it is better to use the cumulant generating function

$$k_n(x_1,..,x_s) = \ln f_n(x_1,..,x_s), \tag{11}$$

which yields the cumulants $K_n(r_1,..,r_s)$:

$$k_n(x_1,..,x_s) = \sum_{\substack{r_1,..,r_s=0\\ \Sigma_{j=1}^{s} r_j>0}} \prod_{\ell=1}^{s} (x_\ell^{r_\ell}/r_\ell!)\, K_n(r_1,..,r_s). \tag{12}$$

For the first few cumulants, some new notation is introduced:

$$K_r(j;n) = K_n(0,..,0,r,0,..,0), \quad r \text{ at the } j\text{-th position},$$

$$K_{r_1r_2}(j_1j_2;n) = K_n(0,..,0,r_1,0,..,0,r_2,0,..,0), \quad r_t \text{ at the } j_t\text{-th position, } t=1,2. \tag{13}$$

and similarly for cumulants with more than two nonzero arguments. The first few cumulants are then given in terms of the moments (for which the notation of eq. (13) is introduced analogously), by

$$K_1(j;n) = M_1(j;n), \quad \text{(average)},$$

$$K_2(j;n) = M_2(j;n)-\{M_1(j;n)\}^2, \quad \text{(variance)}, \tag{14}$$

$$K_{11}(j_1j_2;n) = M_{11}(j_1j_2;n)-M_1(j_1;n)M_2(j_2;n), \quad \text{(covariance), etc.}$$

The recursion relation for the cumulant generating function, which follows from eqs. (10) and (11) as

$$k_n(x_1,..x_s) = m \ln\int \exp\{k_{n-1}(x_1\lambda_1/\lambda_o,...,x_s\lambda_s/\lambda_o)\}p(\underline{E})d\underline{E}, \tag{15}$$

now leads to recursion relations for the cumulants. Defining the averages

$$\beta(r_1,..,r_s) = m \prod_{\ell=1}^{s} (\lambda_\ell/\lambda_o)^{r_\ell} p(\underline{E})d\underline{E}, \tag{16}$$

and using for the first few of these a notation similar to eq. (13), the recursion relations up to third order are seen to be given as:

$$K_1(j;n) = \beta_1(j)K_1(j;n-1),$$

$$K_1(j;0) = \begin{cases} 1 & \text{if a field of type } j \text{ is present at the surface,} \\ 0 & \text{if this is not the case.} \end{cases} \tag{17}$$

$$K_2(j;n) = \beta_2(j)K_2(j;n-1)+\{K_1(j;n-1)\}^2[\beta_2(j)-m^{-1}\beta_1(j)^2],$$

$$K_2(j;0) = 0, \tag{18}$$

$$K_3(j;n) = \beta_3(j)K_3(j;n-1)+K_1(j;n-1)K_2(j;n-1)[3\beta_3(j)-3m^{-1}\beta_1(j)\beta_2(j)]+$$

$$+\{K_1(j;n-1)\}^3[\beta_3(j)-3m^{-1}\beta_1(j)\beta_2(j)+2m^{-2}\beta_1(j)^3], \tag{19}$$

$$K_3(j;0) = 0,$$

$$K_{11}(j_1j_2;n) = \beta_{11}(j_1j_2)K_{11}(j_1j_2;n-1)+K_1(j_1;n-1)K_1(j_2;n-1)[\beta_{11}(j_1j_2)-$$

$$-m^{-1}\beta_1(j_1)\beta_1(j_2)], \tag{20}$$

$$K_{11}(j_1j_2;0) = 0,$$

$$K_{12}(j_1j_2;n) = \beta_{12}(j_1j_2)K_{12}(j_1j_2;n-1)+K_1(j_2;n-1)K_{11}(j_1j_2;n-1)\times$$

$$\times[2\beta_{12}(j_1j_2)-2m^{-1}\beta_1(j_2)\beta_{11}(j_1j_2)]+K_1(j_1;n-1)K_2(j_2;n-1)[\beta_{12}(j_1j_2)-$$

$$-m^{-1}\beta_1(j_1)\beta_2(j_2)]+K_1(j_1;n-1)\{K_1(j_2;n-1)\}^2[\beta_{12}(j_1j_2)-m^{-1}\beta_1(j_1)\beta_2(j_2)-$$

$$-2m^{-1}\beta_1(j_2)\beta_{11}(j_1j_2)+2m^{-2}\beta_1(j_1)\beta_1(j_2)^2],\ K_{12}(j_1j_2;0) = 0, \tag{21}$$

$$K_{111}(j_1j_2j_3;n) = \beta_{111}(j_1j_2j_3)K_{111}(j_1j_2j_3;n-1)+K_1(j_1;n-1)K_{11}(j_2j_3;n-1)$$

$$[\beta_{111}(j_1j_2j_3)-m^{-1}\beta_1(j_1)\beta_{11}(j_2j_3)]+(\text{two similar terms with } j_1j_2j_3 \text{ cyclically permuted})+$$

$$+K_1(j_1;n-1)K_1(j_2;n-1)K_1(j_3;n-1)[\beta_{111}(j_1j_2j_3)-m^{-1}\beta_1(j_1)\beta_{11}(j_2j_3)-$$

$$-m^{-1}\beta_1(j_2)\beta_{11}(j_1j_3)-m^{-1}\beta_1(j_3)\beta_{11}(j_2j_3)+2m^{-2}\beta_1(j_1)\beta_2(j_2)\beta_1(j_3)],$$

$$K_{111}(j_1j_2j_3;0) = 0. \tag{22}$$

The reason for explicitly writing down the complicated third-order equations (19), (21) and (22) is, that the general structure of the recursion relations shows up clearly. This can be formalized as follows: define the notion of a <u>partition</u> of a sequence of $s$ numbers $(r_1,..,r_s)$ as a set $L$ of sequences $(\ell_1,..,\ell_s)$ with the property, that

$$\sum_{\substack{\text{sequences}\\ \text{from } L}} \ell_i = r_i, \quad \text{for} \quad i=1,..,s. \tag{23}$$

If two partitions $L_1$ and $L_2$ are given, a partial order can be defined by: $L_1 \le L_2$ holds, if the sequences of $L_1$ can be combined into disjoint classes $C(L_1)$, of which there are as many as there are sequences in $L_2$, such that

$$\sum_{\substack{\text{sequences}\\ \text{from } C(L_1)}} \ell_i(L_1) = \ell_i(L_2), \quad \text{for all} \quad i. \tag{24}$$

If this is possible, $L_1$ will be called <u>finer</u> than $L_2$. In terms of partitions, the cumulant recursion relations have the following general structure, which is easily checked against the explicit eqs. (17-22): In the equation for $K_n(r_1,..,r_s)$, all products of the form

$$\prod_{\substack{\text{sequences}\\ \text{from } L}} K_{n-1}(\ell_1,..,\ell_s) \tag{25}$$

occur for all partitions $L$ of $(r_1,..,r_s)$. The coefficient of a term such as eq. (25) is a linear combination of products

$$\prod_{\substack{\text{sequences}\\ \text{from } L_1}} \beta(\ell_1',..,\ell_s'), \tag{26}$$

where all partitions $L_1$, such that $L$ is finer than $L_1$, occur.

For the solutions of the recursion relations for the cumulants, this is easily seen to imply

$$\lim_{n\to\infty} K_n(r_1,..,r_s) = 0 \quad \text{iff} \quad |\beta(L)| = \prod_{\substack{\text{sequences}\\ \text{from } L}} |\beta(\ell_1,..,\ell_s)| < 1 \tag{27}$$

holds for every partition $L$ of $(r_1,..,r_s)$; the limit does not exist as soon as $|\beta(L)|>1$ holds for some $L$. Since this changes the distribution function for the magnetization qualitatively, a phase transition of type $(r_1,..,r_s)$ is said to occur for

$$\max_{L} \prod_{\substack{\text{sequences}\\ \text{from } L}} |\beta(\ell_1,..,\ell_s)| = 1. \tag{28}$$

This phase transition type will be studied in some detail for two special distributions $p(\underline{E})$ in the next two sections.

## 10.2. The diluted bond case.

The distribution of energy values for the diluted bond case is

$$p(E_1,..,E_s) = p \prod_{t=1}^{s} \delta(E_t,E_t^{(o)}) + (1-p) \prod_{t=1}^{s} \delta(E_t,0), \tag{1}$$

so that there is a probability $p$ for interaction with energy parameters $\underline{E}^{(o)}$ and a probability $1-p$ for no interaction at all. Eq. (1.16) shows, that the $\beta$'s are then simply given as

$$\beta(r_1,..,r_s) = mp \prod_{\ell=1}^{s} (\lambda_\ell/\lambda_o)^{r_\ell}, \tag{2}$$

where the $\lambda$'s are the eigenvalues of $\underline{\underline{\Omega}}$ for the interaction with $\underline{E}^{(o)}$.

Therefore, $\beta(L)$ as given in eq. (1.27) becomes

$$\beta(L) = (mp)^{|L|} \prod_{\ell=1}^{s} (\lambda_\ell/\lambda_o)^{r} , \tag{3}$$

where $|L|$ is the number of sequences in the partition $L$. Since the finest partition has sequences consisting of one entry 1 and all other entries 0, the criterion (1.28) for a phase transition of type $(r_1,..,r_s)$ is:

$$|\prod_{\ell=1}^{s} (mp\lambda_\ell/\lambda_o)^{r_\ell}|=1, \tag{4}$$

unless $mp<1$ holds, in which case no transition of any type is possible. For $mp=1$, the inequality $|\lambda_\ell|<\lambda_o$, which holds at all temperatures $\neq 0$, shows, that for the critical value $p_c=m^{-1}$, phase transitions are at most possible at $T=0$.

The critical value $p_c=m^{-1}$ is also the critical bond percolation probability for the Cayley branch. This problem is defined as follows: let a bond be present with probability $p$, absent with probability $1-p$. Let $P_n(p)$ be the probability of finding an unbroken path of bonds extending from the top spin to the boundary. Then the structure of the Cayley branch entails the recursion relation

$$P_n(p) = 1-\{1-pP_{n-1}(p)\}^m, \quad P_o(p)=1. \tag{5}$$

Percolation is said to occur if the probabilipy $P_n(p)$ does not go to zero in the limit $n\to\infty$. By linearizing eq. (5), it is easily seen that percolation occurs only for $p>p_c=m^{-1}$. It is not surprising, that no phase transitions occur in the diluted bond case if no unbroken path exists between the top spin (of which the field-distribution function is considered) and the surface (where the field is applied).

Returning now to eq. (4) for the case $m^{-1}<p<1$, one sees, that the following critical hyperplanes exist:

(i) All phase transitions of types $(0,..,0,r_k,0,..,0)$ for $r_k=1,2,..,$ occur at

$$mp\lambda_k/\lambda_o|=1; \tag{6}$$

(ii) All phase transitions of types $(0,..,0,r_{k_1},0,..,0,r_{k_2},0,..,0)$ for $r_{k_1},r_{k_2}=1,2,..,$ occur at

$$|m^2p^2\lambda_{k_1}\lambda_{k_2}/\lambda_o{}^2|=1; \tag{7}$$

etc. The phase transitions of eq. (6) describe the appearance of a distribution with a nonvanishing average; these are, therefore, analogous to the phase transitions on a homogeneous Cayley branch (which are, at least formally, also given by eq. (6) with $p$ set equal to $1$). Since all other cumulants also diverge, this analogy is not perfect, however; in particular, it becomes difficult to make the distinction between symmetry-breaking and dimensionality-changing phase transitions, since in both cases the distribution function now changes dramatically.

Eq. (7) shows, that also nontrivial correlations between different types of eigenvectors may occur, even if these do not all propagate. Such phase transitions have no analogues in the homogeneous case. The same holds true for the higher order correlations. In the following, these points are made more explicit for Potts models and for models with two-graph MI's.

a) Potts models. Since these have only one eigenvalue different from $\lambda_o$, the only phase transitions are those described by eq. (6). The phase transition points are then given by eqs. (9.2.1) with $m$ replaced by $mp$:

$$\omega_f = \frac{mp-1}{mp+M-1}\ , \text{ for } \ p>m^{-1}, \quad \omega_{af} = \frac{mp+1}{mp-M+1}\ , \text{ for } \ p>m^{-1}(M-1). \tag{8}$$

The requirement for the antiferromagnetic case can only be fulfilled for a value of $p$ less than $1$ if $m>M-1$ holds, which is the condition for the existence of this transition on the homogeneous Cayley branch.

b) $S(M_1)\sim S(M_2)$ models. These models have two eigenvalues $\neq\lambda_o$ given by eqs. (9.3.2) and (9.3.5). There are, therefore, six possible phase transitions:

$$\begin{aligned}
&\text{phase } 1 \text{ boundary}: && mp\lambda_1=\lambda_o;\\
&\text{phase } \overline{1} \text{ boundary}: && mp\lambda_1=-\lambda_o;\\
&\text{phase } 2 \text{ boundary}: && mp\lambda_2=\lambda_o;\\
&\text{phase } \overline{2} \text{ boundary}: && mp\lambda_2=-\lambda_o;\\
&\text{phase } 12 \text{ boundary}: && m^2p^2\lambda_1\lambda_2=\lambda_o^2;\\
&\text{phase } \overline{12} \text{ boundary}: && m^2p^2\lambda_1\lambda_2=-\lambda_o^2.
\end{aligned}\tag{9}$$

Figs. 1(a)-(f) show the critical $\omega_c$ as obtained from eqs. (9) and the explicit expressions of the eigenvalues for the model with $S(3)\sim S(2)$

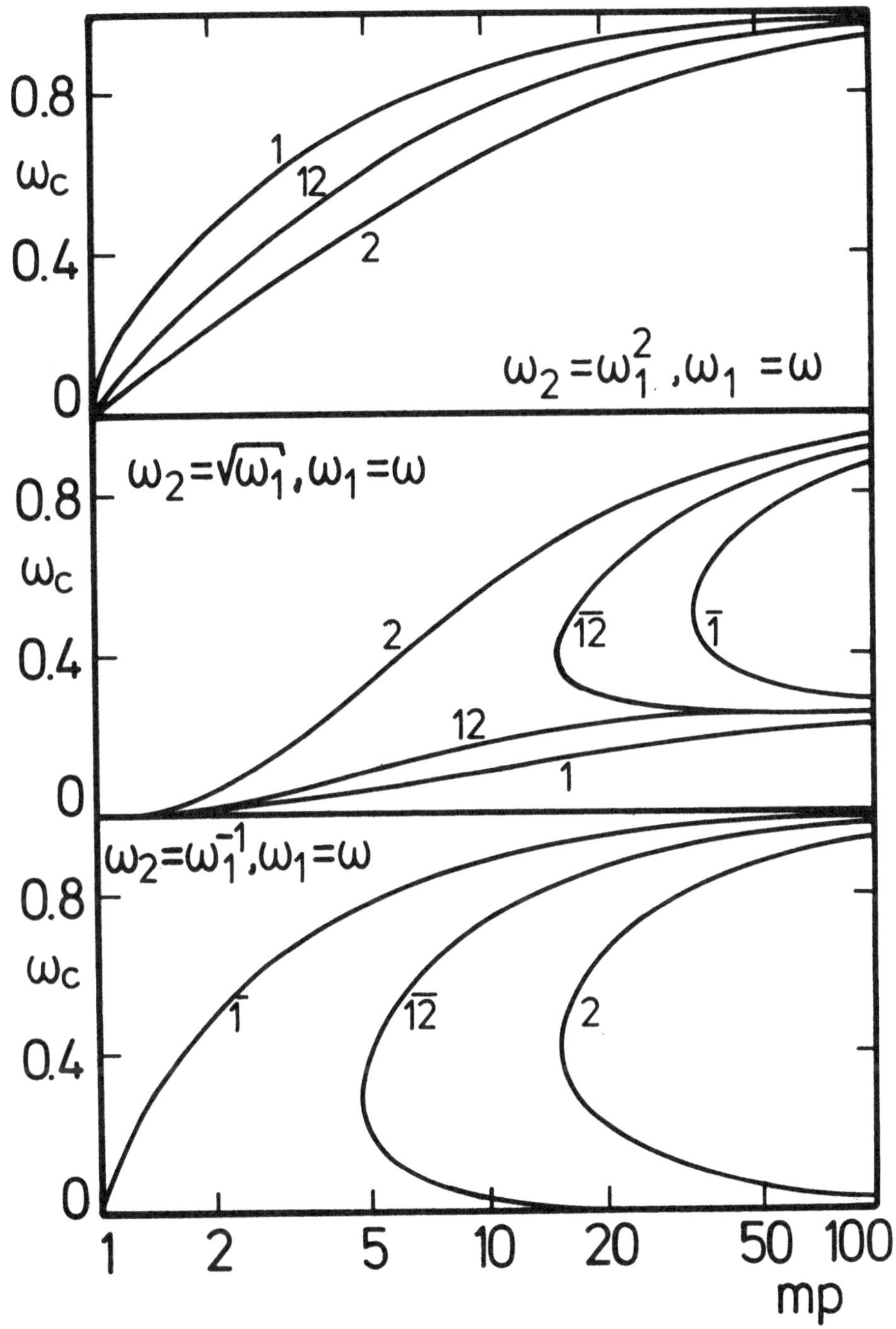


Fig. 1(a)-(c). Phase transitions for the diluted bond $S(3)\sim S(2)$ model on a Cayley branch;for three different ratios of $\omega_1$ and $\omega_2$. The phases are as in eqs. (9), see also the text.

symmetry for six different ratios of $\omega_1$ and $\omega_2$. Note the reentrant behaviour, which many of these curves imply for a thermodynamic path.

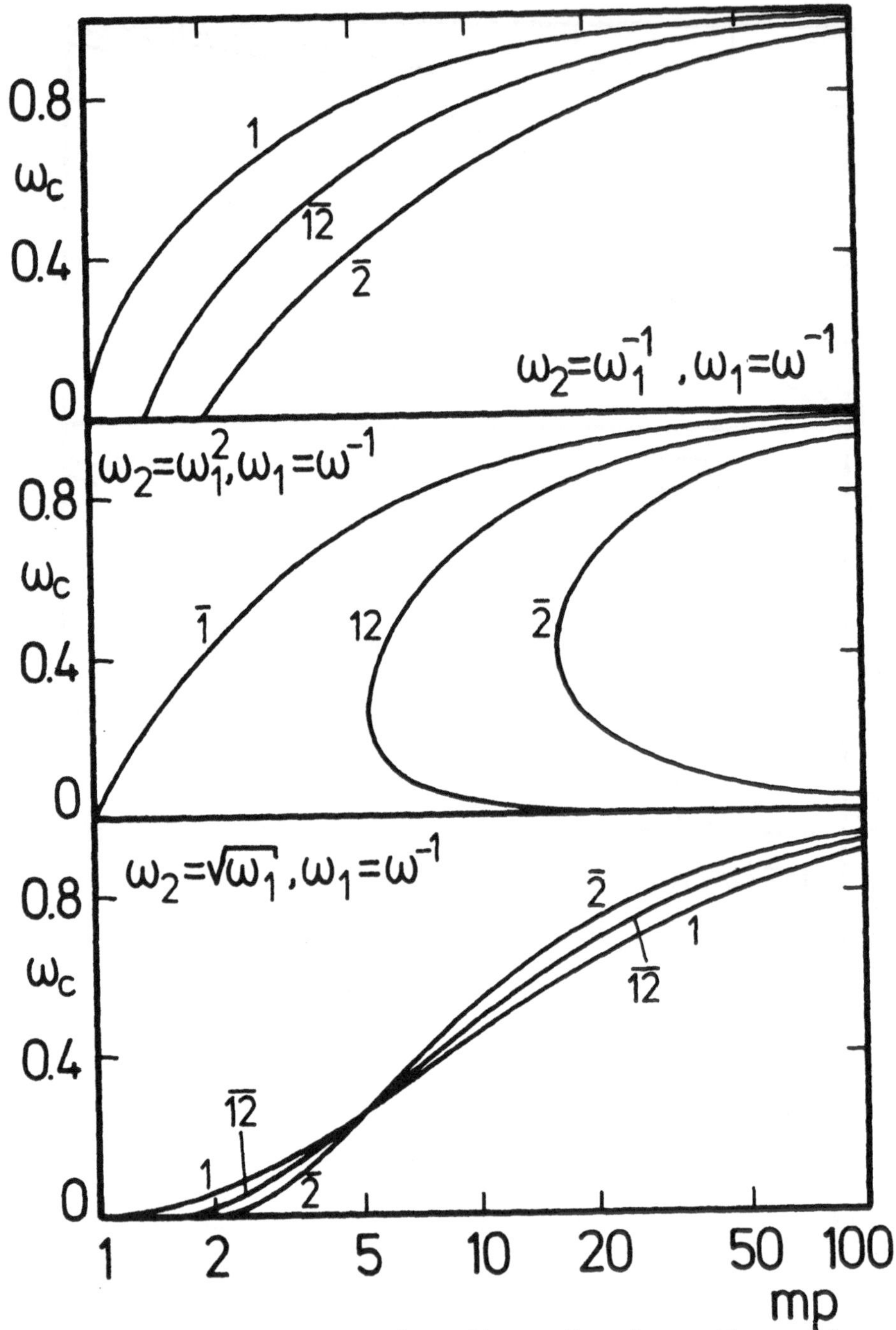


Fig. 1(d)-(e). Same as the previous figure for three other ratios of $\omega_1$ and $\omega_2$.

c) $G(G_9)$. As an example of a primitive group with a two-graph MI, $G(G_9)$ has been chosen. Its two relevant eigenvalues are given in eqs. (7.6.38).

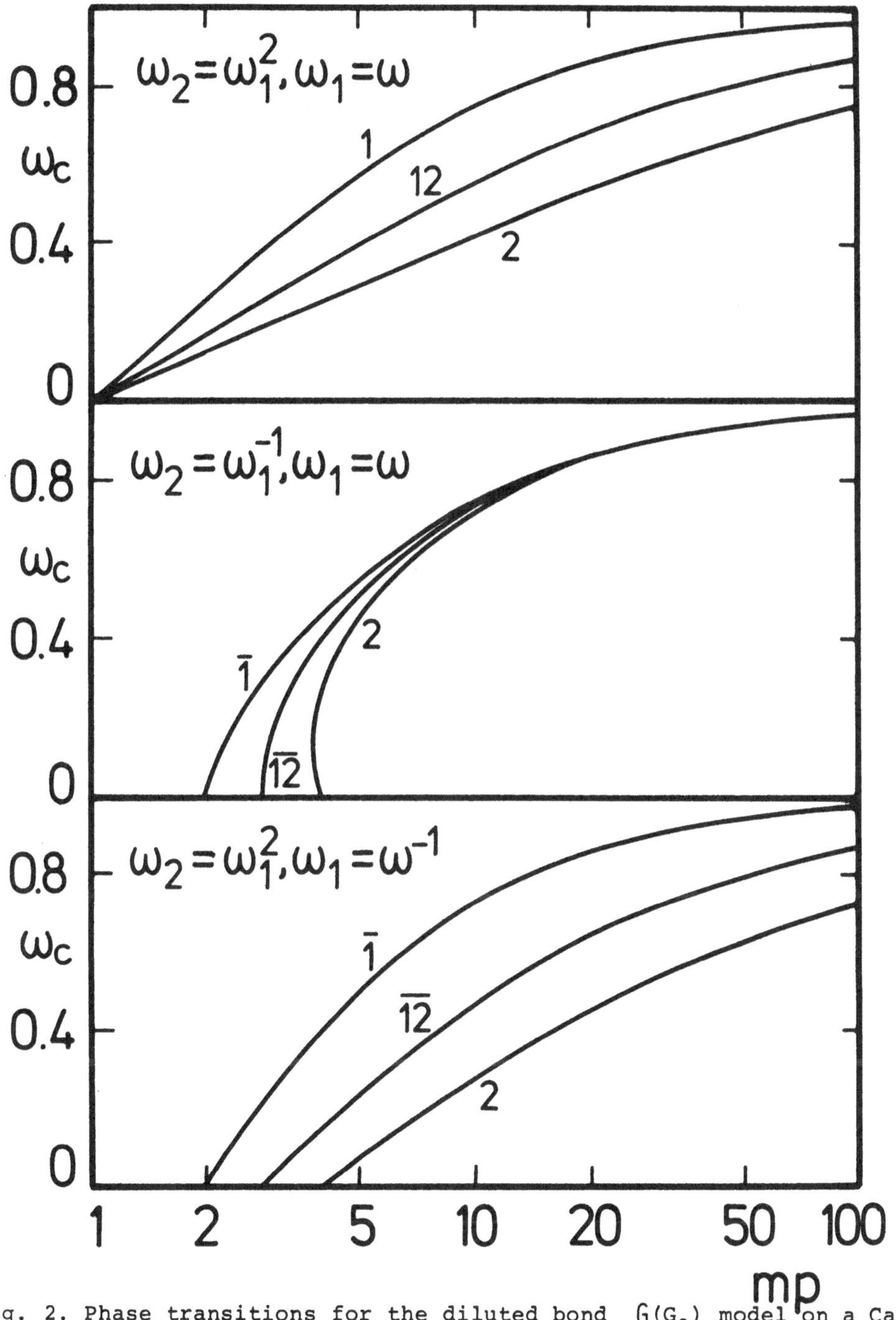


Fig. 2. Phase transitions for the diluted bond $G(G_9)$ model on a Cayley branch for three different ratios of $\omega_1$ and $\omega_2$.

For the phase transitions, given by eqs. (9) again, see Fig. 2; the symmetry with respect to the interchange of $\omega_1$ and $\omega_2$ implies, that only half of the number of possibilities for the imprimitive case suffices to give an impression of the different cases.

## 10.3. The spin glass case.

In this section, the distribution function $p(\underline{E})$ is taken to be

$$p(\underline{E}) = p \prod_{t=1}^{s} \delta(E_t - E_t^{(o)}) + (1-p) \prod_{t=1}^{s} \delta(E_t + E_t^{(o)}), \quad E_t^{(o)} \geqq 0, \tag{1}$$

so that all energies are either all positive (probability $p$) or all negative (probability $1-p$). Now there is no simple way to compare the absolute values of $(\lambda_k/\lambda_o)$ for positive and negative energies, so that a special model has to be chosen in order to proceed. Only the Ising and Potts models will be considered in some detail.

The Ising model. Since there is only one eigenvalue ratio, one has

$$(\lambda_1/\lambda_o)(E) = \frac{1-\omega}{1+\omega} = -(\lambda_1/\lambda_o)(-E), \quad \omega = \exp{-\beta E}, \quad E>0. \tag{2}$$

Therefore, the $\beta$'s are given as (see eq. (1.16)):

$$\beta(2k) = m\{(1-\omega)/(1+\omega)\}^{2k}, \quad \beta(2k+1) = m(2p-1)\{(1-\omega)/(1+\omega)\}^{2k+1}. \tag{3}$$

A phase transition of type $2k$ occurs, by eq. (1.28), for

$$\max \left| \prod_{i=1}^{r} \beta(2\ell_i) \prod_{j=1}^{s} \beta(2m_i+1) \right| = 1, \tag{4}$$

where the maximum has to be taken over all possible values of $r$, $s$, of the $\ell_i$ and $m_i$, subject to the restriction

$$\sum_{i=1}^{r} (2\ell_i) + \sum_{j=1}^{s} (2m_i+1) = 2k. \tag{5}$$

Insertion of eqs. (3) into eqs. (4), noting also eq. (5), gives

$$\max_{r,s} m^r (m|2p-1|)^s \left(\frac{1-\omega}{1+\omega}\right)^{2k} = 1. \tag{6}$$

Several cases must be distinguished:

(i) for $m|2p-1| \leqq 1$, $r$ should be as large as possible, $s$ as small as possible; this yields $r=k$, $s=0$:

$$\beta(2)=1 \quad \text{for} \quad (m-1)/2m \leq p \leq (m+1)/2m. \tag{7}$$

(ii) for $m|2p-1|\geqq 1$ and for a fixed value of $r$, $s$ should be as large as possible; this is the case for $\ell_i=1$, all $i$, and $m_i=0$, all $i$, so that eq. (5) becomes

$$2r+s=2k. \tag{8}$$

Using this to eliminate $s$, eq. (6) becomes

$$\max_r m^{2k-r}(2p-1)^{2k-2r}\left(\frac{1-\omega}{1+\omega}\right)^{2k}=1. \tag{9}$$

Now two subcases follow: (iia) $m(2p-1)^2\geqq 1$ : $r$ should be minimal, i.e., zero:

$$|\beta(1)|=1, \quad p\geqq\frac{\sqrt{m}+1}{2\sqrt{m}} \quad \text{or} \quad p\leqq\frac{\sqrt{m}-1}{2\sqrt{m}}. \tag{10}$$

(iib) $m(2p-1)^2\leqq 1$ : $r$ should be maximal, i.e., equal to $k$:

$$\beta(2)=1, \quad \frac{\sqrt{m}-1}{2\sqrt{m}}\leq p \leq\frac{\sqrt{m}+1}{2\sqrt{m}} . \tag{11}$$

Combination of eqs. (7), (10) and (11) yields the following overall picture: in the interval of eq. (11), all cumulants $K_{2k}$ diverge at $\beta(2)=1$; outside of this interval, they diverge at $|\beta(1)|=1$.

Now consider the phase transitions of type $2k+1$; eqs. (4) and (5) are replaced by

$$\max\left|\prod_{i=1}^{r}\beta(2\ell_i)\prod_{j=1}^{s}\beta(2m_j+1)\right|=1; \quad \sum_{i=1}^{r}(2\ell_i)+\sum_{j=1}^{s}(2m_j+1)=2k+1. \tag{12}$$

The same analysis as for the even case now yields: all odd moments diverge at $|\beta(1)|=1$ outside of the interval of eq. (11); inside this interval, $K_{2k+1}$ diverges for

$$|\beta(1)|\{\beta(2)\}^k=1. \tag{13}$$

Using the explicit forms of $\beta(1)$ and $\beta(2)$ as given by eq. (3), the results can be stated as follows:

(a) All moments diverge for $\omega<\omega_{c_1}$, given as

$$\omega_{c_1} = \frac{m|2p-1|-1}{m|2p-1|+1}; \tag{14}$$

these critical lines do not enter the interval $(m+1)/2m\geq p\geq(m-1)/2m$.

(b) In the interval $(\sqrt{m}-1)/2\sqrt{m} \leq p\leq (\sqrt{m}+1)/2\sqrt{m}$, all <u>even</u> cumulants diverge for $\omega<\omega_{c_2}$, which is given by

$$\omega_{c_2} = (\sqrt{m}-1)/(\sqrt{m}+1); \tag{15}$$

all odd moments with $\ell \geq k$ diverge below $\omega_{c_{2k+1}}$ (k=1,2,..):

$$\omega_{c_{2k+1}} = \frac{\{m^{k+1}|2p-1|\}^{\frac{1}{2k+1}}-1}{\{m^{k+1}|2p-1|\}^{\frac{1}{2k+1}}+1} . \tag{16}$$

These phase transition lines are shown in Fig. 1 for a Cayley branch with m=2. Here "Order" means the divergence of all cumulants, whereas

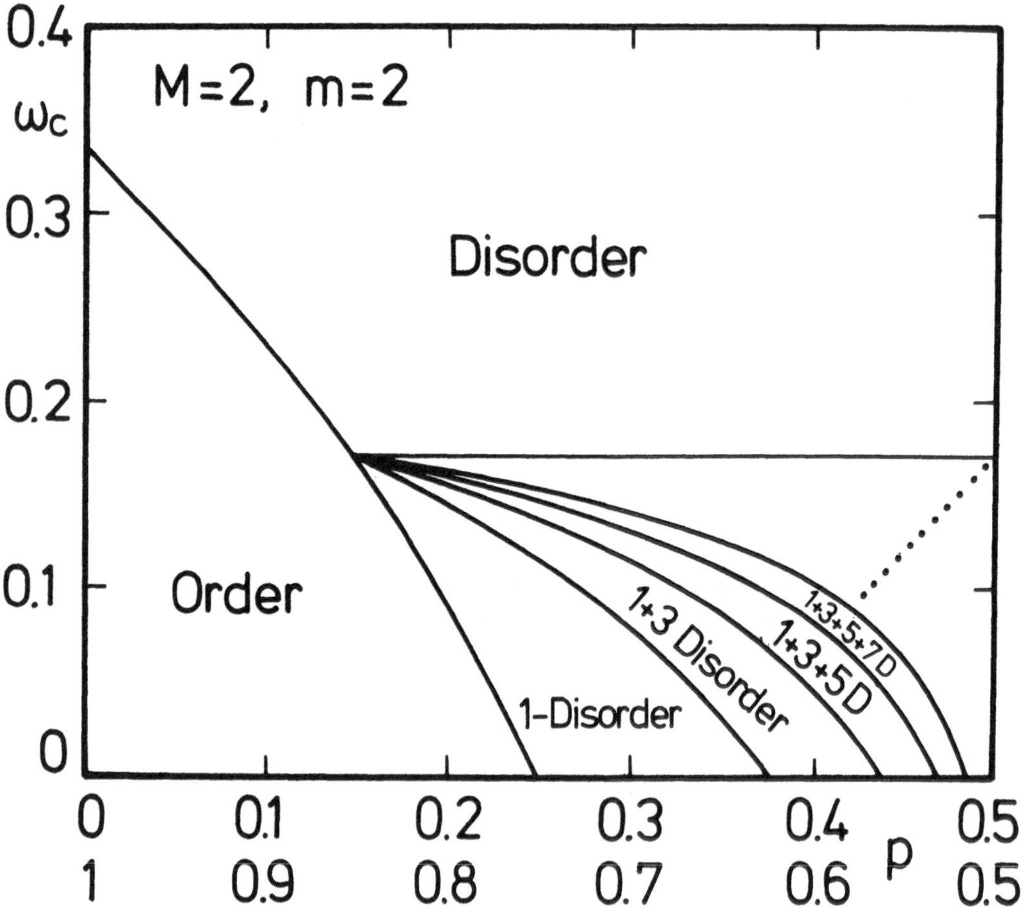


Fig. 1. Phase transitions of the Ising spin glass model on a Cayley branch with m=2. The different phases are explained in the text.

"Disorder" means, that all cumulants approach zero in the limit $n\to\infty$; "1+3+..+(2k-1) D" means, that all cumulants except the first, third,

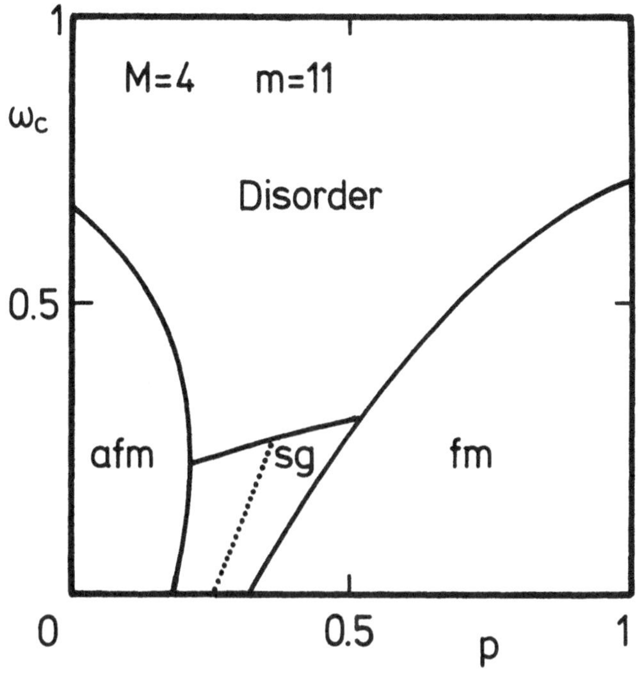


Fig. 1. Phase diagram of the spin glass model for the 4-state Potts model on a Cayley branch with $m=11$. The different phases are: fm for $\beta(1)>1$; afm for $\beta(1)<-1$; sg for $\beta(2)>1$. The dotted line is the "pure spin glass" curve of eq. (19) for this case.

which takes the value 0 for an a spin and 1 for a b spin. The interaction for a pair of spins may then be written in the form:

$$\Omega_{\mu_1\mu_2}(i,j) = (1-\mu_1)(1-\mu_2)\Omega_{aa}(i,j) + \{\mu_1(1-\mu_2)+\mu_2(1-\mu_1)\}\Omega_{ab}(i,j) + \\ + \mu_1\mu_2\Omega_{bb}(i,j). \tag{1}$$

For a special configuration $\Psi$ of the spins, eq. (1.6) is now replaced by:

$$\alpha_k^{(n)}(\Psi) = (\lambda_k/\lambda_o)_{aa}(1-\mu)\sum_{t=1}^{m}(1-\mu_t)\alpha_k^{(n-1)}(\Psi_t') + \\ + (\lambda_k/\lambda_o)_{ab}\sum_{t=1}^{m}\{\mu(1-\mu_t)+(1-\mu)\mu_t\}\alpha_k^{(n-1)}(\Psi_t') + \\ + (\lambda_k/\lambda_o)_{bb}\mu\sum_{t=1}^{m}\mu_t\alpha_k^{(n-1)}(\Psi_t'). \tag{2}$$

Here $\mu$ is the spin-type variable for the top spin of the n-generation branch, whereas the $\mu_t$ are these variables for the top spins of the m

..., (2k-1)-th diverge. Note, that the $\omega_{c_{2k+1}}$ -curves approach, for $k\to\infty$, $\omega_{c_2}$ everywhere except at the point $p=0.5$. The part of the phase diagram, where $K_2$ diverges, whereas the average $K_1\to 0$, is called the <u>spin glass phase</u> ([1]). As shown above, there are still different phase transitions within this phase.

The Potts model for $M\neq 2$. For this case, the eigenvalue ratio is

$$(\lambda_1/\lambda_o)(E) = (1-\omega)/[1+(M-1)\omega], \quad (\lambda_1/\lambda_o)(-E) = -(1-\omega)/(\omega+M-1). \quad (17)$$

The $\beta$-parameters are then, from eq.(1.16),

$$\begin{aligned} \beta(2k) &= m(1-\omega)^{2k}\{p[1+(M-1)\omega]^{-2k}+(1-p)[\omega+M-1]^{-2k}\}, \\ \beta(2k+1) &= m(1-\omega)^{2k+1}\{p[1+(M-1)\omega]^{-2k-1}-(1-p)[\omega+M-1]^{-2k-1}\}. \end{aligned} \quad (18)$$

As in the case of the Ising model, the following two phase transitions are found:
(i) $|\beta(1)|=1$ is the boundary of the completely ordered phase, i.e., a phase, in which all cumulants diverge.
(ii) for $\beta(2)=1$, but $|\beta(1)|<1$, the boundary of a spin glass phase, in which all even cumulants diverge.
Inside this spin glass phase, there are again still more phase transition lines, separating parts of the phase diagram, in which different numbers of odd order cumulants diverge. The line $p=0.5$, on which only even order moments or cumulants diverge for the Ising model, is here replaced by a curve given by:

$$\omega = \frac{pM-1}{(1-p)M-1}, \quad (19)$$

as follows from the second of eqs. (18) for $k=0$. In Fig. 2, the two phase transition boundaries (i) and (ii) above are plotted for the 4-state Potts model on a Cayley branch with $m=11$, see the next page.

## 10.4. Site disorder.

In this section, the randomness is not introduced by bonds with a distribution of interactions, but to a mixture of different types of spins being present. For conciseness, the presence of only two types of spins, called a and b spins, will be treated in what follows. These spins are distinguished by an extra variable $\mu$ at every vertex, (n-1)-generation branches to which it is connected. Eq. (2) leads to two coupled recursion relations for the quantities $\alpha_k^{(n)}(a)=(1-\mu)\alpha_k^{(n)}(\Psi)$ and $\alpha_k^{(n)}(b)=\mu\alpha_k^{(n)}(\Psi)$:

$$\alpha_k^{(n)}(a) = (\lambda_k/\lambda_o)_{aa}(1-\mu)\sum_{t=1}^{m}\alpha_k^{(n-1)}(a) + (\lambda_k/\lambda_o)_{ab}(1-\mu)\sum_{t=1}^{m}\alpha_k^{(n-1)}(b),$$

$$\alpha_k^{(n)}(b) = (\lambda_k/\lambda_o)_{ab}\mu\sum_{t=1}^{m}\alpha_k^{(n-1)}(a) + (\lambda_k/\lambda_o)_{bb}\mu\sum_{t=1}^{m}\alpha_k^{(n-1)}(b). \tag{3}$$

Eqs. (3) reduce markedly in the site-diluted case: if a spin of type b is no spin at all, i.e., has no interaction with any type of spin, then $\alpha_k^{(n)}(\Psi)=\alpha_k^{(n)}(a)$ and $\alpha_k^{(n)}(b)=0$, so that only one recursion remains:

$$\alpha_k^{(n)}(\Psi) = (\lambda_k/\lambda_o)_{aa}(1-\mu)\sum_{t=1}^{m}\alpha_k^{(n-1)}(\Psi_t'). \tag{4}$$

If the concentration of spins of type a is p, $p=\langle 1-\mu\rangle$, then eq. (4) is equivalent to eq. (1.6) for a diluted bond model, see Section 2 as well. Therefore, the diluted site and bond problems are equivalent for Cayley branches and all results of Section 2 are also valid for the site-diluted case. In particular, the critical site percolation probability, defined as the concentration $p_c$ of spins of type a such, that for $p>p_c$ an unbroken chain of a spins extends from the top spin to the surface, equals $m^{-1}$, see Section 2.

Returning now to the general case, the averaging of eqs. (3) is seen to lead to a matrix recursion:

$$\begin{pmatrix}\langle\alpha_k^{(n)}(a)\rangle \\ \langle\alpha_k^{(n)}(b)\rangle\end{pmatrix} = m\begin{pmatrix} p(\lambda_k/\lambda_o)_{aa} & p(\lambda_k/\lambda_o)_{ab} \\ (1-p)(\lambda_k/\lambda_o)_{ab} & (1-p)(\lambda_k/\lambda_o)_{bb}\end{pmatrix}\begin{pmatrix}\langle\alpha_k^{(n-1)}(a)\rangle \\ \langle\alpha_k^{(n-1)}(b)\rangle\end{pmatrix}, \tag{5}$$

where $\langle 1-\mu\rangle=p$ again. A complete analysis of the cumulants for a general spin model is rather complicated and will not be presented here; it is, however, rather easy to show, that, for a model with only one eigenvalue, i.e., for the Ising and Potts model, all cumulants of $\alpha_k^{(n)}(\Psi)$ diverge in the limit $n\to\infty$ as soon as the average of this quantity diverges; by eq. (5), this is the case iff the largest eigenvalue of the matrix in this equation (in absolute value) is larger than 1.

For the Ising model, the matrix of eq. (5) takes the form of eq. (6) on the next page; here $\omega_{11}$, $\omega_{12}$, $\omega_{22}$, are the Boltzmann factors for aa, ab and bb spin-spin interactions, respectively. From this, the largest eigenvalue is easily found.

$$m\begin{bmatrix} p(1-\omega_{11})/(1+\omega_{11}) & p(1-\omega_{12})/(1+\omega_{12}) \\ (1-p)(1-\omega_{12})/(1+\omega_{12}) & (1-p)(1-\omega_{22})/(1+\omega_{22}) \end{bmatrix} \quad (6)$$

The eigenvalues of the matrix (6) have been used to obtain the sample phase diagrams of Fig. 1. In Fig. 1(a), the critical value $\omega_c=\omega_{11}=\omega_{22}$ is shown as a function of the concentration $p$ of type $a$ spins for several values of $\omega'=\omega_{12}$ for a Cayley branch with $m=3$. This figure is typical for the case, that both of the diagonal interactions are ferromagnetic. The structure of the matrix (6) implies, that there is no difference in the phase diagram if $\omega'=\omega_{12}$ is replaced by its corresponding antiferromagnetic value $\omega_{12}^{-1}$. Fig. 1(b) shows the critical curves for several values of $\omega'$ in case $\omega_c=\omega_{11}=\omega_{22}^{-1}$ holds; this is the typical phase diagram for the case that one of the diagonal interactions is ferromagnetic and the other one antiferromagnetic. For Potts models with $M\neq 2$, the phase diagrams look similar, but are somewhat asymmetric.

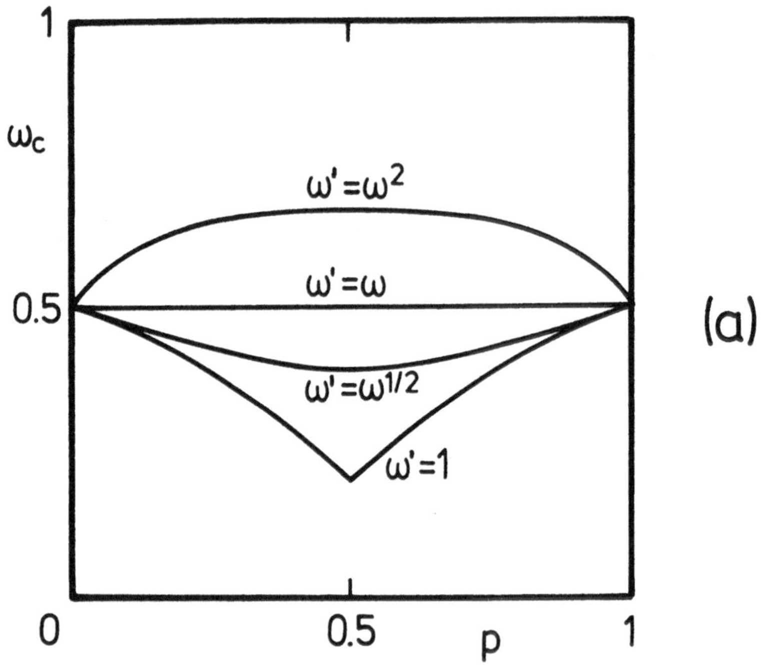


Fig. 1(a). Phase diagram for the site-disordered Ising model on a Cayley branch with m=3. The diagonal interactions are equal and ferromagnetic; if both are replaced by their corresponding antiferromagnetic values, the phase diagram does not change. This is also the case, if the off-diagonal interaction is replaced by its antiferromagnetic analogue, see the text.

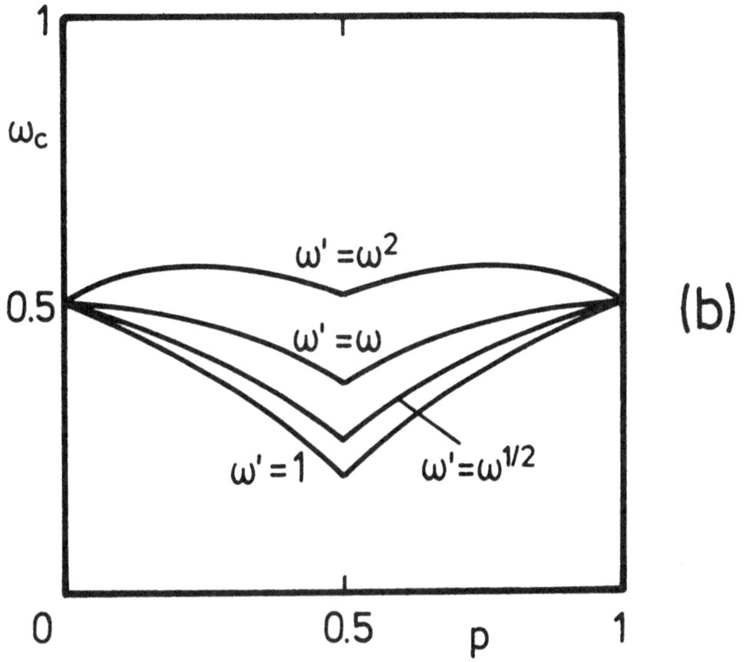


Fig. 1(b). Phase diagram for the site-disordered Ising model on a Cayley branch with m=3. Here one of the diagonal interactions is ferromagnetic, the other one is antiferromagnetic of equal strength.

REFERENCE.

(1). S.F. Edwards and P.W. Anderson, J. Phys. F 5 (1975) 965.

GENERAL REFERENCES.

It is not possible to give here a more or less complete bibliography of the vast subject area touched upon in this chapter. Instead, the reader is referred to two recent reviews on disordered systems:
Ill-Condensed Matter, Les Houches 1978, eds. R. Balian, R. Maynard and G. Toulouse (North-Holland Publ. Co.,Amsterdam, 1978),
Proceedings Rome 1981, Int. Conference on Disordered Systems and Localization, eds. C. Castellani, C. Di Castro and L. Peliti, Lecture Notes in Physics Vol. 149 (Springer-Verlag, Berlin, Heidelberg, New York, 1981).
For percolation problems, good introductions are:
D. Stauffer, Physics Reports 54 (1979) 1
for the general theory and some of the articles from
Percolation Structures and Processes, eds. G. Deutscher, R. Zallen and J. Adler, Annals of the Israel Physical Society, Vol. 5 (1983).
The present chapter is a realization for the Cayley branch of a program proposed recently by
B. Derrida, Physics Reports 103 (1984) 29.
The phase boundaries for the first and second cumulant transitions for spin glass case of the Ising model have also been derived by

T. Horiguchi and T. Morita, J. Phys. A 13 (1980) L71.
who used a homogeneous field. Earlier work for the dilute case was performed by
J. Heinrichs, Phys. Rev. B 19 (1979) 3788,
and for random external fields by
C.E.T. Goncalves da Silva, J. Phys. C 12 (1979) L219.

# 11. SPIN MODELS ON RECURSIVE BOND GRAPH SEQUENCES AND FRACTALS.

## 11.1. Fractals: Koch curves and the Sierpinski gasket.

In this section, the Koch curves, defined as recursive bond graph sequences in Section 8.5, and the Sierpinski gasket, defined as a triangular-site-recursive graph sequence in the same section, are studied. A simple argument shows, that these cannot support phase transitions if spin models are defined on them, except possibly at $T=0$: in both cases, the cutting of a finite number of bonds results in a number of disconnected graphs, even in the thermodynamic limit, and one expects, that graphs with such a finite cutting set cannot sustain long-range order. Below, this is shown explicitly for the Koch curve of Fig. 8.5.1(a) and for the Sierpinski gasket, Fig. 8.5.2, in both cases for the Ising model only.

a) Koch curve. For the Koch curve of Fig. 8.5.1(a), the recursion relation, eq. (8.3.15), for a recursive bond graph sequence, reads:

$$Z_n(i,j) = \sum_{k,\ell,m=1}^{M} Z_{n-1}(i,k)Z_{n-1}(k,\ell)Z_{n-1}(\ell,m)Z_{n-1}(m,j)\Omega'(k,m),$$
$$Z_o(i,j) = \Omega_o(i,j). \qquad (1)$$

Here $\underline{\underline{\Omega}}'$ is the interaction matrix for the noniterated bond. For the Ising model, eq. (1) is easily transformed into a recursion relation for the ratio $x_n$ defined by:

$$x_n = Z_n(i,j)/Z_n(k,k), \quad i\neq j, \quad i,j,k=1,2. \qquad (2)$$

This ratio is independent of $i\neq j$ and $k$ by symmetry; it satisfies

$$x_n = \frac{2(1+\omega')x_{n-1}(1+x_{n-1}{}^2)}{1+2(1+2\omega')x_{n-1}{}^2+x_{n-1}{}^4}, \quad x_o=\omega. \qquad (3)$$

The only fixed points of this recursion relation are $x=0$ and $x=1$, as follows also from the general arguments in subsection 8.3.2. The high-temperature fixed point $x=1$ is stable, whereas the low-temperature fixed point $x=0$ is unstable, see Fig. 1. This shows, that there is no phase transition, except at $T=0$: the Koch curve behaves as a one-dimensional graph, even though its graph fractal dimension is larger than 1.

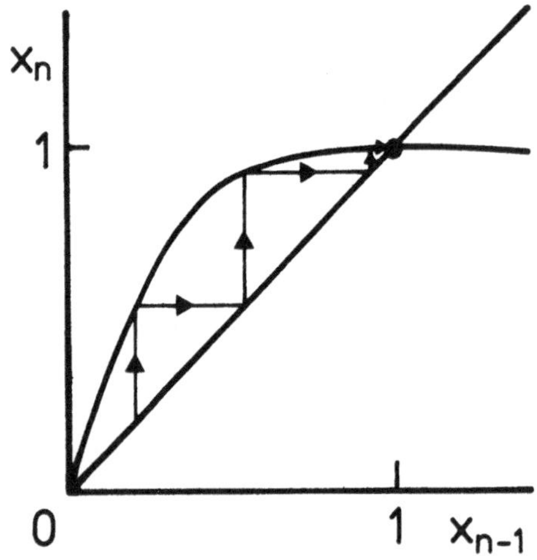


Fig. 1. The recursion relation, eq. (3), for the Ising model on the Koch curve of Fig. 8.5.1(a), showing the absence of phase transitions for $T \neq 0$.

b) Sierpinski gasket. The recursion relation for the partition function $Z_n(i,j,k)$ with the three spins at the corners of a triangle fixed, follows from Fig. 8.5.2 as

$$
\begin{aligned}
Z_n(i,j,k) &= \sum_{\ell,m,h}^{M} Z_{n-1}(i,\ell,m)\,Z_{n-1}(j,\ell,h)\,Z_{n-1}(k,m,h), \\
Z_o(i,j,k) &= \Omega(i,j)\,\Omega(j,k)\,\Omega(i,k).
\end{aligned} \tag{4}
$$

For the Ising model, the symmetry implies, that $Z_n(i,j,k)$ can take only two values:

$$
\begin{aligned}
A_n &= Z_n(1,1,1)=Z_n(2,2,2), \\
B_n &= Z_n(1,1,2)=Z_n(1,2,1)=Z_n(2,1,1)=Z_n(1,2,2)=Z_n(2,1,2)=Z_n(2,2,1).
\end{aligned} \tag{5}
$$

The ratio $x_n=B_n/A_n$ satisfies the recursion relation

$$
x_n = \frac{3x_{n-1}^3+4x_{n-1}^2+x_{n-1}}{4x_{n-1}^3+3x_{n-1}^2+1}, \quad x_o = \omega^2. \tag{6}
$$

Again, $x=0$ and $x=1$ are the only fixed points, the high-temperature one being stable and the low-temperature one marginal, see Fig. 2. There is, therefore, no phase transition, except possibly at $T=0$; since the fixed point is marginal here, a more detailed study would be necessary to establish this.

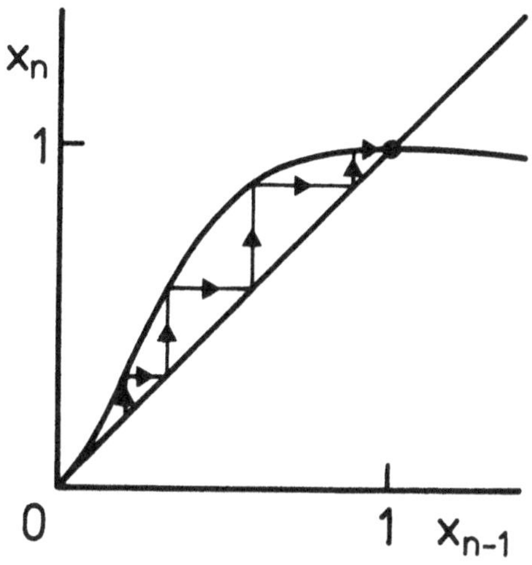


Fig. 2. The recursion relation, eq. (6), for the Ising model on the Sierpinski gasket of Fig. 8.5.2, showing the absence of phase transitions for $T \neq 0$.

In contrast to the above, the Sierpinski carpets, see, e.g., Fig. 8.5.3, are expected to show nontrivial behaviour. This has been found in an approximate treatment only ([1]), since no exact renormalization equations have been found for these fractals, which do not fit in with the general theory presented here. For this reason, these are not considered here in detail.

## 11.2. A self-dual diamond hierarchical lattice.

As an example of a nontrivial recursive bond graph sequence, the self-similar sequence $\{D_n\}$ built from the diamond of Fig. 8.2.1(d) is considered in this section ([2]). This sequence consists of planar graphs, which are nearly self-dual, see Fig. 1, where the graph $D_2$ is shown together with its dual. In fact, the duality relation (7.1.26)

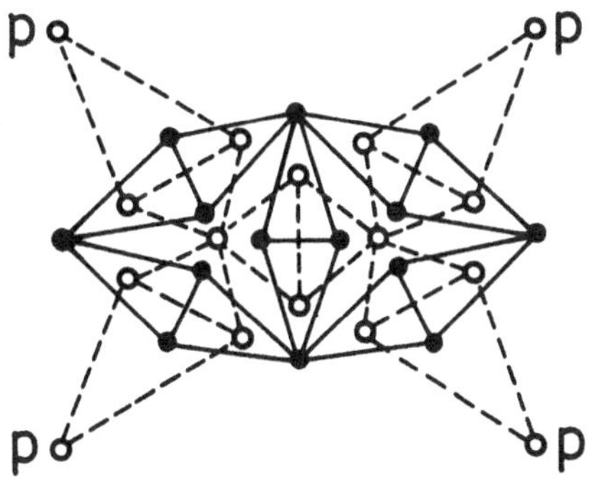


Fig. 1. The graph $D_2$, solid edges and vertices, and its dual, open vertices and broken edges. The four vertices marked P are one and the same vertex of the dual graph.

reads, for $D_n$,

$$Z(\Omega,D_n) = \sqrt{M}\ (\lambda_o/\sqrt{M})^{5^n} Z(\tilde{\Omega},\tilde{D}_n), \tag{1}$$

where $\tilde{D}_n$ is obtained from $D_n$ by the merging of its two blue vertices (which corresponds to the identification of the four vertices marked P in Fig. 1). Here use has been made of the numbers of vertices and edges for a self-similar recursive bond graph sequence, eqs. (8.2.9), with $m'=5$, $\Delta m'=0$, $f'=4$, to yield

$$E(D_n) = E_y(D_n) = 5^n, \quad V(D_n) = (5^n+3)/2. \tag{2}$$

The sequence $\{D_n\}$ is, therefore, nontrivial with $\delta'=1$, $\varepsilon'=2$, see eqs. (8.2.11,12).

The partition functions occurring in eq. (1) can be expressed in terms of the partition functions $Z_n(i,j)$ and $\tilde{Z}_n(i,j)$ for fixed blue spins, corresponding to interactions $\Omega(i,j)$ and $\tilde{\Omega}(i,j)$, both on $D_n$, by

$$Z(\Omega,D_n) = \sum_{i,j=1}^{M} Z_n(i,j), \quad Z(\tilde{\Omega},\tilde{D}_n) = \sum_{i=1}^{M} \tilde{Z}_n(i,i). \tag{3}$$

From eq. (1), the free energies $\gamma$ and $\tilde{\gamma}$ per spin then are related by

$$\gamma = \ln(\lambda_o^2/M) + \tilde{\gamma} \tag{4}$$

in the thermodynamic limit; this is the same relation as derived for hypercubic lattices in Section 7.6. Since the average number of bonds per vertex is $\varepsilon'=2$, the sequence $\{D_n\}$ models the square lattice. Fixed points of the basic recursion relation for $\{D_n\}$,

$$Z_n(i,j) = \sum_{k,\ell}^{M} Z_{n-1}(i,k)Z_{n-1}(i,\ell)Z_{n-1}(k,\ell)Z_{n-1}(k,j)Z_{n-1}(\ell,j), \tag{5}$$

will, therefore, give rise to phase diagrams, which respect the duality restrictions derived in Section 7.6, at least for models with a symmetry group $\mathcal{G}$, which contains an Abelian, regular subgroup, and for which $\mathcal{G}$ is self-dual. Models, for which one or the other of these requirements is not satisfied, can, of course, also be studied on $\{D_n\}$; this will, however, not be attempted in this section.

The recursion relation, eq. (5), can be normalized in such a way, that the diagonal elements all become equal to 1; this is then an effective interaction matrix $\underline{\underline{\Omega}}_n$:

$$\Omega_n(i,j) = \frac{\Sigma_{k,\ell}\ \Omega_{n-1}(i,k)\,\Omega_{n-1}(i,\ell)\,\Omega_{n-1}(k,\ell)\,\Omega_{n-1}(k,j)\,\Omega_{n-1}(\ell,j)}{\Sigma_{k,\ell}\ \Omega_{n-1}(1,k)\,\Omega_{n-1}(1,\ell)\,\Omega_{n-1}(k,\ell)\,\Omega_{n-1}(k,1)\,\Omega_{n-1}(\ell,1)}, \tag{6}$$

to be started with some initial interaction $\Omega_o(i,j)$. The mapping R: $\underline{\underline{\Omega}}_{n-1} \to \underline{\underline{\Omega}}_n$ can be considered as a mapping of the s-dimensional vector space of $\omega$-parameters $\omega_1,\ldots,\omega_s$, on which $\underline{\underline{\Omega}}$ depends, into itself. It is easily seen, that R commutes with the duality transformation D of Section 7.6, since the diagram of Fig. 2 is commutative:

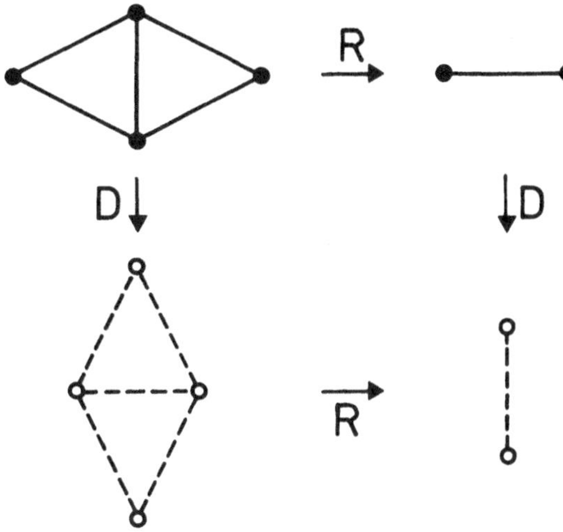


Fig. 2. Commutative diagram of the renormalization R and of the duality transformation D; edges and vertices as in Fig. 1.

The same holds for other duality transformations, obtained by combining D with a symmetry of the partition function. Therefore, R leaves all self-dual hyperplanes invariant, so that there must be fixed points of the renormalization transformation on these. In the following, some specific models are considered.

a) Ising and Potts models. The recursion relation, eq. (6), reduces to a single recursion relation for the off-diagonal elements of $\underline{\underline{\Omega}}_n$, which will be denoted by $x_n$:

$$x_n = f_M(x_{n-1}), \quad x_o=\omega, \quad f_M(x) = \frac{2x^2+2x^3+5(M-2)x^4+(M-2)(M-3)x^5}{1+2(M-1)x^3+(M-1)x^4+(M-1)(M-2)x^5}. \tag{7}$$

It is not difficult to show, that the equation $x=f_M(x)$ has three real solutions: these are the stable high- and low-temperature fixed points $x=1$ and $x=0$ and one unstable fixed point, $x_c$, which is exactly the value for $\omega$, for which the Potts model is self-dual:

$$x_c = (1+\sqrt{M})^{-1}. \tag{8}$$

The function $f_3(x)$ is shown in Fig. 3 as an example.

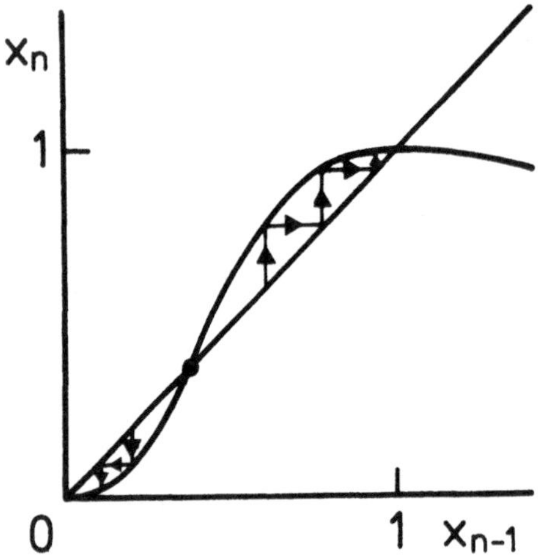


Fig. 3. The recursion relation for the off-diagonal effective Boltzmann factor for the 3-state Potts model on $\{D_n\}$, showing the unique phase transition at the self-dual fixed point.

The asymptotic behaviour of $f_M(x)$ is different for different values of $M$: for $M=2$ or 3, the function approaches 0 for $x\to\infty$; this implies, that there is also an antiferromagnetic phase transition point at a value $x_{af}$, such that

$$x_c = f_M(x_{af}) \tag{9}$$

holds. The values for $M=2,3$ are 4.0143.. and 6.0265.., respectively. For $M\geqq 4$, $f_M(x)$ approaches the value $(M-3)/(M-1)$ as $x\to\infty$ and it is monotonically decreasing for $x>1$. Since one has

$$x_c \leqq (M-3)/(M-1), \quad \text{equality iff } M=4, \tag{10}$$

there is no antiferromagnetic phase transition for $M\neq 2$ or 3. This is in complete agreement with the Cayley branch calculation in subsection 9.2.1, since $\varepsilon'=2$ corresponds to a Cayley branch with $m=3$ and the criterion for the presence of an antiferromagnetic phase transition is

$$M > m+1=4, \quad M=4 \text{ is marginal.} \tag{11}$$

This evidence strongly suggests, that the Potts model on a square lattice has an antiferromagnetic phase transition at a finite temperature for $M=2$ or 3, at $T=0$ for $M=4$ and no such transition for $M>4$.

b) $\mathcal{S}(2)\sim\mathcal{S}(2)=\mathcal{D}(4)$. For this model, the $\underline{\underline{\Omega}}_n$ contain two parameters, $x_n$, corresponding to the z=1-graph of the MI, and $y_n$, corresponding to the graph with z=2. The recursion relations for these quantities are found as

$$x_n=f_1(x_{n-1},y_{n-1}),\quad y_n=f_2(x_{n-1},y_{n-1}),$$

$$f_1(x,y)=\frac{2x^2+2y^3+8xy^3+2xy^4+2y^4}{1+2x^3+x^4+2xy^4+4x^2y^3+2y^4+4y^3}, \tag{12}$$

$$f_2(x,y)=\frac{2y^2+2y^3+6x^2y^2+4xy^3+2x^2y^3}{1+2x^3+x^4+2xy^4+4x^2y^3+2y^4+4y^3}.$$

The fixed points of these recursion relations are:
(i) the stable fixed points (attractors) $(x,y)=(0,0)$, $(1,0)$ and $(1,1)$;
(ii) the repellor $(x,y)=(\frac{1}{3},\frac{1}{3})$, corresponding to the four-state Potts model critical point;
(iii) the saddle points $I=(\sqrt{2}-1,0)$, $\tilde{I}=(1,\sqrt{2}-1)$ and $I_1=(3-2\sqrt{2},\sqrt{2}-1)$, all corresponding to points, where the model factorizes into a product of factors, one of which at least is a critical Ising model point.
These points are all shown in the phase diagram of Fig. 4, together

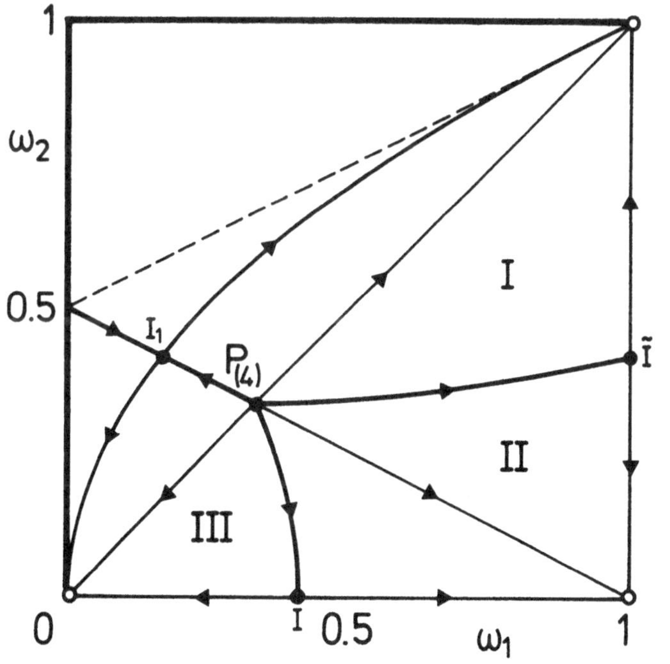


Fig. 4. The phase diagram of the $\mathcal{D}(4)$ model from the diamond renormalization. The solid lines are the phase transition curves, separating the phases I, II and III, see also the text.

with some of the flows or trajectories of the recursion relation, i.e., curves left invariant by R, such as the self-dual line $\omega_1+2\omega_2=1$, and the Ising-like curves $\omega_2=0$, $\omega_1=1$ and $\omega_1=\omega_2^2$. The nature of the fixed points as given above, is determined by linearizing eqs. (12) in the neighbourhood of such a fixed point:

$$\begin{pmatrix}\Delta x\\ \\ \Delta y\end{pmatrix}' = \begin{pmatrix}(\partial f_1/\partial x) & (\partial f_1/\partial y)\\ \\ (\partial f_2/\partial x) & (\partial f_2/\partial y)\end{pmatrix}\begin{pmatrix}\Delta x\\ \\ \Delta y\end{pmatrix}; \tag{13}$$

here all partial derivatives are evaluated at the fixed point and the vector on the left-hand-side is the (small) deviation from this, on which R maps the original small deviation. There are now three possibilities:

(i) both eigenvalues of the matrix in eq. (13) are, in absolute value, less than 1: this is a stable fixed point or attractor;

(ii) both eigenvalues are, in absolute value, larger than 1: this is a repellor, or completely unstable fixed point;

(iii) one eigenvalue larger, the other smaller than 1 (absolutely): this is a saddle point: the fixed point is attractive in the direction of the eigenvector corresponding to the smaller eigenvalue, repulsive in the other eigenvector direction.

In the above classification, the marginal fixed points, for which one or both of the eigenvalues are equal to 1 in absolute value, have not been mentioned, since these do not occur for any of the models in this section.

The phase transition lines are, by definition, those on which the recursion, eq. (12), goes towards an unstable fixed point or stays at such a point; since there are three stable fixed points for the present model, there are three basins of attraction and then, consequently, also three phase transition curves separating these. One of them is the self-dual line for $\omega_2 \geq \frac{1}{3}$; the other two phase transition lines must start at the critical Potts model point P(4) and correspond to trajectories ending at the saddle points I and $\tilde{I}$; these trajectories are dual to each other, see Fig. 4. This phase diagram is very similar to the one obtained for the same model on the Cayley branch in the ferromagnetic region, see Fig. 7.6.1(a). Therefore, the three phases marked I, II and III in Fig. 4 can be identified as the disordered phase, the phase with residual $\mathcal{K}(4)$-symmetry and the completely ordered phase, respectively. For other models with $\mathcal{S}(M_1)\sim\mathcal{S}(M_2)$ as symmetry group, similar results confirming the Cayley branch calculations of Section 7.6 are also found by the present renormalization prescription.

c) $\mathbb{D}(5)$ and $\mathbb{G}(G_9)$. For the model with $\mathbb{D}(5)$ symmetry, recursions for $x_n$ and $y_n$, defined as the values of $\omega_1$ and $\omega_2$ for the matrix $\underline{\underline{\Omega}}_n$ for this model, follow as in eqs. (12), but with $f_1$ and $f_2$ replaced by

$$f_1'(x,y) = \frac{2x^2+2x^3+y^4+4xy^3+4x^3y+6x^2y^2+6x^2y^3}{1+2x^4+2y^4+4x^3+4y^3+2x^4y+2xy^4+4x^2y^3+4x^3y^2}\ , \quad f_2'(x,y)=f_1'(y,x)\,. \tag{14}$$

As stable fixed points, only the low- and high-temperature fixed points (0,0) and (1,1) are found, whereas the 5-state Potts model critical point P(5) is again a repellor. It follows, that there must be two saddle points $S_1$ ans $S_2$ on the self-dual line; these are shown in Fig. 5 as well. These saddle points are now not so, that the model reduces here in some way; in fact, the trajectories of the recursion are not the curves $\omega_1=\omega_2{}^a$ and $\omega_2=\omega_1{}^a$ with a such, that these pass through the saddle points. (a=4.60461..).

It is concluded, that the whole self-dual line is the unique phase transition line separating the disordered high-temperature phase from the completely ordered low-temperature one. This conclusion was also reached tentatively from duality alone in Section 7.6 and supported by the Cayley branch calculation in Section 9.3.

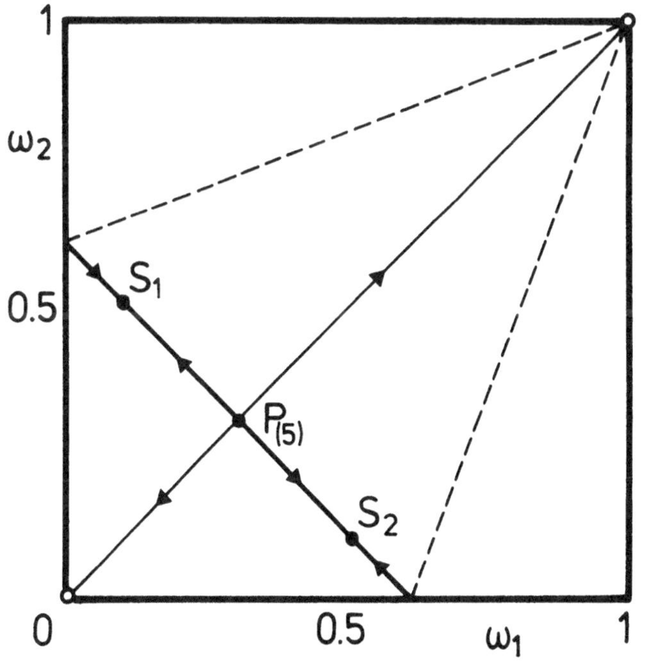


Fig. 5. The phase diagram of the $\mathbb{D}(5)$ model from the diamond renormalization. The stable and unstable fixed points are described in the text; the solid (self-dual) line is the phase transition line.

For $\mathcal{G}(G_9)$, similar conclusions are reached, see Fig. 6. Here, however, the saddle points on the self-dual line, $P_1(3)$ and $P_2(3)$, do correspond to the critical 3-state Potts model critical points on the trajectories $\omega_1=\omega_2^2$ and $\omega_2=\omega_1^2$. The high- and low-temperature fixed points are again the only attractors and $P(9)$, the 9-state Potts model critical point, is the only repellor.

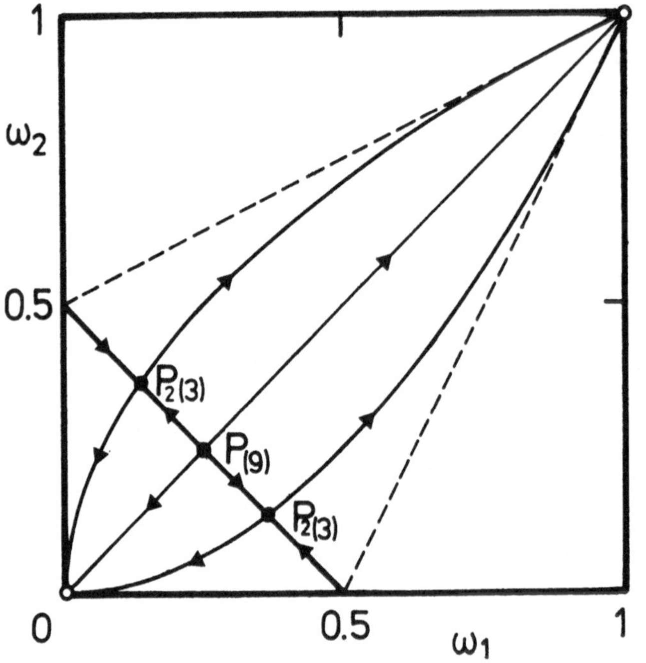


Fig. 6. The phase diagram of the $\mathcal{G}(G_9)$ model, as obtained from the self-dual diamond renormalization. The fixed points and some of the trajectories are shown; the unique phase transition line is the (solid) self-dual line.

d) $\mathcal{S}(2)\otimes\mathcal{S}(2)$. Here, there are three recursion relations for the three parameters $\omega_1$, $\omega_2$ and $\omega_3$ of $\underline{\underline{\Omega}}_n$. Instead of giving these explicitly, the results for the fixed points and the trajectories will be discussed using Fig. 7.6.2 for the phase diagram obtained by duality:
(i) There are five stable fixed points: $(1,1,1)$ is the high-temperature fixed point; its basin of attraction is the disordered phase; the three stable fixed points $(1,0,0)$, $(0,1,0)$ and $(0,0,1)$ have basins of attraction corresponding to the three different unbroken $\mathcal{S}(2)$-symmetries found in the Cayley branch calculation, Section 9.4; finally, the low-temperature fixed point $(0,0,0)$ attracts the completely ordered phase.
(ii) The 4-state Potts critical point $P(4)$ is the only repellor.
(iii) The points $I_1$, $I_2$, $I_3$, $D_1$, $D_2$ and $D_3$ are saddle points, which

attract in two directions and repel in the third one.
(iv) The points $C_1$, $C_2$ and $C_3$ are saddle points with two repulsive directions and one attractive one.
The phase diagram contains two phase transition planes; these touch each other along those parts of the self-dual lines, which extend in the directions $P(4)C_i$, $i=1,2,3$, so forming three isolated "pockets" of unbroken $\mathcal{S}(2)$ symmetry.

Similar conclusions can be drawn for other self-dual models with three graphs in their MI's; in all cases, the qualitative conclusions reached by means of the Cayley branch calculations of Section 9.4 are supported by the present renormalization group calculations. The same is true for the non-self-dual models; for these, however, it is simpler to use the (non-self-dual) diamond lattice of Fig. 8.2.1(b). The recursion relation is here:

$$z_n(i,j) = \left\{\sum_{k=1}^{M} z_{n-1}(i,k)\,z_{n-1}(k,j)\right\}^2 . \tag{15}$$

This is known as the (approximate) Migdal-Kadanoff ([3]) renormalization group. The qualitative results obtained by means of thes recursion for the non-self-dual models are again in agreement with the Cayley branch calculations of Sections 9.3-5 and will not be given here explicitly ([4]).

## 11.3. Other recursive bond graph sequences: the Potts model.

It is obvious from Section 8.2, that there is an enormous variety of recursive bond graph sequences, even if only the self-similar ones are considered. In this section, the influence of the non-existence of one or both of the "trivial" fixed points, i.e., the high-temperature fixed point with all $\omega_i=1$ and the low-temperature one with all $\omega_i=0$, will be studied for some simple sequences. Throughout, only the Potts model (and its special case, the Ising model) will be considered, since this gives one-dimensional recursion relations.

### 11.3.1. Absence of the high-temperature fixed point.

It has been shown in subsection 8.3.2, that the high-temperature fixed point is absent if there is a path of "normal", i.e., noniterated, bonds in every graph $H_n$ in the store of a recursive bond graph sequence. Therefore, the simplest, nontrivial sequence is the self-similar one built up from the graph of Fig. 8.2.1(c), see also Fig. 8.2.2(a) for the first construction step. Calling the Boltzmann factor matrix for the noniterated bond $\underline{\underline{\Omega}}'$, the recursion relation for the partition function with fixed states i and j of the spins at the blue vertices is given by:

$$z_n(i,j) = \Omega'(i,j) \sum_{k=1}^{M} z_{n-1}(i,k) z_{n-1}(k,j), \quad z_o(i,j)=\Omega(i,j). \tag{1}$$

Specializing this to the Potts model, the renormalized off-diagonal element of $\underline{\underline{\Omega}}_n$ is given as

$$x_n = \omega' \frac{2x_{n-1} + (M-2)x_{n-1}^2}{1+(M-1)x_{n-1}^2}, \quad x_o=\omega. \tag{2}$$

Consider first the case $M=2$, the Ising model. Then the fixed points of the recursion relation (2) are the low-temperature fixed point at $x_0=0$ and, for $\omega'>\frac{1}{2}$, also the fixed point $x_1=\sqrt{2\omega'-1}$. This latter fixed point is stable as soon as it exists. Therefore, the model has no phase transition for $\omega'\leq\frac{1}{2}$: for every initial value $\omega$, the sequence $\{x_n\}$ converges to 0. For $\omega'>\frac{1}{2}$, $\{x_n\}$ converges to $x_1$ for all $\omega\neq 0$; there is a T=0 phase transition now. The resulting phase diagram is shown in Fig. 1. The fixed point $x_1$ plays the role of the absent high-temperature fixed point, but only for $\omega'$ sufficiently large.

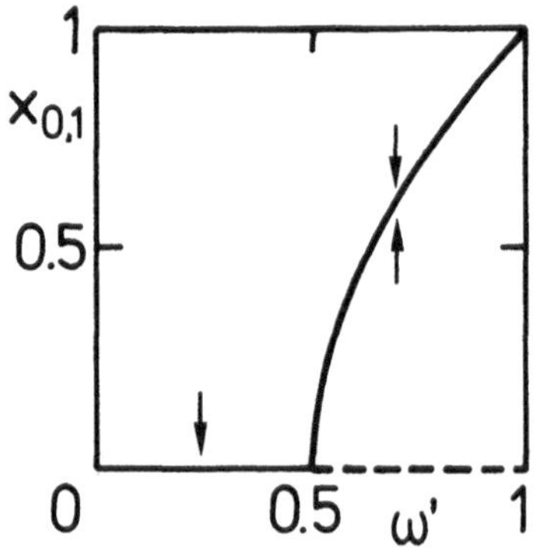


Fig. 1. Phase diagram of the Ising model on a recursive bond graph sequence without high-temperature fixed point. Solid curves are stable fixed points, broken curves unstable ones; the arrows indicate the direction, in which the recursion goes in different regions of the phase diagram.

For $M\neq 2$, the situation is slightly more complicated. For $\omega'<\omega_c$, given by

$$\omega_c = \{2\left(4(M-1)^2+(M-1)(M-2)^2\right)^{\frac{1}{2}}-4(M-1)\}/(M-2)^2, \tag{3}$$

there is still only the stable low-temperature fixed point. For $\omega_c\leq\omega'<\frac{1}{2}$, there are three fixed points, $x_0$ and $x_{1,2}$:

$$x_{1,2} = \{(M-2)\omega'\pm\left[(M-2)^2\omega'^2-4(M-1)(1-2\omega')\right]^{\frac{1}{2}}\}/2(M-1). \tag{4}$$

Of these, $x_0$ and the largest of $x_{1,2}$ are stable. For $\omega' \geq \frac{1}{2}$, the fixed points are $x_0$ and the largest of $x_{1,2}$, the latter being stable only. As an example, the phase diagram for $M=8$ is shown in Fig. 2. The phase transitions not at $T=0$ (for $\omega' > \frac{1}{2}$) are on the curve connecting $\omega_c$ and $\frac{1}{2}$. This is akin to the high-field phase transitions found for Potts models on Cayley branches in subsection 9.2.4.

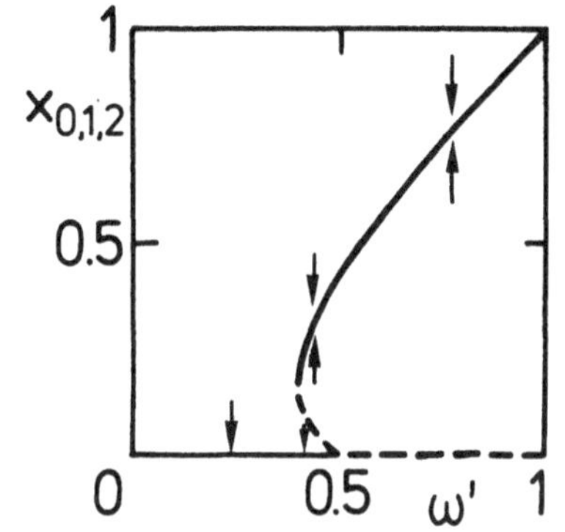


Fig. 2. Phase diagram of the 8-state Potts model on a recursive bond graph sequence without high-temperature fixed point. The curves are as in Fig. 1.

11.3.2. Absence of the low-temperature fixed point.

The low-temperature fixed point is absent, if there is no path of yellow (iterated) edges connecting the two blue vertices of each $H_n$. The simplest self-similar, nontrivial recursive bond graph sequence with this property can be built up from the graph of Fig. 3 below, where also the first iteration is shown. The recursion relation for such a

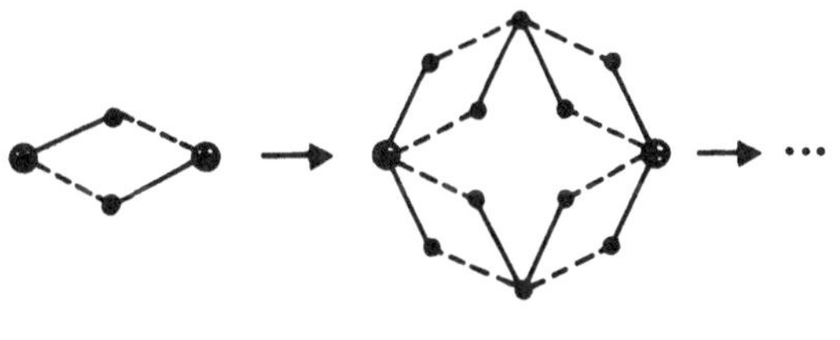

Fig. 3. The first two steps of the construction of the recursive bond graph sequence without low-temperature fixed point considered in the present subsection.

sequence is:

$$Z_n(i,j) = \{\sum_{k=1}^{M} \Omega'(i,k) Z_{n-1}(k,j)\}\{\sum_{k=1}^{M} Z_{n-1}(i,k)\Omega'(k,j)\}, \quad Z_0(i,j)=\Omega(i,j). \tag{5}$$

For a model with a completely permissible symmetry group, the indices $i$ and $j$ can be exchanged in the second term to yield the simpler recursion relation:

$$z_n(i,j) = \left[\sum_{k=1}^{M} \Omega'(i,k)\, z_{n-1}(k,j)\right]^2, \quad z_o(i,j)=\Omega(i,j). \tag{6}$$

This amounts to an exchange of a normal and a yellow edge in Fig. 3; iteration of this yields another recursive graph sequence similar to a Cayley branch with $m=2$. (Other values of $m$ can also be simulated in this way by connecting the blue vertices by $m$ instead of 2 copies of the unit consisting of a yellow and a normal edge in series.) This is shown in Fig. 4. It is, therefore, to be expected, that the recursion eq. (6) is similar to a Cayley branch recursion. For the Potts model,

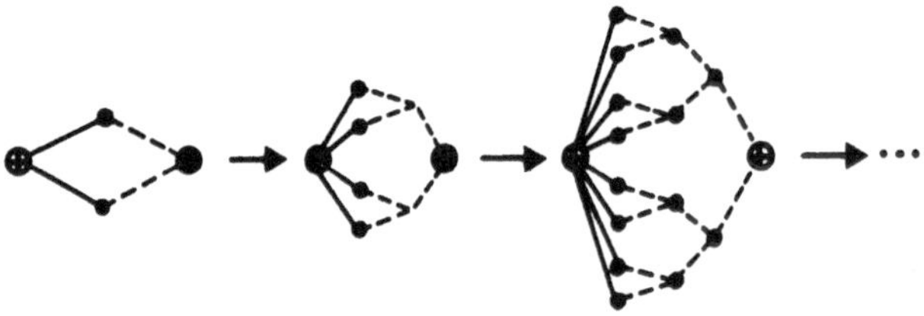

Fig. 4. This recursive bond graph sequence is equivalent to the one of Fig. 3 for models with completely permissible symmetry group.

this is easily seen to be the case, since eq. (6) then becomes, for the off-diagonal element of $\underline{\underline{\Omega}}_n$,

$$x_n = \left(\frac{\omega'+x_{n-1}+(M-2)\,\omega' x_{n-1}}{1+(M-1)\,\omega' x_{n-1}}\right)^2, \quad x_o=\omega, \tag{7}$$

which is easily recognized as being the same as eq. (9.2.13) for the Potts model on a Cayley branch with $m=2$, provided all $x(i)$ are set equal. For $M=2$, in particular, the Ising model recursion, eq. (9.2.6), is recovered. All results of subsections 9.2.3 and 9.2.4 are then here valid as well upon replacement of the term "zero-field fixed point" by "high-temperature fixed point". This translation will be given in what follows:

(i) For $M=2$, the recursion, eq. (7), converges to the stable high-temperature fixed point for $\omega'$ in the range $[\frac{1}{3},3]$. For $0 \leq \omega' < \frac{1}{3}$, this is unstable and $\{x_n\}$ converges to the solution $x_1<1$ of

$$\omega'^2 x^2-(1-2\omega'-\omega'^2)\,x+\omega'^2=0, \tag{8}$$

see Fig. 5. For $\omega'>3$, the recursion approaches a two-point limit cycle consisting of $x_1$ and $x_1^{-1}$ belonging to $\omega'^{-1}$.

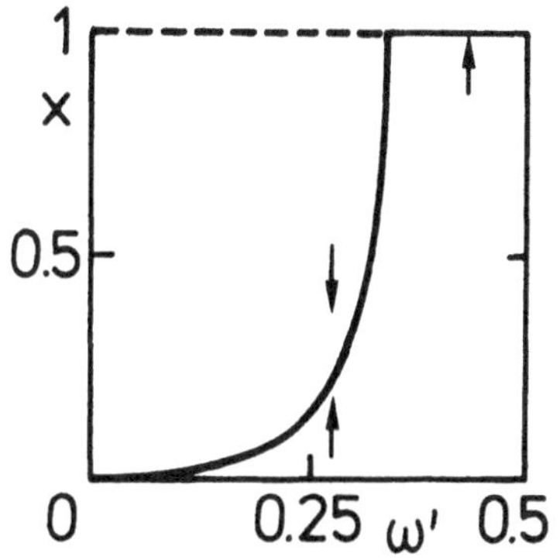


Fig. 5. Phase diagram of the Ising model on the recursive bond graph sequence of Figs. 3 or 4 without low-temperature fixed point. Curves as in Fig. 1.

(ii) For $M\neq 2$, the situation is slightly more complex because of the high-field phase transitions treated in subsection 9.2.4. One has the following regions: (a) $\omega'\geq\omega_c=(1+2\sqrt{M-1})^{-1}$ (see eq. (9.2.18) for $r=M-1$); here $x=1$ is the only (stable) fixed point; (b) $(M+1)^{-1}\leq\omega'<\omega_c$: the smallest solution of

$$(M-1)^2\omega'^2x^2-\{1-2\omega'-(2M-3)\omega'^2\}x+\omega'^2=0 \tag{9}$$

is a stable fixed point as is $x=1$; the other solution of eq. (9) is unstable; (c) $\omega<(M+1)^{-1}$: here the smallest solution of eq. (9) is stable, $x=1$ is unstable. The resulting phase diagram is shown in Fig. 6 for the case $M=5$. Figs. 5 and 6 are very similar to Figs. 1 and 2 with the roles of the high- and low-temperature fixed points exchanged.

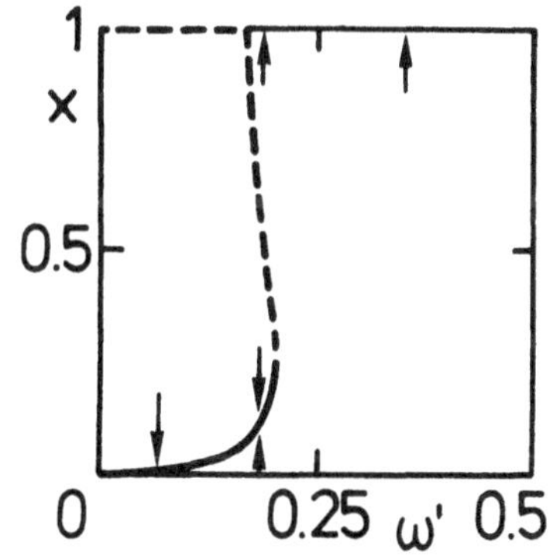


Fig. 6. As Fig. 5 for the five-state Potts model.

### 11.3.3. Absence of both high-and low-temperature fixed points.

In the previous two subsections, only one of the two "trivial" fixed points was absent; in these cases, the unstable fixed point from the case with both trivial fixed points present, becomes stable at sufficiently high or low temperatures. For the Potts models with $M \neq 2$, there is also an "in-between region", in which the phase diagram is very similar to the case with both trivial fixed points present: an unstable fixed point sandwiched between two stable ones. This suggests, that for cases, in which both of the trivial fixed points are absent, only one new, stable fixed point will be found. For the simple self-similar recursive bond graph sequences built up from the graphs of Fig. 7, for which there are, obviously, no trivial fixed points, only one stable fixed point is indeed found for the Potts model for all values of $M$. There are, therefore, no phase transitions on these pseudo-lattices.

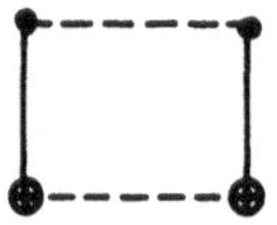

(a)

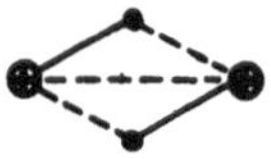

(b)

Fig. 7. Two building blocks for self-similar recursive bond graph sequences without trivial fixed points. There are no phase transitions for the Potts models on the corresponding pseudo-lattices.

REFERENCES.


(1). Y. Gefen, A. Aharony and B.B. Mandelbrot, J. Phys. A 17 (1984) 1277.
(2). M. Kaufman and R.B. Griffiths, Phys. Rev. B 24 (1981) 496.
(3). A. A. Migdal, Zh. Eksp. Teor. Fiz. 69 (1975) 1457; [Sov. Phys. JETP 42 (1976) 743].
L.P. Kadanoff, Ann. Phys. (N.Y.) 100 (1976) 359.
(4). See, e.g., V.L. Balton, G.M. Carneiro, M.E. Pol and N. Zagury, J. Phys. A 17 (1984) 2119.

## 12. GAUGE MODELS ON PLAQUETTE BRANCHES.

### 12.1 General formulation.

q-Plaquette branches with branching ratio $m$ are defined (see the end of Section 8.2) as the self-similar recursive bond graph sequences obtained from a primitive graph $H(q,m)$, which consists of $m$ polygons with $q$ edges, which all have exactly one normal edge (noniterated) in common, the remaining $m(q-1)$ edges being yellow (iterated); the blue vertices are at the end points of the noniterated edge. The basic recursion relations for gauge models with regular group $\mathcal{R}$ on such graphs have been given as eqs. (8.3.39-41) in subsection 8.3.3. For the q-plaquette branch with branching ratio $m$, these become, for a Higgs field on the surface only, (the tildes are dropped in this chapter)

$$\rho_n(r_1) = \frac{\left[\sum_{r_2,\ldots,r_q\varepsilon\mathcal{R}} \Omega_C\left(\prod_{i=1}^{q} r_i\right)\prod_{i=2}^{q}\rho_{n-1}(r_i)\right]^m}{\sum_{r_1\varepsilon\mathcal{R}}\left[\sum_{r_2,\ldots,r_q\varepsilon\mathcal{R}} \Omega_C\left(\prod_{i=1}^{q} r_i\right)\prod_{i=2}^{q}\rho_{n-1}(r_i)\right]^m}, \quad \rho_o(r_1)=A(r_1)/\{\sum_{r'\varepsilon\mathcal{R}} A(r')\}. \tag{1}$$

Here use has been made of the fact, that $\rho_n(r)=\rho_n(r^{-1})$ holds, since this holds for $n=0$, see the definition of the Higgs field in Section 7.7. Also, all plaquettes are traversed in the same direction, so that $r_1$ can always be taken as the first element in the product of group elements, which is the argument of the Boltzmann factor $\Omega_C(.)$. This Boltzmann factor is a class function on the group $\mathcal{R}$.

A stability analysis of the field-free fixed point

$$\rho^{(o)}(r) = M^{-1}, \quad \text{all } r\varepsilon\mathcal{R},\ M=|\mathcal{R}|, \tag{2}$$

in the same vein as performed in Section 9.1, immediately confirms Elitzur's theorem (see Section 7.7 and subsection 8.3.3); setting

$$\rho_n(r) = M^{-1}+\delta_n(r), \tag{3}$$

so that the normalization implies

$$\sum_{r\varepsilon\mathcal{R}}\delta_n(r) = 0, \tag{4}$$

the numerator of the right-hand-side of eq. (1) becomes, up to first order in $\underline{\delta}$ :

$$\sum_{r_2,\ldots,r_q \in R} \Omega_C\left(\prod_{i=1}^{q} r_i\right)\{M^{-q+1}+M^{-q+2}\sum_{i=2}^{q}\delta_{n-1}(r_i)\}. \tag{5}$$

Since $q \geq 3$ holds, every term in the sum contains at least one free summation:

$$\sum_{r_k \in R} \Omega_C\left(\prod_{i=1}^{q} r_i\right) = \sum_{r \in R} \Omega_C(r) = \lambda_o, \tag{6}$$

so that the other summations imply, for eq. (1),

$$\rho_n(r_1) = \rho^{(o)}(r_1) + (\text{terms at least quadratic in } \underline{\delta}_{n-1}). \tag{7}$$

Therefore, $\delta_n(r)=0$ for all $n \neq 0$, and the field-free fixed point is always stable.

For practical calculations, it is expedient to introduce another normalized distribution by

$$\sigma_n(r) = [\rho_n(r)]^{\frac{1}{m}}/\{\sum_{r' \in R}[\rho_n(r')]^{\frac{1}{m}}\}. \tag{8}$$

In terms of this distribution, eq. (1) becomes

$$\sigma_n(r_1) = \frac{\sum_{r_2,\ldots,r_q \in R} \Omega_C\left(\prod_{i=1}^{q} r_i\right) \prod_{i=2}^{q} \sigma_{n-1}(r_i)^m}{\lambda_o\left(\sum_{r' \in R}\sigma_{n-1}(r')^m\right)^{q-1}}, \quad \sigma_o(r)=A(r)^{\frac{1}{m}}/\{\sum_{r' \in R} A(r')^{\frac{1}{m}}\}. \tag{9}$$

Clearly, $\sigma^{(o)}(r)=M^{-1}$ (all $r \in R$) is again a stable fixed point of this recursion.

In case that $R$ is an Abelian group $A$, eq. (9) can be made more explicit; writing the group $A$ additively as in Section 7.1, the ratio

$$x_n(\underline{a}) = \sigma_n(\underline{a})/\sigma_n(\underline{0}) \tag{10}$$

can be defined for all $\underline{a} \in A$. Eq. (9) then gives as recursion for these ratios the expression:

$$x_n(\underline{a}) = \frac{\sum\limits_{\underline{a}_2,\ldots,\underline{a}_q \varepsilon A} \Omega_C(\underline{a}+\underline{a}_2+\ldots+\underline{a}_q) \prod\limits_{i=1}^{q} x_{n-1}(\underline{a}_i)^m}{\sum\limits_{\underline{a}_2,\ldots,\underline{a}_q \varepsilon A} \Omega_C(\underline{a}_2+\ldots+\underline{a}_q) \prod\limits_{i=1}^{q} x_{n-1}(\underline{a}_i)^m}, \quad x_o(\underline{a}) = \sigma_o(\underline{a})/\sigma_o(\underline{0}). \tag{10}$$

Now the Fourier expansion of $\Omega_C(\underline{w})$ as given in eqs. (7.2.18,19) is inserted into eq. (10); the result is

$$x_n(\underline{a}) = \frac{\sum\limits_{\underline{t}\varepsilon A} \tilde{\Omega}(\underline{t}) \left[\sum\limits_{\underline{a}'\varepsilon A} \chi_{\underline{a}'}(\underline{t})\ x_{n-1}(\underline{a}')^m\right]^{q-1} \chi_{\underline{t}}(\underline{a})}{\sum\limits_{\underline{t}\varepsilon A} \tilde{\Omega}(\underline{t}) \left[\sum\limits_{\underline{a}'\varepsilon A} \chi_{\underline{a}'}(\underline{t})\ x_{n-1}(\underline{a}')^m\right]^{q-1}}. \tag{11}$$

But now here several such Fourier expansions occur, since the function

$$y_{n-1}(\underline{a}) = x_{n-1}(\underline{a})^m \tag{12}$$

is, just as $\Omega_C(\underline{a})$, a function on $A$ with $y(\underline{a})=y(-\underline{a})$ and $y(\underline{0})=1$:

$$x_n(\underline{a}) = \frac{\sum\limits_{\underline{t}\varepsilon A} \tilde{\Omega}(\underline{t})\ \tilde{y}_{n-1}(\underline{t})^{q-1}\ \chi_{\underline{t}}(\underline{a})}{\sum\limits_{\underline{t}\varepsilon A} \tilde{\Omega}(\underline{t})\ \tilde{y}_{n-1}(\underline{t})^{q-1}}. \tag{13}$$

Now this is again a Fourier transform of a function on $A$ with the correct properties; therefore, eq. (13) can be written in terms of the eigenvalues $\lambda_{\underline{a}}$ of such functions, see eq. (7.1.6), as

$$x_n(\underline{a}) = \frac{\lambda_{\underline{a}}\left(\frac{\lambda_{\underline{t}}(\omega_{\underline{s}})}{\lambda_o(\omega_{\underline{s}})}\left[\frac{\lambda_{\underline{t}}(x_{n-1}(\underline{u}))^m}{\lambda_o(x_{n-1}(\underline{u}))^m}\right]^{q-1}\right)}{\lambda_o\left(\frac{\lambda_{\underline{t}}(\omega_{\underline{s}})}{\lambda_o(\omega_{\underline{s}})}\left[\frac{\lambda_{\underline{t}}(x_{n-1}(\underline{u}))^m}{\lambda_o(x_{n-1}(\underline{u}))^m}\right]^{q-1}\right)}. \tag{14}$$

Formally, setting $q=2$ in this equation gives also the expression for the recursion relation of the corresponding spin model on a Cayley branch with branching ratio $m$.

## 12.2. The gauge Ising model.

For this model, $\mathcal{R}$ is simply the Abelian group $\mathcal{S}(2)=\{0,1\}$. A function on this group, which can be Fourier transformed, has the simple form $f(0)=1$, $f(1)=f$, so that its eigenvalues are

$$\lambda_o(f(t)) = 1+f, \quad \lambda_1(f(t)) = 1-f. \tag{1}$$

With these expressions inserted, eq. (1.14) yields a single recursion relation for $x_n=x_n(1)$:

$$x_n = f_{qm}(x_{n-1},\omega), \quad x_o=\{A(1)/A(0)\}^{\frac{1}{m}}, \tag{2}$$

$$f_{qm}(x,\omega) = \left\{1-\left(\frac{1-\omega}{1+\omega}\right)\left(\frac{1-x^m}{1+x^m}\right)^{q-1}\right\}\left\{1+\left(\frac{1-\omega}{1+\omega}\right)\left(\frac{1-x^m}{1+x^m}\right)^{q-1}\right\}^{-1}. \tag{3}$$

In order to find the fixed points, which satisfy $x=f_{qm}(x,\omega)$, the cases $q$=odd and $q$=even have to be distinguished, since one has

$$\begin{aligned} f_{qm}(x,\omega) &= f_{qm}(x^{-1},\omega), \quad \text{for } q=\text{odd}, \\ f_{qm}(x,\omega) &= \{f_{qm}(x^{-1},\omega)\}^{-1}, \quad \text{for } q=\text{even}. \end{aligned} \tag{4}$$

In addition, one always has the symmetry:

$$f_{qm}(x,\omega^{-1}) = \{f_{qm}(x,\omega)\}^{-1}. \tag{5}$$

The fixed points are now given as follows:

(i) $q$=odd; there is a critical $\omega_c$, such that one has:

(ia) for $\omega_c<\omega<\omega_c^{-1}$, the only fixed point is the stable one at $x=1$;

(ib) for $\omega<\omega_c$, there are three fixed points, $x_o=1$ and $x_1<x_2<1$; of these, $x_o$ and $x_1$ are stable, $x_2$ is unstable;

(ic) for $\omega>\omega_c^{-1}$, the fixed points are $x_o$ and $x_1^{-1}>x_2^{-1}>1$, of which $x_o$ and $x_1^{-1}$ are stable and $x_2^{-1}$ unstable.

(id) for $\omega=\omega_c$ or $\omega=\omega_c^{-1}$, the two nontrivial fixed points coincide and are marginal.

(ii) $q$=even; there is again a critical $\omega_c$, but now the cases are:

(iia) for $\omega>\omega_c$, there is only the stable fixed point $x_o$;

(iib) for $\omega<\omega_c$, there are five fixed points, $x_o$ and $x_1<x_2<1<x_2^{-1}<x_1^{-1}$, of which $x_o$, $x_1$ and $x_1^{-1}$ are stable, the other two unstable;
(iic) for $\omega=\omega_c$, there are three fixed points $x_1=x_2<1<x_2^{-1}=x_1^{-1}$, of which $x_o=1$ is, of course, still stable, whereas the other two are marginal.

The critical values $\omega_c$ and $x_c$ for the marginal fixed points can be obtained in all cases from the two requirements

$$x=f_{qm}(x,\omega) \quad \text{and} \quad 1=(\partial/\partial x)f_{qm}(x,\omega). \tag{6}$$

This yields for $x_c$ the two solutions of the equation

$$m(q-1)x_c^{m-1}=(1-x_c^{2m})/(1-x_c^{2}); \tag{7}$$

the value(s) of $\omega_c$ then follow from

$$\omega_c = \left\{\left(\frac{1-x_c^{m}}{1+x_c^{m}}\right)^{q-1} - \frac{1-x_c}{1+x_c}\right\}\left\{\left(\frac{1-x_c^{m}}{1+x_c^{m}}\right)^{q-1} + \frac{1-x_c}{1+x_c}\right\}^{-1}. \tag{8}$$

From the above and from more detailed consideration of the case $\omega>\omega_c^{-1}$ for q=even, the following picture for the flows of the recursion relation, eqs. (2) and (3), is obtained:

| | | |
|---|---|---|
| q odd, $\omega_c<\omega<\omega_c^{-1}$ | : $\{x_n\}$ | converges to 1 for every initial value, denoted by x in the rest of this table. |
| q odd, $\omega=\omega_c$ | : $\{x_n\}$ | converges to 1 for $x_c<x<x_c^{-1}$, to $x_c$ for $x\leqq x_c$ and for $x\geqq x_c^{-1}$. |
| q odd, $\omega<\omega_c$ | : $\{x_n\}$ | converges to 1 for $x_2<x<x_2^{-1}$, to $x_1$ for $x<x_2$ and for $x>x_2^{-1}$; it converges to $x_2$ for $x=x_2$ and for $x=x_2^{-1}$. |
| q odd, $\omega=\omega_c^{-1}$ | : $\{x_n\}$ | converges to 1 for $x_c<x<x_c^{-1}$, to $x_c^{-1}$ for $x\leqq x_c$ and for $x\geqq x_c^{-1}$. |
| q odd, $\omega>\omega_c^{-1}$ | : $\{x_n\}$ | converges to 1 for $x_2<x<x_2^{-1}$, to $x_1^{-1}$ for $x>x_2^{-1}$ and for $x<x_2$; it converges to $x_2^{-1}$ for $x=x_2^{-1}$ and for $x=x_2$. |
| q even, $\omega_c<\omega<\omega_c^{-1}$ | : $\{x_n\}$ | converges to 1 for all x. |
| q even, $\omega=\omega_c$ | : $\{x_n\}$ | converges to 1 for $x_c<x<x_c^{-1}$, to $x_c$ for $x\leqq x_c$ and to $x_c^{-1}$ for $x\geqq x_c^{-1}$. |
| q even, $\omega<\omega_c$ | : $\{x_n\}$ | converges to 1 for $x_2<x<x_2^{-1}$, to $x_1$ for $x<x_2$, to $x_1^{-1}$ for $x>x_2^{-1}$, and stays |

at $x_2$ or at $x_2^{-1}$, if it starts there.

q even, $\omega=\omega_c^{-1}$ : $\{x_n\}$ converges to 1 for $x_c<x<x_c^{-1}$ and to a two-point limit cycle, consisting of $x_c$ and $x_c^{-1}$, for $x\leqq x_c$ and for $x\geqq x_c^{-1}$.

q even, $\omega>\omega_c^{-1}$ : $\{x_n\}$ converges to 1 for $x_2<x<x_2^{-1}$ and to the two-point limit cycle $(x_1,x_1^{-1})$ for $x<x_2$ and for $x>x_2^{-1}$. $(x_2,x_2^{-1})$ is then obtained, if $x$ is one of these two points.

A typical phase diagram for odd $q$ is shown in Fig. 1 for $m=3$, $q=3$, i.e., a triangular plaquette tree; in Fig. 2, an example is given of

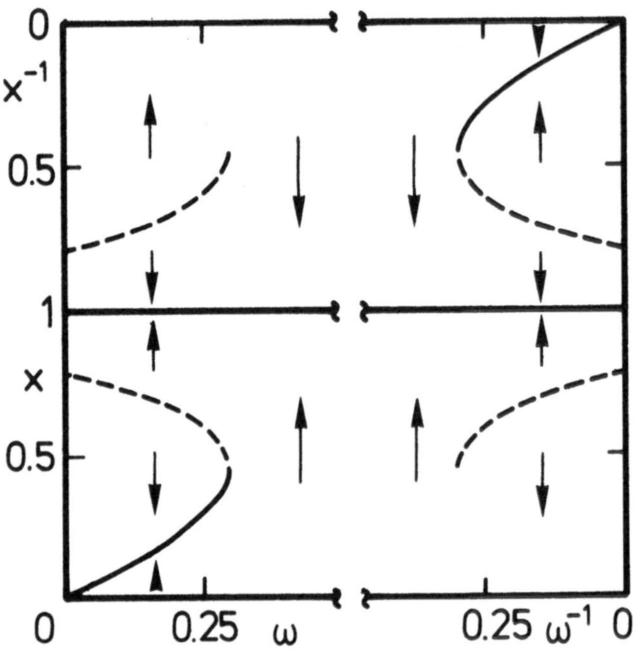


Fig. 1. Phase diagram of the gauge Ising model on a triangular plaquette tree with branching ratio 3. The solid curves are the stable fixed points, the broken curves are the unstable ones, except for the ones in the upper left and lower right corners, where they are lines of points that are mapped on the unstable fixed points by one recursion step. These broken lines are all phase transitions, since they separate parts of the phase diagram, in which the recursion goes to different stable fixed points; these are marked by arrows.

the case of even $q$: $q=4$, i.e., a square plaquette tree, and $m=5$. This last value has been chosen so, that this pseudo-lattice has exactly as many plaquettes per bond as the four-dimensional hypercubic lattice. The value of $\omega_c$ is here 0.436.., in tolerable agreement with the exact value $\sqrt{2}-1=0.414..$ found by duality in Section 7.6.

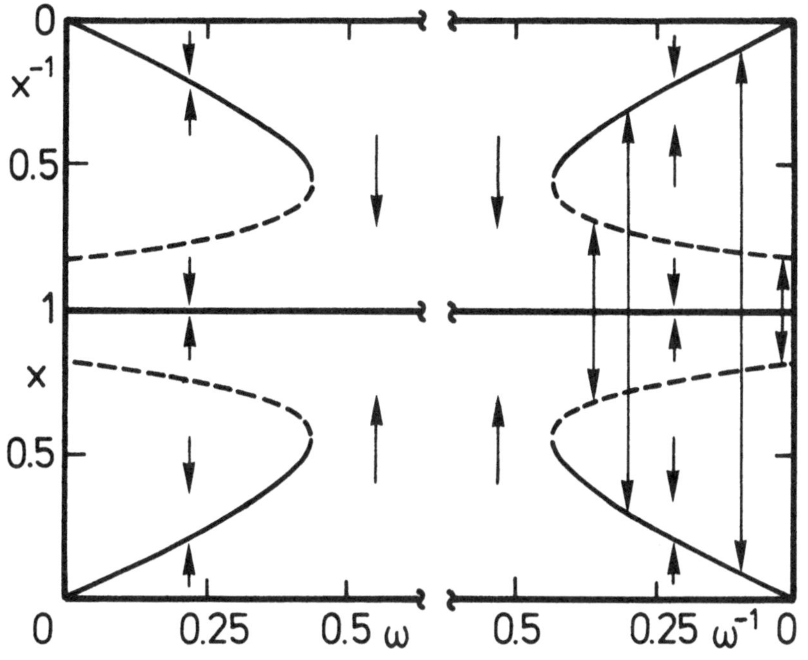


Fig. 2. Phase diagram of the gauge Ising model on a square plaquette branch with m=5, modelling the four-dimensional hypercubic lattice. Solid lines are stable fixed points or stable two-point limit cycles, these latter have extra arrows. Broken lines are unstable fixed points or limit cycles; these are always phase transition lines. The short arrows indicate to which fixed point or limit cycle the recursion goes.

## 12.3. Gauge Potts models.

For a Potts model on $M$ letters, the group $R$ can be chosen as any Abelian group with $M$ elements. The function $f(\underline{a})$ on this group is, if it is properly normalized, given as

$$f(\underline{0}) = 1, \quad f(\underline{a}) = f \quad \text{for all} \quad \underline{0}\neq\underline{a}\in A, \tag{1}$$

so that its Fourier transform is given by the eigenvalues

$$\lambda_o = 1+(M-1)f, \quad \lambda_{\underline{a}} = 1-f \quad \text{for all} \quad \underline{0}\neq\underline{a}\in A. \tag{2}$$

If also the initial function $x_o(\underline{a})$ has the structure of eq. (1), the model will be called the gauge Potts model. In this case, eq. (1.14) gives a single recursion for all $x_n(\underline{a})$, since these are necessarily all equal to $x_n$ for $\underline{a}\neq\underline{0}$:

$$x_n = \frac{1 - \frac{1-\omega}{1+(M-1)\omega}\left(\frac{1-x_{n-1}^{\ m}}{1+(M-1)x_{n-1}^{\ m}}\right)^{q-1}}{1+(M-1)\frac{1-\omega}{1+(M-1)\omega}\left(\frac{1-x_{n-1}^{\ m}}{1+(M-1)x_{n-1}^{\ m}}\right)^{q-1}} . \tag{3}$$

Note, that this does indeed reduce to eq. (11.3.7) for $q=2$, $m=2$, and the present $x_n$ replaced by its m-th root, in accordance with the remark on Cayley branches following eq. (1.14) above.

The recursion relation (3) has, for $M\neq 2$, either only the fixed point 1 (for $\omega>\omega_c$) or three fixed points $x_1<x_2<1$ for $\omega<\omega_c$, of which $x_1$ and 1 are stable and $x_2$ is unstable, or, for $\omega=\omega_c$, the two fixed points 1 and $x_c$, the latter being marginal. There are, therefore, less different possibilities than for the gauge Ising model:

| | | | |
|---|---|---|---|
| q arbitrary, $\omega>\omega_c$ | : | $\{x_n\}$ | converges to 1 for every value of x. |
| q odd, $\omega=\omega_c$ | : | $\{x_n\}$ | converges to 1 for $x_c<x<x_c^{-1}$, to $x_c$ for $x\leqq x_c$ and for $x\geqq x_c^{-1}$. |
| q odd, $\omega<\omega_c$ | : | $\{x_n\}$ | converges to 1 for $x_2<x<x_2^{-1}$, to $x_1$ for $x<x_2$ and for $x>x_2'$, to $x_2$ for $x=x_2$ and for $x=x_2'$. |
| q even, $\omega=\omega_c$ | : | $\{x_n\}$ | converges to 1 for $x>x_c$, to $x_c$ for $x\leqq x_c$. |
| q even, $\omega<\omega_c$ | : | $\{x_n\}$ | converges to 1 for $x>x_2$, to $x_1$ for $x<x_2$ and stays at $x_2$ for $x=x_2$. |

Here $x'$ is the value $>1$, which is mapped by one recursion step onto $x<1$; such a value can exist only for odd q; it does, in general, not exist for all $\omega\leqq\omega_c$.

Two sample phase diagrams are shown as Figs. 1 and 2, both for the three-state Potts model; as an example of the case q odd, $q=3$ and $m=3$ have been chosen again (Fig. 1), whereas q even is again represented by the case $q=4$, $m=5$, as model of the four-dimensional hypercubic lattice (Fig. 2). In this latter case, the phase transition starts at $\omega_c=0.404..$, which is not in as good agreement with the exact value of 0.366.. found from duality in Section 7.6, as was the case for the gauge Ising model of the previous section. This is a general trend: with increasing M, the discrepancy between the exact values from duality and the approximate ones from the $q=4$, $m=5$ pseudo-lattice becomes larger. This result has been obtained from exact calculations for $\omega_c$ and $x_c$, which proceed analogous to eqs. (2.7) and (2.8).

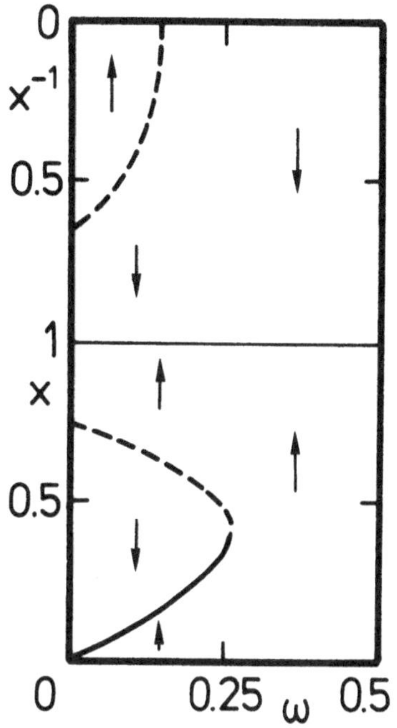


Fig. 1. Phase diagram of the three-state gauge Potts model on a triangular plaquette tree with m=3. Solid curves are stable fixed points, the broken curve in the lower half of the phase diagram is the unstable fixed point and the broken curve in the upper part is x', corresponding to this unstable fixed point. Both of these broken curves are phase transition lines, since they separate regions of the phase diagram, in which the recursion goes to different stable fixed points; this is indicated by arrows.

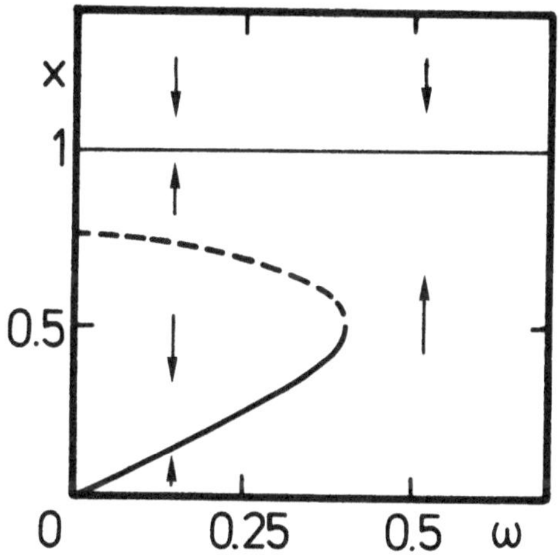


Fig. 2. Phase diagram of the 3-state Potts gauge model on a square plaquette branch with m=5. Solid curves are again stable fixed points, the broken curve is the unstable one, which is also the phase transition line. The arrows have their usual meaning.

## 12.4. The gauge $C(4)$ model.

For models with more than one energy parameter or Boltzmann factor, the recursion relations become very complicated; here only the simplest case of a $C(4)$ model will be treated, and then only in a superficial manner. The group $C(4)$ is generated by an element a with 4a=0; therefore, a properly normalized vector, which can be Fourier transformed, is, for example,

$$1,\ \omega_1,\ \omega_2,\ \omega_1, \tag{1}$$

since such a function must take on the same value for a and for -a= 3a. The eigenvalues are

$$\lambda_o=1+2\omega_1+\omega_2,\quad \lambda_1=\lambda_3=1-\omega_2,\quad \lambda_2=1+\omega_2-2\omega_1. \tag{2}$$

Setting $x_n(1)=x_n(3)=x_n$ and $x_n(2)=y_n$, eq. (1.14) gives rise to two coupled recursion relations:

$$x_n = f_1(x_{n-1},y_{n-1}),\quad y_n = f_2(x_{n-1},y_{n-1}),$$

$$f_1(x,y)=D^{-1}\left\{1-\frac{1+\omega_2-2\omega_1}{1+\omega_2+2\omega_1}\left(\frac{1+y^m-2x^m}{1+y^m+2x^m}\right)^{q-1}\right\},$$

$$f_2(x,y)=D^{-1}\left\{1+\frac{1+\omega_2-2\omega_1}{1+\omega_2+2\omega_1}\left(\frac{1+y^m-2x^m}{1+y^m+2x^m}\right)^{q-1}-2\frac{1-\omega_2}{1+\omega_2+2\omega_1}\left(\frac{1-y^m}{1+y^m+2x^m}\right)^{q-1}\right\}, \tag{2}$$

$$D=1+\frac{1+\omega_2-2\omega_1}{1+\omega_2+2\omega_1}\left(\frac{1+y^m-2x^m}{1+y^m+2x^m}\right)^{q-1}+2\frac{1-\omega_2}{1+\omega_2+2\omega_1}\left(\frac{1-y^m}{1+y^m+2x^m}\right)^{q-1}.$$

Needless to say, this is so complicated, that it is of numerical use only; nonetheless, a number of special cases are extracted rather easily from these equations. These will be briefly considered in the following, except for the case $\omega_1=\omega_2$, $x_n=y_n$ (all n), which is the four-state Potts gauge model of the previous section.

(i) For $\omega_1=\omega_2=\omega$, $x_n=1$ (for all n) is a solution of eqs. (2) for <u>even</u> q-values; there only remains one recursion relation for $y_n$: this turns out to be the same as eq. (3.3) for the four-state Potts model, upon replacement of $x_n$ by $y_n^{-1}$. Therefore, there are again three fixed points for $\omega<\omega_c=0.436..$

(ii) $y_n=1$ for all n, is always a solution of eqs. (2). There only remains a recursion relation for $x_n$, which is of the form of eq. (2.3) for the gauge Ising model with an effective interaction parameter:

$$\omega_{eff} = 2\omega_1/(1+\omega_2). \tag{3}$$

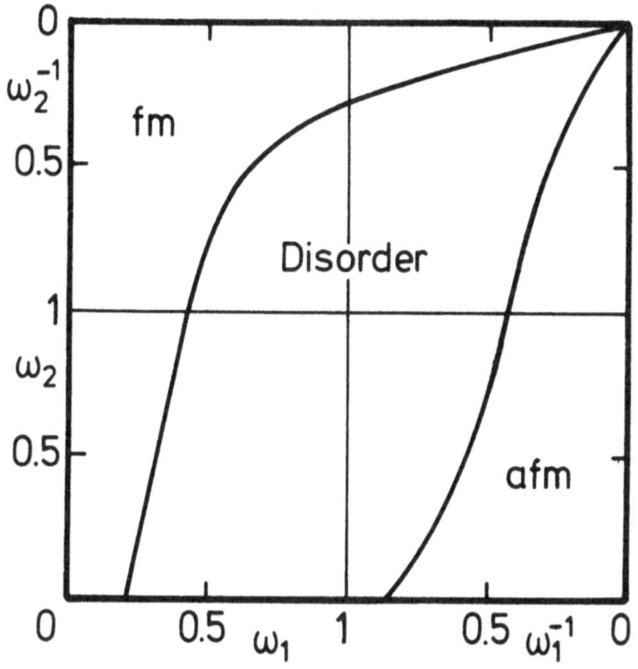


Fig. 1. Phase diagram of the gauge $\mathcal{C}(4)$ model for the special case (ii), see the text.

Therefore, everything said for the gauge Ising model holds here too, with this effective interaction parameter. The resulting phase diagram is shown in Fig. 1 for the case $q=4$, $m=5$. Here "disorder" means, that 1 is the only stable fixed point, "fm" indicates, that another stable fixed point can be reached for suitable initial values and "afm" indicates a region, where a stable two-point limit cycle can be reached in this way.

(iii) For $\omega_1=0$, $x_n=0$ for all $n$, is a solution; the remaining recursion relation for $y_n$ has the gauge Ising form, eq. (2.3), with $\omega=\omega_2$.

(iv) For $\omega_2=\omega_1^2$, $y_n=x_n^2$ (all $n$) is a solution; the recursion for $x_n$ has the Ising form with $\omega=\omega_1$.

Cases (iii) and (iv) are two of the three cases, in which the corresponding spin model, $\mathcal{S}(2)\sim\mathcal{S}(2)$, factorizes; the third case is a special instance of case (ii).

From the above special cases, it is clear, that the complete four-dimensional phase diagram for the present model is very complex. This will, therefore, not be pursued further here. In the next section, a simpler approach is briefly described.

## 12.5. The Bethe approximation for gauge models.

As noted in subsection 11.3.2, there exists a recursive bond graph sequence such, that the renormalization prescription for the interaction on this is equal to the field-renormalization procedure for a Cayley branch. It may, therefore, be expected, that a pseudo-lattice can be found, such that the renormalization of the plaquette interaction on this lattice is equal to the Higgs field renormalization prescription on a plaquette branch. This is indeed the case for square plaquettes: consider a "cubic branch" built up from three-dimensional cubes, $m$ of which have a face in common; call this common face blue, the $5m$ others yellow and build up the branch as in Chapter 8: the n-generation branch is obtained by replacing each of the $5m$ yellow faces of the basic cube by the blue faces of as many (n-1)-generation branches. Directing the edges of the basic cube as in Fig. 1, a recursion relation for the partition function $Z_n(a_1,a_2,a_3,a_4)$ with fixed spins on the edges of the

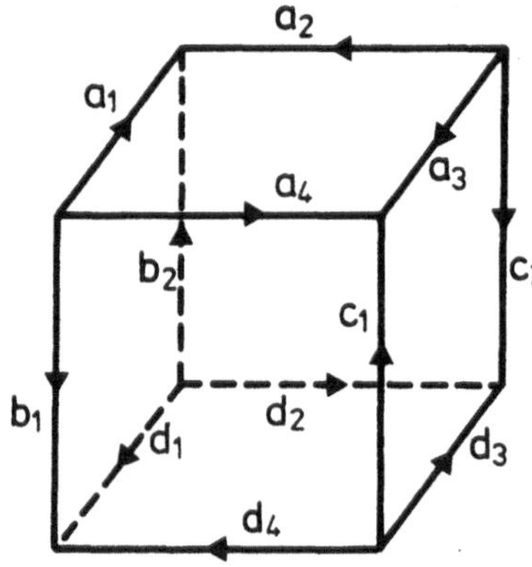


Fig. 1. The basic building block of the cubic branch; the spins on the top face are kept fixed, see eq. (1). There are $m$ copies of this cube, all with the same top=blue face.

blue plaquette is easily obtained as

$$Z_n(a_1a_2^{-1}a_3a_4^{-1}) = \Omega_C(a_1a_2^{-1}a_3a_4^{-1})\Big\{\sum_{\substack{b_1,b_2,c_1,c_2,\\ d_1,d_2,d_3,d_4\in R}} Z_{n-1}(a_1b_2^{-1}d_1b_1^{-1})$$

$$Z_{n-1}(a_3c_1^{-1}d_3c_2^{-1})\,Z_{n-1}(a_2b_2^{-1}d_2c_2^{-1})\,Z_{n-1}(a_4c_1^{-1}d_4b_1^{-1})$$

$$Z_{n-1}(d_1d_4^{-1}d_3d_2^{-1})\Big\}^m, \quad Z_o(a_1a_2^{-1}a_3a_4^{-1})=\Omega_o(a_1a_2^{-1}a_3a_4^{-1}). \tag{1}$$

Here use has been made of the fact, that these partition functions can only depend in the way shown on their arguments, due to the form of the initial value.

For the case, that $\mathcal{R}$ is Abelian, eq. (1) gives a somewhat simpler recursion relation for the ratio

$$y_n(\underline{a}) = Z_n(\underline{a})/Z_n(\underline{0}): \tag{2}$$

$$y_n(\underline{a}) = \Omega_C(\underline{a}) \frac{\left\{\lambda_{\underline{a}}\left[\frac{\lambda_{\underline{t}}(y_{n-1}(\underline{u}))}{\lambda_o(y_{n-1}(\underline{u}))}\right]^5\right\}^m}{\left\{\lambda_o\left[\frac{\lambda_{\underline{t}}(y_{n-1}(\underline{u}))}{\lambda_o(y_{n-1}(\underline{u}))}\right]^5\right\}^m} \tag{3}$$

Defining new parameters by

$$x_n(\underline{a}) = \frac{\lambda_{\underline{a}}(y_n(\underline{u}))}{\lambda_o(y_n(\underline{u}))}, \tag{4}$$

eq. (3) takes the form:

$$x_n(\underline{a}) = \frac{\lambda_{\underline{a}}\left\{\frac{\lambda_{\underline{t}}(\tilde{\omega}_{\underline{s}})}{\lambda_o(\tilde{\omega}_{\underline{s}})}\left[\frac{\lambda_{\underline{t}}(x_{n-1}(\underline{u}))^5}{\lambda_o(x_{n-1}(\underline{u}))^5}\right]\right\}^m}{\lambda_o\left\{\frac{\lambda_{\underline{t}}(\tilde{\omega}_{\underline{s}})}{\lambda_o(\tilde{\omega}_{\underline{s}})}\left[\frac{\lambda_{\underline{t}}(x_{n-1}(\underline{u}))^5}{\lambda_o(x_{n-1}(\underline{u}))^5}\right]\right\}^m}, \quad x_o(\underline{a}) = \tilde{\Omega}_o(\underline{a}). \tag{5}$$

The case $m=3$ yields the proper number of plaquettes per cube, so as to model the four-dimensional hypercubic lattice; eq. (5) is then formally the same as eq. (1.14) for $q=4$, also applicable in order to model this lattice. Note, however, the occurrence of the dual $\tilde{\omega}$-parameters in the present case. All results of the preceding sections for the case $q=4$, $m=5$ can now easily be translated to give the results for the cubic branch with $m=3$; this will not be done explicitly.

The fixed points of eq. (5) may also be used to obtain free energies in a Bethe-like approximation as detailed for spin systems in Section 8.4. Such a calculation for the gauge $\mathcal{C}(4)$ model has recently been performed ([1]). The results are in excellent qualitative and, maybe surprisingly, quantitative agreement with Monte Carlo calculations. It is not known, why this approximation works so much better for gauge systems than it does for the corresponding spin models. This should be an interesting question for future research.

REFERENCE.

([1]). J.B. Zuber, Nucl. Phys. B 235 [FS 11] (1984) 435.

GENERAL REFERENCES.

A different type of plaquette tree (or branch) has been studied by A. Maritan and A.L. Stella, J. Phys. A 16 (1983) L157.
The Bethe approximation for the gauge Ising model is given by C. Itzykson, R.B. Pearson and J.B. Zuber, Nucl. Phys. B 220 [FS 8] (1983) 415.
Ref. ([1]) above actually gives a very general description of this approximation, valid also for nonabelian gauge models.

# 13. CRITICAL EXPONENTS FOR SYSTEMS ON SELF-SIMILAR PSEUDO-LATTICES.

## 13.1. Introductory remarks.

In Chapters 9-12, pseudo-lattices have been used mainly to derive results,which have relevance for real lattices as well: symmetry breaking patterns and qualitative pictures of renormalization flows should be independent of the sequence of graphs, which defines the thermodynamic limit. Recursively defined graph sequences make these problems soluble, so that one naturally has recourse to them to obtain first ideas as to the shape of phase diagrams.

In the present chapter, the pseudo-lattices are taken more seriously: the nature of the singularities associated with unstable fixed points (or limit cycles) is studied. Here, one may expect differences from the real world to occur: the fact, that a recursively defined graph sequence does not converge to a unique infinite graph should result in effects not found for the same systems on real lattices. This is indeed the case at first glance: the phase transitions are mostly of continuous order, as first found by Müller-Hartmann and Zittartz for the case of a Cayley tree in a (vanishingly small) homogeneous field [1]. From these also, however, information for systems on real lattices can be obtained; there are strong indications, that this type of phase transition does occur for spin systems with a continuous symmetry group on low-dimensional lattices [2].

## 13.2. General expressions for the critical exponents.

For a self-similar recursive graph sequence, the free energy of a (spin or gauge) model can be written in the form:

$$\gamma(\underline{x}_o) = \gamma_o \sum_{k=0}^{\infty} B(\underline{x}_k)/N^k, \tag{1}$$

where $\gamma_o$ is a constant, $N$ is the number of (n-1)-generation graphs needed to build the n-generation graph (as in Section 8.5) and the quantity $B(\underline{x}_k)$ depends on a vector of recursively defined numbers:

$$\underline{x}_k = R(\underline{x}_{k-1}), \quad \underline{x}_o \text{ given.} \tag{2}$$

Examples are quoted in Section 8.3: eq. (8.3.8) (site recursion) and eq. (8.3.23) (bond recursion) are applicable to spin systems, whereas eq. (8.3.41) is of the same form for a general gauge model. Such an equation can also be deduced for the cubic branches introduced in Section 12.5.

Eq.(1) has the following obvious, but important, consequence:

$$\gamma(\underline{x}_o) = \gamma_o \sum_{k=0}^{K-1} B(\underline{x}_k)/N^k + N^{-K}\gamma(R^k(\underline{x}_o)) \tag{3}$$

holds for any natural number $K$. In the neighbourhood of an unstable fixed point, denoted by $\underline{x}$, eq. (2) takes the linearized form

$$\underline{\delta}_k = \underline{\underline{L}}\underline{\delta}_{k-1} \tag{4}$$

upon setting $\underline{x}_k = \underline{x} + \underline{\delta}_k$; here $\underline{\underline{L}}$ is a matrix, at least one of the eigenvalues of which is larger than 1 in absolute value, since $\underline{x}$ is unstable. Let $\underline{\hat{e}}$ be a unit eigenvector of $\underline{\underline{L}}$ belonging to such an eigenvalue $\lambda$. Then eq. (3) can be rewritten: setting

$$\underline{x}_k = \underline{x} + \underline{\hat{e}}\delta_k \tag{5}$$

and

$$\gamma(\underline{x} + \underline{\hat{e}}\delta) = \gamma(\underline{x}, \underline{e}; \delta), \tag{6}$$

one obtains:

$$\gamma(\underline{x}, \underline{e}; \delta_o) = \gamma_o \sum_{k=0}^{K-1} B(\underline{x}_k)/N^k + N^{-K}\gamma(\underline{x}, \underline{e}; \lambda^K \delta_o). \tag{7}$$

Since the finite sum in eq. (7) must be an analytic function of $\delta_o$, the singular part $\gamma_{sing}(\delta_o)$ of $\gamma(\underline{x}, \underline{e}; \delta_o)$ satisfies:

$$\gamma_{sing}(\delta) = N^{-K}\gamma_{sing}(\lambda^K \delta). \tag{8}$$

Assuming a power-law singularity of the form

$$\gamma_{sing}(\delta) \simeq |\delta|^{\tau}, \tag{9}$$

eq. (8) gives, for <u>all</u> values of $K$, the critical exponent $\tau$ as

$$\tau = \ln N / \ln|\lambda|. \tag{10}$$

This is the fundamental equation for the critical exponent; some more care has to be taken in case $\tau$ is equal to a natural number. If this is $\ell_o$, the leading singularity acquires a logarithmic correction [1,3]:

$$\gamma_{sing}(\delta) \simeq A|\delta|^{\ell_o} + B\delta^{\ell_o} \ln|\delta|. \tag{11}$$

## 13.3. Cayley branches: phase transitions of continuous order.

For the Cayley branches with field on the boundary studied in Chapter 9, the zero-field fixed point is unstable against perturbations in the eigenspace of the eigenvalue $\lambda_k$ of the interaction matrix $\underline{\underline{\Omega}}$, as soon as the temperature is low enough, so that

$$m|\lambda_k/\lambda_o| \geq 1 \tag{1}$$

holds. Eq. (2.10) yields the critical exponent for this direction as

$$\tau_k = (\ln m)(\ln m + \ln|\lambda_k/\lambda_o|)^{-1}. \tag{2}$$

This is a continuous function of the temperature: for $T \to 0$, the ratio $\lambda_k/\lambda_o$ approaches 1, so that

$$\lim_{T \to 0} \tau_k(T) = 1 \tag{3}$$

holds; for nonzero temperatures, $\tau_k(T)$ is an increasing function of $T$, which finally diverges at the phase transition, given by equality in eq. (1).

For the Potts model, there ia also an unstable fixed point unequal to the zero-field fixed point for slightly higher temperatures, see subsection 9.2.3, Figs. 9.2.3 and 9.2.4. In Fig. 1, the critical exponent $\tau$ for the three-state Potts model on a Cayley branch with $m=2$ is plotted as a function of $\omega$; for $\omega<0.25$, this is a plot of the function

$$\tau(\omega) = (\ln 2)\{\ln 2 + \ln(1-\omega)/(1+2\omega)\}^{-1}, \tag{4}$$

in accordance with eq. (2); for $0.25<\omega<(1+2\sqrt{2})^{-1}$, the exponent has been calculated numerically. Note the symmetry around $\omega=0.25$ and the much faster divergence at $(1+2\sqrt{2})^{-1}$. It is to be noted, that the minimum value of $\tau$ above 0.25 is about 47: this is an extremely weak singularity !

Phase transitions of continuous order have first been discovered for the Ising model on a Cayley branch in a homogeneous field [1]. In

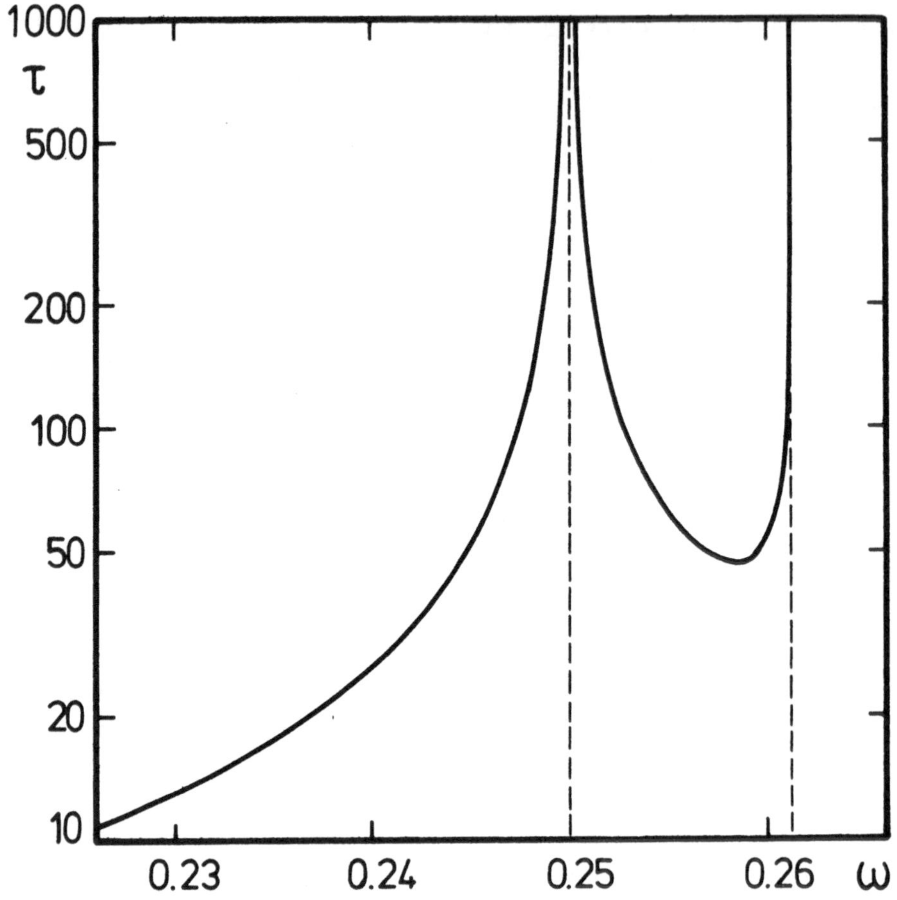


Fig. 1. Plot of the critical exponent $\tau$ for the three-state Potts model on a Cayley branch with $m=2$. Below $\omega=0.25$, the phase transition is of the small-field type, for higher temperatures of the high-field type.

this case, the derivation is not as simple as in Section 2, since in eq. (2.1) the quantity B now depends on $\underline{x}_o$ and on $\underline{x}_k$. A careful analysis ([4]) then yields corrections to the simple power law behaviour, which are oscillatory. Since this type of analysis makes essential use of the distribution of the zeroes of the partition function in the complex plane, it can only be performed for the Ising model, for which one knows, that these zeroes all lie on the unit circle, see subsection 9.2.2. For this reason, a restriction to boundary fields has been made here for the general case.

## 13.4. Critical exponents for spin and gauge Potts models on recursive bond graph sequences.

The examples of systems described by the title of this section are the same as those treated in Chapters 11 and 12.

(i) Koch curve. The recursion relation for the Potts model on this pseudo-lattice, eq. (11.1.3), has an unstable fixed point at $\omega=0$. Taking $\omega$ and $\omega'$ as independent variables, the critical exponent equation (2.10) with $N=4$ and $\lambda=2(1+\omega')$ yields

$$\tau = (2\ln 2)\{\ln 2 + \ln(1+\omega')\}^{-1} . \tag{1}$$

The transition is again of continuous order: it interpolates between a second order ($\tau=2$) transition at $\omega'=0$ and a first order one ($\tau=1$) at $\omega'=1$. As a function of the temperature only, both $\omega$ and $\omega'$ must approach 1 as $T\to 0$; interpreted in this way, there is a second order phase transition at $T=0$ only.

(ii) Self-dual diamond hierarchical lattice. This recursive bond graph sequence has been treated at length in Section 11.2. For the Potts model, the recursion relation, eq. (11.2.7), together with eq. (2.10), gives a fixed value of the critical exponent at the critical, self-dual coupling $\omega_c=(1+\sqrt{M})^{-1}$. Since $N=5$, this is

$$\tau = (\ln 5)\left[\ln\{(5M^{\frac{3}{2}}+18M+21M^{\frac{1}{2}}+8)/(M^{\frac{3}{2}}+8M+15M^{\frac{1}{2}}+8)\}\right]^{-1} \tag{2}$$

This is about $2\frac{2}{3}$ for the Ising model, $M=2$, and decreases below 2 for $M\geqq 6$. For large $M$, the value 1 is slowly approached ($\tau=1.085..$ for the 1000-state Potts model).

(iii) Absence of trivial fixed points. A recursive bond graph sequence with only the low- but not the high-temperature fixed point present has been introduced in subsection 11.3.1. This low-temperature fixed point becomes unstable for $\omega'$, which measures the strength of the noniterated interaction, larger than 0.5. The critical exponent is, since $N=2$, given as

$$\tau = (\ln 2)/(\ln 2\omega'), \quad \omega'\geqq 0.5. \tag{3}$$

This is again a phase transition of continuous order, but with the value 1 (first order transition) approached for $\omega'\to 1$ and the divergence at $\omega'=0.5$.

The absence of the low-temperature fixed point only has been shown

in subsection 11.3.2 to lead to Cayley branch recursion relations, so that the continuous order critical exponents of Section 3 are recovered for this case. Since in the absence of both the high- and low-temperature fixed points, only one stable fixed point has been found in subsection 11.3.3, there now only remains the plaquette branch:
(iv) Gauge models on plaquette branches. Along the unstable fixed point lines of the phase diagrams of Chapter 12, phase transitions of continuous order occur. The critical exponent diverges at the marginal fixed point $\omega_c, x_c$ and decreases for $\omega \to 0$; for all models studied, however, the limit of $\tau$ for $\omega \to 0$ is not equal to 1. The minimal value for the gauge Potts model varies somewhat with q, m and M, but is generally within the range $2.5<\tau<3$. This implies, that the cubic branches of Section 12.5 also show phase transitions of this type.

In conclusion, all recursive bond graph sequences lead to phase transitions of continuous order if they contain noniterated bonds; the order of the transition is a function of the interaction parameters for these bonds. In the cases of the diamond hierarchical lattices, there are fixed critical exponents, since these do contain yellow, i.e., iterated, bonds only. The values found above for the Potts model are, however, very different from the real lattice ones: it is known ([5]), that the Potts model on a square lattice has a continuous (higher than first order) phase transition for M=2, 3 and 4, and a first order one for M>4. Obviously, different renormalization prescriptions (which may not be realizable as exact recursions for a pseudo-lattice) are needed to give useful results for the critical exponents, see,e.g., the articles in ([6]).

REFERENCES.

([1]). E. Müller-Hartmann and J. Zittartz, Phys. Rev. Lett. 33 (1974) 893; Z. Phys. B 22 (1975) 59.
([2]). J. Zittartz, Z. Phys. B 23 (1976) 55, 63.
([3]). M. Kaufman and D. Andelman, Phys. Rev. B 29 (1984) 4010.
([4]). E. Müller-Hartmann, Z. Phys. B 27 (1977) 161.
([5]). R.J. Baxter, J. Phys. C 6 (1973) L445.
([6]). Phase Transitions and Critical Phenomena, Vol. 6, eds. C. Domb and M.S. Green (Academic Press, London, New York, San Francisco, 1976).

# APPENDIX: TWO EXTENSIONS OF DISCRETE CLASSICAL SPIN MODELS.

## A.1. The Potts model for noninteger M and percolation.

In Section 10.2, bond percolation on a Cayley branch has been defined. In this appendix, a different definition of bond percolation, valid for an arbitrary graph sequence, will be given. This definition is based on the extension of the Potts model to noninteger values of the number of states $M$. To see how such an extension is possible, consider the partition function of the M-state Potts model on a graph $G=(V,E,I)$ in an external field provided by a ghost spin (see Section 7.5):

$$Z = \sum_{\substack{i_v=1,..,M \\ v\varepsilon V}} \prod_{e\varepsilon E} \Omega(i_{v_1(e)}, i_{v_2(e)}) \prod_{v\varepsilon V} A(i_v), \tag{1}$$

$$\Omega(i,j) = \omega+(1-\omega)\delta(i,j), \quad A(i) = a+(1-a)\delta(i,1). \tag{2}$$

Insertion of eqs. (2) into eq. (1) gives an expression for the partition function, which, upon introduction of variables $\lambda_e=0$ or $1$ for all edges, takes the form

$$Z = \sum_{\substack{i_v=1,..,M \\ v\varepsilon V}} \sum_{\substack{\lambda_e=0,1 \\ e\varepsilon E}} \prod_{e\varepsilon E} [\omega^{1-\lambda_e}\{(1-\omega)\delta(i,j)\}^{\lambda_e}] \prod_{v\varepsilon V} \{a+(1-a)\delta(i_v,1)\}. \tag{3}$$

For a particular configuration $\{\lambda_e\}$, the graph $G$ splits up in a number $N_c$ of connected components $G_i=(V_i,E_i,I_i)$ upon deletion of the edges with $\lambda_e=0$; for each of these connected components, there is a factor in the partition function of the form:

$$\sum_{\substack{i_v=1,..,M \\ v\varepsilon V_i}} \prod_{e\varepsilon E_i} \{(1-\omega)\delta(i_{v_1(e)}, i_{v_2(e)})\} \prod_{v\varepsilon V_i} \{a+(1-a)\delta(i_v,1)\}. \tag{4}$$

In this expression, the summation over the spin variables may be performed to give:

$$(1-\omega)^{|E_i|}\{1+(M-1)a^{|V_i|}\}, \tag{5}$$

so that eq. (3) becomes:

$$Z = \sum_{\substack{\lambda_e=0,1\\ e\in E}} \prod_{e\in E} \{\omega^{1-\lambda_e}(1-\omega)^{\lambda_e}\} \prod_{i=1}^{N_c}\{1+(M-1)a^{|V_i|}\}. \tag{6}$$

Now $1-\omega$ may be interpreted as the probability $p$ with which a bond is present, provided that $\omega\leqq 1$ holds, i.e., in the ferromagnetic region. One can then define a "percolation average" by

$$\langle X\rangle_p = \sum_{\substack{\lambda_e=0,1\\ e\in E}} \prod_{e\in E} \{p^{\lambda_e}(1-p)^{1-\lambda_e}\}\, X(\{\lambda_e\}). \tag{7}$$

In this notation, the partition function <u>is</u> a percolation average:

$$Z = \langle \prod_{i=1}^{N_c}\{1+(M-1)a^{|V_i|}\}\rangle_p. \tag{8}$$

This can now obviously be defined for all real $M$, so that the goal of extending the Potts model has been achieved.

In order to see what the above has to do with percolation, the limit $M\to 1$ is studied; setting $M=1+\varepsilon$, eq. (8) reads, to first order in $\varepsilon$,

$$Z = 1+\varepsilon\langle \sum_{i=1}^{N_c} a^{|V_i|}\rangle_p. \tag{9}$$

From this, the free energy $\gamma=|V|^{-1}\ln Z$ follows as

$$\gamma(\varepsilon) = \varepsilon|V|^{-1}\langle \sum_{i=1}^{N_c} a^{|V_i|}\rangle_p. \tag{10}$$

In case there is no external field, this gives

$$\lim_{\varepsilon\to 0} \gamma(\varepsilon)/\varepsilon = |V|^{-1}\langle N_c\rangle_p, \tag{11}$$

which is the average (per vertex) of the number of disconnected graphs in which $G$ splits up if its bonds are present with probability $p$. The first derivative with respect to the field gives the relation

$$\lim_{\varepsilon\to 0} \frac{1}{\varepsilon}\left.\frac{\partial\gamma}{\partial a}\right|_{a=1} = |V|^{-1}\langle \sum_{i=1}^{N_c}|V_i|\rangle_p = 1. \tag{12}$$

This result may, however, be different from 1 if the thermodynamic limit $|V|\to\infty$ is taken, <u>before</u> the fields are allowed to go to zero, $a\to 1$. This corresponds, for the spin model, to the appearance of a spontaneous magnetization; in percolation theory, this implies

$$\lim_{|V|\to\infty} |V|^{-1} < \sum_{i=1}^{N_c} |V_i| >_p < 1, \tag{13}$$

which shows, that an infinite cluster or infinite connected graph is present in the thermodynamic limit. The value $\omega_c$, for which this first occurs, gives the critical percolation probability $p_c'=1-\omega_c$.

It is easy to see, that Cayley branches (or even recursive site graph sequences in general) in a homogeneous magnetic field never show a spontaneous magnetization; this follows immediately from an expansion of the free energy, eq. (8.3.8), in terms of a small external field. Also, the phase transition being of continuous order, see Section 13.3, rules out the possibility of a spontaneous magnetization. Therefore, all these graph sequences have $p_c'=1$; clearly, the "infinite cluster" criterion is not equivalent to the "top-to-bottom" criterion of Section 10.2.

For the square lattice, $\omega_c$ for the Potts model is given by eq. (7.6.21), derived from duality. In the limit $M\to1$, this yields the exact percolation threshold, $p_c'=\frac{1}{2}$, for this lattice.

## A.2. A model with an infinite symmetry group.

There is, in principle, no difficulty to formulate the definitions of permissible and completely permissible groups, of interactions and maximal interactions, etc., for cases, where the spin models have infinitely many degrees of freedom. No systematic study, such as has been presented in the first six chapters for the case of finite groups, has been performed for the infinite group case thus far. Here, a simple example only will be treated; the questions of permissibility and symmetry breaking are briefly discussed.

Suppose, that the state of a spin is given by an angle $\phi\varepsilon[0,2\pi)$, and let the energy function be invariant with respect to rotations of both spins over the same angle:

$$E(\phi_1,\phi_2) = E(\phi_1+\phi,\phi_2+\phi) \quad \text{for all} \quad \phi\varepsilon[0,2\pi). \tag{1}$$

Requiring, as in Section 1.2, that the energy function is symmetric with respect to spin exchange, this implies, that the full symmetry group of the interaction contains, in addition to the rotations, which form the Lie group U(1), the reflection $\sigma:\phi\to-\phi$. This resulting permissible symmetry group is, in a sense, the limit of the dihedral group for $M\to\infty$. The energy function is now a function only of the absolute value of the difference of the two arguments:

$$E(\phi_1,\phi_2) = f(|\phi_1-\phi_2|). \tag{2}$$

Eq. (2) implies, that the group is completely permissible. The (common) eigenvectors and the eigenvalues of a Boltzmann factor operator $\Omega(|\phi|)$ are given as

$$\int_0^{2\pi} \Omega(|\phi-\phi_1|)\ \mu_k(\phi_1)\ d\phi_1 = \lambda_k\ \mu_k(\phi), \tag{3}$$

$$\mu_k(\phi) = \exp(\pm ik\phi), \quad k=0,1,2,\ldots , \tag{4}$$

$$\lambda_k = \int_0^{2\pi} \Omega(|\phi'|)\ \exp(\pm ik\phi')\ d\phi'. \tag{5}$$

All eigenvalues except $\lambda_o$ are doubly degenerate. From eq. (5), the phase transitions on a Cayley tree with its boundary in a field follow as in Section 9.1:

$$m|\lambda_k/\lambda_o|=1. \tag{6}$$

Eq. (4) shows, that the pure phases $k$ and $\bar{k}$ have a residual k-th order cyclic symmetry.

For the special case, that the Boltzmann factors are given by

$$\Omega(|\phi|) = \exp(-\beta J \cos \phi), \tag{7}$$

the model is known as the XY model. In this case, the eigenvalues are modified Bessel functions:

$$\lambda_k = 2\pi\ I_k(-\beta J). \tag{8}$$

It follows from this, that the highest temperature phase transition points, as given by eq. (6), are for the entrance in the $1$ and $\bar{1}$ phases. Therefore, there is then no residual symmetry below the phase transition temperature.

GENERAL REFERENCES.

The extension of the Potts model to noninteger values of $M$ and the connection with percolation in the limit $M\to 1$ are due to
P.W. Kasteleyn and C.M. Fortuin, J. Phys. Soc. Jp. Suppl. 26 (1969) 11,
C.M. Fortuin and P.W. Kasteleyn, Physica 57 (1972) 536.
The limit as $M\to 0$ is also of interest:
M.J. Stephen, Phys. Lett. 56A (1976) 149.

This extension can also be used to classify those graph sequences, on which the free energy of a Potts model is an analytic function of the temperature in the absence of external fields:
H. Moraal, Z. Phys. B 45 (1982) 237.
Another extension of the Potts model is the random cluster model of Kasteleyn and Fortuin; this has been extensively studied on a Cayley branch in a set of homogeneous external fields:
H.G. Baumgärtel and E. Müller-Hartmann, Z. Phys. 46 (1982) 227.
The difference between the two definitions of percolation for a Cayley branch has also been discussed carefully by
L. Billard and P. Villemain, J. Phys. A 13 (1980) 1335.
The occurrence of a percolative phase transition without the formation of an infinite cluster has been discussed in the context of correlated site-bond percolation by
F. Delyon, B. Souillard and D. Stauffer, J. Phys. A 14 (1981) L243.
In this type of percolation, a cluster is defined not only by the presence of connecting bonds, but it is also required, that the spins at the vertices, which pair interact in the standard way, are all in states from a particular subset of the set S of all states. This problem was first considered by
A. Coniglio and W. Klein, J. Phys. A 13 (1980) 2774,
for the Ising model case. On a Cayley branch, it can be solved for any spin model with a permissible symmetry group:
H. Moraal, Physica 122A (1983) 313.

The XY model on a Cayley branch has been studied in some detail by
H. Moraal, Physica 105A (1981) 472.
The Heisenberg model gives very similar results:
H. Moraal, Physica 85A (1976) 457.
These results are very different from the exact results known for two dimensions for these models: there is no spontaneous symmetry breaking possible by the theorem of
N.D. Mermin and H. Wagner, Phys. Rev. Lett. 17 (1966) 1133.
This is due to the high (infinite) graph fractal dimensionality of the Cayley branches.
It has been suggested, that the phase transitions for the XY and Heisenberg models in two dimensions is of continuous order:
J. Zittartz, Z. Phys. B 23 (1976) 55, 63, (XY model),
C. Schönfelder, Ph. D. Dissertation, University of Cologne (1980, unpublished), (classical Heisenberg model).
For the more general n-vector models, see
J. Zittartz, Z. Phys. B 31 (1978) 63, 79, 89.

There are, of course, also quantum mechanical analogues of classical spin systems, such as the finite-spin Heisenberg model. It is also possible, to translate all models into quantum mechanical ones by using the Hamiltonian formalism based on a highly anisotropic square lattice, see, e.g.,
J. B. Kogut, Rev. Mod. Phys. 51 (1979) 659.
For permissible groups containing a regular,Abelian subgroup, this formalism has been presented in some detail by
M. Marcu, A. Regev and V. Rittenberg, J. Math. Phys. 22 (1981) 2740.
It is not very difficult, to extend this treatment to arbitrary groups.

SUBJECT INDEX

# Lecture Notes in Physics

Vol. 195: Trends and Applications of Pure Mathematics to Mechanics. Proceedings, 1983. Edited by P.G. Ciarlet and M. Roseau. V, 422 pages. 1984.

Vol. 196: WOPPLOT 83. Parallel Processing: Logic, Organization and Technology. Proceedings, 1983. Edited by J. Becker and I. Eisele. V, 189 pages. 1984.

Vol. 197: Quarks and Nuclear Structure. Proceedings, 1983. Edited by K. Bleuler. VIII, 414 pages. 1984.

Vol. 198: Recent Progress in Many-Body Theories. Proceedings, 1983. Edited by H. Kümmel and M. L. Ristig. IX, 422 pages. 1984.

Vol. 199: Recent Developments in Nonequilibrium Thermodynamics. Proceedings, 1983. Edited by J. Casas-Vázquez, D. Jou and G. Lebon. XIII, 485 pages. 1984.

Vol. 200: H.D. Zeh, Die Physik der Zeitrichtung. V, 86 Seiten. 1984.

Vol. 201: Group Theoretical Methods in Physics. Proceedings, 1983. Edited by G. Denardo, G. Ghirardi and T. Weber. XXXVII, 518 pages. 1984.

Vol. 202: Asymptotic Behavior of Mass and Spacetime Geometry. Proceedings, 1983. Edited by F. J. Flaherty. VI, 213 pages. 1984.

Vol. 203: C. Marchioro, M. Pulvirenti, Vortex Methods in Two-Dimensional Fluid Dynamics. III, 137 pages. 1984.

Vol. 204: Y. Waseda, Novel Application of Anomalous (Resonance) X-Ray Scattering for Structural Characterization of Disordered Materials. VI, 183 pages. 1984.

Vol. 205: Solutions of Einstein's Equations: Techniques and Results. Proceedings, 1983. Edited by C. Hoenselaers and W. Dietz. VI, 439 pages. 1984.

Vol. 206: Static Critical Phenomena in Inhomogeneous Systems. Edited by A. Pękalski and J. Sznajd. Proceedings, 1984. VIII, 358 pages. 1984.

Vol. 207: S. W. Koch, Dynamics of First-Order Phase Transitions in Equilibrium and Nonequilibrium Systems. III, 148 pages. 1984.

Vol. 208: Supersymmetry and Supergravity/Nonperturbative QCD. Proceedings, 1984. Edited by P. Roy and V. Singh. V, 389 pages. 1984.

Vol. 209: Mathematical and Computational Methods in Nuclear Physics. Proceedings, 1983. Edited by J. S. Dehesa, J. M. G. Gomez and A. Polls. V, 276 pages. 1984.

Vol. 210: Cellular Structures in Instabilities. Proceedings, 1983. Edited by J. E. Wesfreid and S. Zaleski. VI, 389 pages. 1984.

Vol. 211: Resonances – Models and Phenomena. Proceedings, 1984. Edited by S. Albeverio, L. S. Ferreira and L. Streit. VI, 369 pages. 1984.

Vol. 212: Gravitation, Geometry and Relativistic Physics. Proceedings, 1984. Edited by Laboratoire "Gravitation et Cosmologie Relativistes", Université Pierre et Marie Curie et C.N.R.S., Institut Henri Poincaré, Paris. VI, 336 pages. 1984.

Vol. 213: Forward Electron Ejection in Ion Collisions. Proceedings, 1984. Edited by K.O. Groeneveld, W. Meckbach and I.A. Sellin. VII, 165 pages. 1984.

Vol. 214: H. Moraal, Classical, Discrete Spin Models. VII, 251 pages. 1984.

Selected Issues from

# Lecture Notes in Mathematics